Theory and Technology of Laser Imaging Based Target Detection

Yihua Hu

Theory and Technology of Laser Imaging Based Target Detection

Yihua Hu
Shanghai Institute of Technical Physics
Chinese Academy of Sciences
Shanghai
China

ISBN 978-981-10-3496-1 ISBN 978-981-10-3497-8 (eBook)
DOI 10.1007/978-981-10-3497-8

Jointly published with National Defense Industry Press, Beijing, China

The print edition is not for sale in China Mainland. Customers from China Mainland please order the print book from: National Defense Industry Press.

Library of Congress Control Number: 2016961341

Printed on acid-free paper

This Springer imprint is published by Springer Nature
The registered company is Springer Nature Singapore Pte Ltd.
The registered company address is: 152 Beach Road, #21-01/04 Gateway East, Singapore 189721, Singapore

Preface

Since May 1960, when the scientist Maiman from the Hughes Research Laboratories in USA invented the fantastic laser, laser has been studied and applied extensively in various fields. Laser has been utilized in irradiation, cutting, welding, CD/DVD, printing and copying, anti-counterfeit, high-speed typesetting, measurement, communication, remote sensing, and even detecting and imaging the moon. As a consequence, laser has shown increasingly obvious influence and benefit.

Laser has been widely applied in the production and living of human. In the remote sensing detection field studied by the author, using laser to detect targets is probably the most precise method for perceiving the appearance of long-distance targets at present. This is because laser is characterized by the high brightness, high collimation, and strong coherence and can actively, real timely, and precisely acquire the three-dimensional (3D) information of the detected targets.

Target detection based on a laser imaging system is a target detecting method applying laser beams as detection media, which are radiated to illuminate targets according to certain spatial distribution law. Then, the data including time delay, intensity, waveform, phase, and polarization of laser echoes reflected from the detected targets are collected. After being processed, these data are presented in images. Afterward, feature extraction and object inversion are conducted based on the images so as to obtain the information of targets, including distance, position, reflection attribute, structure size, and motion feature. By doing so, targets are found, identified, and confirmed. This technology can provide the information, including the range images, gray images, and feature images of targets with high resolution that cannot be obtained using general imaging methods. In addition, characterized by high-resolution, high-measurement accuracy, anti-interference ability, and strong anti-shadowing ability, the laser imaging system for target detection is especially suitable for the detailed detection of targets. In international society, attention has been paid to the application of a laser imaging system in target detection: Since the mid and late 1990s, America has began to research this technology; while at the beginning of the twentieth century, European countries have embarked on the investigation of sea-aero targets by using laser imaging. In contrast, laser imaging is mainly studied from the perspective of laser remote sensing imaging for the earth observation in China so that the laser imaging system for target detection is relatively backward. Since the 1990s, the author has been occupied in the study of laser remote imaging and target detection. He has hosted multiple national research projects, organized, or participated in developing various spaceborne and airborne target detection systems based on laser imaging. Based on a large amount of first-hand data about laser imaging target detection acquired, the author has published a series of research papers. After integrating the data we obtained in recent years, the author writes this book, aiming to explain the basic principles and technology realization of laser imaging in target detection so as to provide helpful guidance in the field. Moreover, the author aims to provide more valuable reference to the researchers who are engaged in the study of target detection based on a laser imaging system and postgraduates who are endeavoring to participate in this field.

According to the implementation process of target detection based on a laser imaging system, this paper illustrates the basic principle, laser imaging, image processing, target detection and localization, and classification and recognition in four parts with ten chapters. The first to the third chapters comprehensively analyze the concept and mechanism of target detection based on a laser imaging system, modulation mechanism of laser, and composition and characteristics of the detection system; the fourth and fifth chapters systematically elaborate the principle and method of information acquisition, data collection, data processing, and laser image generation; the sixth to the ninth chapters focus on the technical realization of the image feature extraction, target feature extraction, target detection and location, target classification, and identification based on laser images for typical targets, while the tenth chapter discusses the source and characteristics of the errors of the target detection based on a laser imaging system.

Since the 1990s, academician Xue Yongqi, from Shanghai Institute of Technical Physics, the Chinese Academy of Science, has given elaborate guidance and support to the author's research work. Moreover, he encouraged and helped the author to write this book. Meanwhile, academicians Ling Yongshun and Gong Zhiben also provided valuable opinions and suggestions for the author's research work and this manuscript. In addition, part of the research work involved in this book was conducted in Shanghai Institute of Technical Physics, the Chinese Academy of Science. A great help has been obtained from researchers Wang Jianyu and Shu Rong as well as other people in the research group of the institute for more than a decade. Professors Lei Wuhu and Hao Shiqi, and assistant professor Zhao Nanxiang as well as other workmates and postgraduates in the research group participated in realizing part of the research work and provided helpful suggestions to this book. In writing this book, Dr. Shi Liang helped in word arrangement and typesetting; Huang Tao, Li Lei, and Xu Shilong participated in collecting data in the early period. Here, the author expresses heartfelt thanks to them in the research process. Furthermore, as numerous literatures were investigated during the writing, here, the author presents gratitude to these authors. The author also acknowledges the support of The Publishing Fund Committee of National Defense Science and

Technology and the reviewers, the experts, and the editors from the National Defense Industry Press in China.

Since this book contains new theories and technologies, some problems are remained to be further studied. Moreover, some deficiencies are inevitably presented owing to different viewpoints. Criticisms are welcome from experts and readers.

Hefei, Anhui, China Yihua Hu

Contents

Chapter 1
Introduction

Target detection based on a Laser imaging system, as a new remote sensing technology developed from laser ranging, laser guidance, laser track-pointing and laser imaging, is another detection method for obtaining target information apart from visible light imaging, infrared imaging and radar imaging. This technology mainly uses lasers with high brightness, directivity and coherence as the light source which are radiated to illuminate targets. Therefore, non-touch detection imaging is performed directly to targets so as to acquire the images of targets. Then, based on laser images of targets, feature extraction and object inversion are conducted so as to directly, quickly and precisely gain the 3D feature information of targets including position, structure and attributes. In this way, targets are found, identified and confirmed, thus satisfying the requirement of real-time target information acquisition with high precision, multiple dimensions, efficiency and accuracy. The technology of applying a laser imaging system in target detection, as a novel detection manner, has gradually become an important method for obtaining the intelligence information of battlefield targets. To help readers accurately understand the concept and outline of the target detection based on a laser imaging system, this chapter introduces the basic concept and principles of the target detection using a laser imaging system.

1.1 Concept of Laser Imaging

1.1.1 Origin of Laser Imaging

Light amplification by stimulated emission of radiation. Compared with ordinary light, laser is characterized by good monochromaticity, strong directivity, high brightness and good coherence. Moreover, it presents strong controllability, long operating range and high precision. Therefore, laser has been extensively applied in

Y. Hu, *Theory and Technology of Laser Imaging Based Target Detection*,
DOI 10.1007/978-981-10-3497-8_1

various fields since it was invented in the 1960s. Currently, there have been many laser application technologies including laser ranging, laser guidance and laser track-pointing.

As one of the earliest and the most mature applied technologies of laser, laser ranging is used to calculate the distance from targets to observation points by measuring the time taken by a laser beam on a round trip between observation points and targets [1]. According to working principle and technical methods, laser ranging can be divided into pulsed ranging, phased laser ranging and interferometry ranging. Though laser ranging is in principle the extension of radar ranging in optical frequency band, it can realize long-distance measurement with high precision, thus meeting the need of high-precision location for targets. At present, laser ranging is studied to develop eye-safe devices which have good atmosphere penetration. Moreover, these devices are supposed to be small and standard with solid components and multi-function. Meanwhile, the measurement range and precision of the existing laser ranging devices are expected to be improved.

Laser guidance is another earliest and most extensively utilized laser application. It is a method by using lasers to obtain guidance information or transmit guidance instructions so that weapons can fly to targets according to certain guidance law [2]. It has advantages of high guidance precision and strong anti-interference ability. Based on different working principles, laser guidance can be separated into laser beam riding guidance and laser homing guidance, while the latter can be further divided into semi-active laser guidance, laser command guidance and active laser radar guidance. The rapid development of solid lasers with high power, laser modulation and transmission technique, laser detector array, and high speed signal processing lays a solid foundation for the development of laser guidance technology, and promotes the application of laser guidance. Moreover, driven by technical progress and actual demand, laser guidance is supposed to develop to active imaging guidance. Meanwhile, laser guidance devices are expected to be more remote, intelligent, universal, miniaturized, multi-functional and compound, and can be used in long ranging distances. With lasers as media for detecting targets, laser pointing is used to detect research, locate and track targets so as to acquire the information including position, angle and velocity of targets with high precision. By making full use of the advantages of lasers including good monochromaticity, strong directivity, high brightness, short wavelength and high spatial resolution, laser track-pointing shows stronger directivity, and higher resolution and tracking precision; moreover, it presents good concealment and is unlikely to be subject to interference. Laser radars are typical laser track-pointing devices which are developed by adding devices for measuring the direction and pitch as well as automatic tracking devices based on laser ranging. Therefore, they present similar principle, structure and function to microwave radars. They can be considered as the extension of microwave radars in laser bands. With the development of the technology, the laser radars for the purpose of pointing currently is developed to be multi-functional ones which are imaginable, intelligent, real time, integrated, standardized and applied in micro optics.

Owing to favorable technical characteristics and successful applications, lasers have gradually presented a wider application prospect. Especially in the fields of information acquisition, target guidance, anti-interference laser detection and guidance, lasers require to obtain more effective information of targets. Based on the existing technical methods for obtaining the target attributes including distance, angle and velocity, adding other methods which can be used to acquire the physical attributes of targets including size, shape and material can improve the ability of tracking and identifying targets. Therefore, laser imaging has been emerged under the promotion of increasing requirement of informatization, demand for stronger detection ability for targets, and development of photoelectric technology.

Laser imaging is a remote sensing imaging method with lasers as the media. By transmitting laser beams with specific parameters through a specific laser source, remotely sensed regions are irradiated in a traversal way so as to receive the laser echoes reflected from the targets using a photoelectric receiving system. Afterwards, these echoes are changed into electronic signals by using a photoelectric detector. Finally, laser images of targets and their background region with information including time delay, gray (energy), waveform, distribution and phase are obtained through electronic signal processing, feature extraction, analog-digital conversion, data processing and image generation. From theoretical discussion, laser images can be acquired by processing data as long as the lasers transmitted have enough transmission power, parameters meet certain need, and the echo signals of the received laser reach a certain signal-to-noise ratio (SNR). Laser images of a target mainly include range images, intensity images, Doppler profiles, range-angle images, etc. According to the characteristic parameters of laser images of a target including size, shape and surface material of targets, people can better understand and master objects detected.

Developed from laser detection technology, laser imaging is a new technology generated by combining laser detection and imaging technology. The former lays the foundation for the transmission and receiving technology of laser imaging, while the latter founds a basis for the whole laser imaging technology. The appearance of laser imaging is closely related to the development of photoelectric detector technology and the discovery of laser in the 1950s and 1960s: the technological development of semiconductor physics in the 1950s promoted the progress of photoelectric detectors and made preparation for the appearance of laser imaging, while the invention of laser in the 1960s made laser imaging come true. Characterized by high brightness, strong directivity, good monochromaticity, strong coherence and short wavelength, lasers show inherent advantages in imaging.

1.1.2 Characteristics of Laser Imaging

Apart from laser imaging, common imaging methods also include visible light imaging, infrared imaging and radar imaging. Thereinto, laser imaging and radar imaging show similar action mechanism and therefore are essentially the same.

From the perspective of technical principle, they can be traced to the same origin and merely different in that laser imaging employs lasers instead of microwaves and millimeter waves as the detection media. Lasers have far shorter wavelengths than microwaves and millimeter waves, as well as narrower beams. Compared with radar imaging, laser imaging presents the following advantages.

(i) The images with high resolution and containing large amounts of information. Compared with microwaves, lasers show shorter wavelengths, higher frequencies, narrower beams, and higher resolution in terms of distance, velocity and angular. The processed target images present high resolution. Moreover, laser imaging cannot only obtain amplitude, frequency, phase and velocity information, but also can acquire information such as waveform and polarization state.

(ii) Working round-the-clock and having strong anti-interference ability. Laser imaging can work all time without the limitation of day and night. Slightly interfered by ground and sky background, laser imaging can work without the interferences of various radio waves in the environment.

(iii) This type of equipments is small-sized and light. As lasers are with shorter wavelengths and the structure size of antennae and the system is small, the devices present small volume and slight mass. With the development of microelectronics and micro-optical technology, laser imaging devices are improved in terms of the solid integrated optics and the volume.

Compared with visible light imaging and infrared imaging, laser imaging is characterized by the following superiorities.

(i) Laser imaging is active and presents strong adaptive capacity. It is not limited by light condition in day or night and slightly affected by background environment. Laser imaging is significantly different from visible light imaging and infrared imaging in that laser imaging can actively image as itself has light source and does not need to depend on external light sources such as sunlight, starlight and airglow. Not influenced by background and temperature difference, laser imaging is able to work on the condition of large clouds and raining days.

(ii) Laser images with higher resolution and conveying larger amounts of information. As lasers present high frequency and narrow beam and spectrum, the resolutions of obtained laser images are higher. Visible light imaging and infrared imaging can merely obtain the two-dimensional images of targets. While laser imaging can gain 3D images which not only include position, velocity, location and attitude, but also contain geometrical shape, surface characteristic, interior structure, motion and the variation of characteristic parameters. Therefore, a larger amount of information can be conveyed in laser images.

Though with various advantages, laser imaging inevitably presents some shortcomings.

(i) It can be easily influenced by atmospheric and meteorological conditions. This is because both atmospheric attenuation and bad weather can reduce the action distance of lasers; moreover, atmospheric turbulence can influence the measurement precision of distance and echo intensity and therefore affect the accuracy of the laser images.
(ii) Detection range can be significantly influenced by lasers. Apart from atmospheric and meteorological conditions, the detection range of laser imaging based detection is also closely related to the power of laser sources. Currently, limited by the generation method of lasers and the production technologies of optoelectronic devices, it is still difficult to produce lasers with large power which can satisfy the requirement of laser imaging.

1.1.3 Types of Laser Imaging

There are various types of laser imaging, and different types present different working principles, system compositions, imaging effects and are always applied to different fields. According to the types of laser sources, there are gas laser imaging including those applying CO_2 as the laser gas, solid laser imaging, semiconductor laser imaging; based on the modulation types of laser signals transmitted, there are continuous wave modulation imaging, pulse modulation imaging and hybrid modulation laser imaging; in terms of the sampling or scanning modes for targets, there are scanning imaging and staring imaging which is also called as non-scanning imaging and array imaging; on the basis of the detection modes of laser echoes, laser imaging can be divided into coherent laser imaging including synthetic aperture imaging, and non-coherent laser imaging which is also called direct laser imaging; etc. Figure 1.1 shows the classification of typical laser imaging.

Although there are various types of laser imaging, the most basic ones include CO_2 laser imaging, diode pump solid state laser (DPSSL) imaging and semiconductor laser imaging. With high radiation efficiency and good atmospheric transmission, CO_2 lasers can easily realize coherent detection and 3D imaging. At early stage, most laser imaging was realized by using CO_2 laser imaging in angle-angle-range (AAR) and angle-angle-Doppler (AAD) modes. However, CO_2 lasers are large and cost much. Moreover, low temperature is needed for HgCdTe detectors for refrigeration, thus restricting the application of CO_2 laser imaging. In recent years, DPSSL imaging has become a research focus as it overcomes the disadvantages of CO_2 laser imaging including large size and high cost and the detector does not need to be refrigerated. It adopts DPSSLs with high repetition frequency and high peak power and APD/PIN detectors with high sensitivity to realize range and intensity imaging with high resolution through direct detection. Semiconductor laser imaging is characterized by small size, light weight, low cost, long service life, high reliability and low power consumption, and adopts

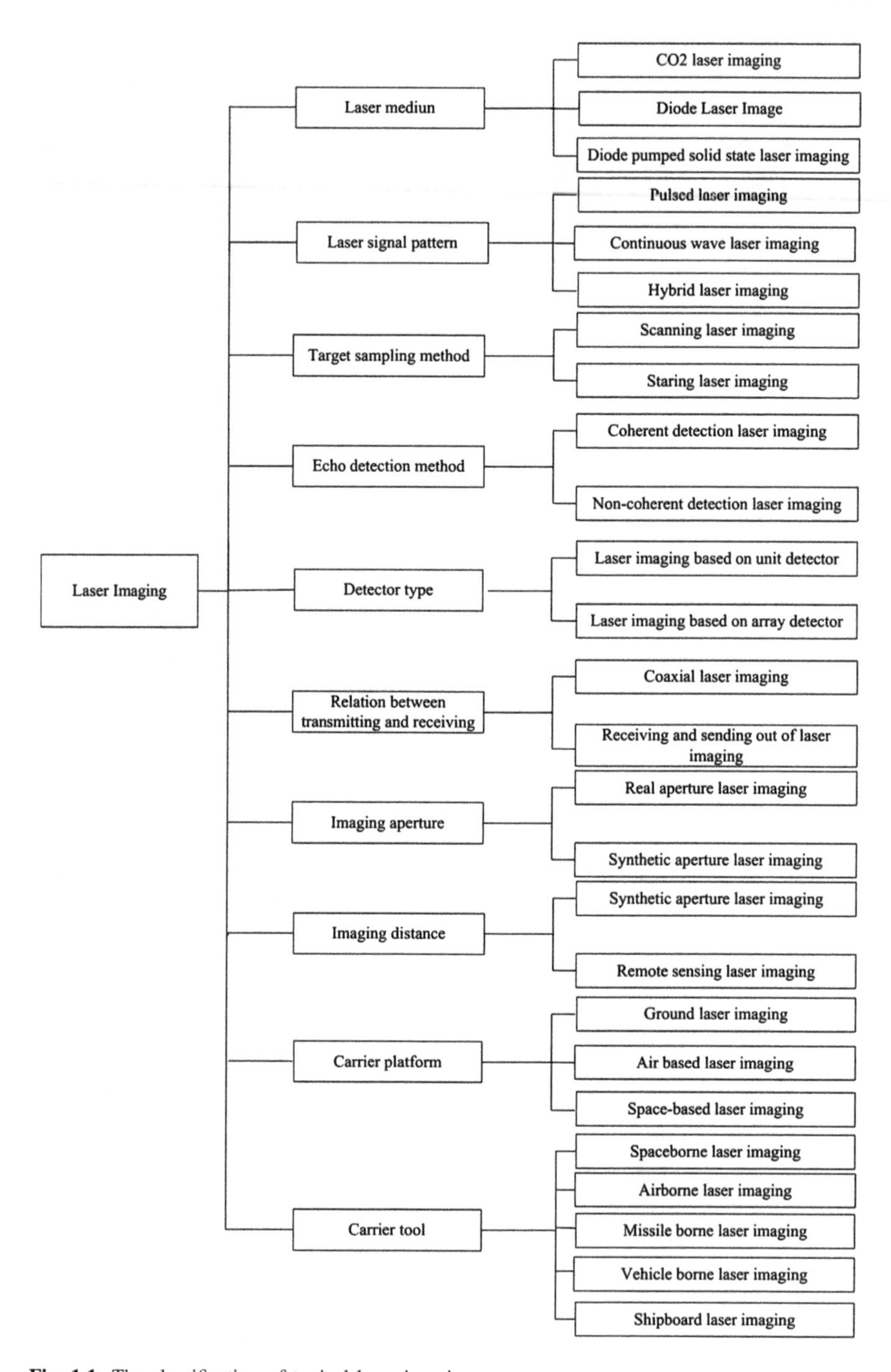

Fig. 1.1 The classification of typical laser imaging

semiconductors or semiconductor laser array. As the detectors are arranged in APD/PIN or APD array, semiconductor laser imaging generally adopts direct detection, thus presenting functions such as ranging and 3D outline imaging of targets.

The remote sensing of laser imaging is mainly used for long-distance imaging, thus having high requirement for the power of lasers. However, based on the current technical level and that in a certain time in the future, lasers with large power and complex signals cannot be produced. Therefore, pulse modulation and direct detection receiving which is also called as non-coherent detection with large power lasers are mostly applied. This is generally called direct laser imaging. The target detection technique based on a laser imaging system in this book is on the basis of this kind of laser imaging. Therefore, the laser imaging in subsequent content basically refers to this kind of direct laser imaging.

Direct laser imaging can generate laser images through receiving pulse signals of laser echoes and acquiring a series of information including pulse delay, pulse energy (gray), pulse waveform and sampling distribution. However, the phase information of laser echoes is not collected. Therefore, the repetition frequency, transmitting power, receiving sensitivity, scanning mode, time delay and waveform of laser pulses require to be measured at a certain precision. In contrast, it is necessary to extract phase, frequency and polarization information of echoed laser signals for coherent laser imaging, or to receive weak laser amplitude or energy signals using heterodyne method for receiving lasers. Compared with the direct one, coherent laser imaging has higher receiving sensitivity and velocity resolution while shows higher requirement for the performance of the system: (1) the coherence of transmitted laser is supposed to be very good; (2) there are intrinsic signals with high stability; (3) more strict requirement for the optical system; (4) the coherence of lasers is easily to be influenced by the transmission path; (5) it is more complicated to process information.

1.2 Concept of Target Detection Based on a Laser Imaging System

1.2.1 Laser Detection and Laser Imaging

All methods which apply lasers to acquire target information are called laser detection. Since lasers are with high brightness, good monochromaticity and strong coherence and directivity, laser detection presents advantages of fast information acquisition, strong anti-interference ability, abundant amount of information obtained and high precision. However, there also exists the shortcoming that the operating range can easily be influenced by atmospheric and meteorological conditions.

Laser detection is a technological means for obtaining target information. The laser imaging introduced in the last section is also a similar means. Therefore, as for as the purpose is concerned, laser detection and laser imaging show certain similar aspects. However, in fact, many types of target information can be acquired through laser detection of targets. It not only includes non-image information such as range, pitch, height, location and velocity, but also contains image information like gray images, range images and Doppler images. Accordingly, from the perspective of the content of target information obtained, laser imaging is a method of laser detection. Moreover, both non-imaging laser methods and laser imaging methods can be used in laser detection. The former include laser ranging, velocity measurement by laser, non-imaging detection of laser radar and target indication of laser non-imaging; and the latter contains laser radar imaging, laser imaging guidance, laser TV detection and laser scanning camera imaging can all be used in laser detection.

In addition, in terms of application, laser detection can be used in various typical scenes with specific names. Laser imaging can be applied to image not only moving targets such as automobiles, aircrafts and ships, but also static targets like city buildings, forests, mountains, farmlands, ditches, and airports. Therefore, when laser imaging is performed to specific targets to obtain their waveform images, gray images and range images, laser imaging and laser detection are combined to generate a new detection technology: applying laser imaging in target detection. This method is designed to acquire target information by processing the obtained laser images of the target.

1.2.2 Definition of the Target Detection Technology Based on Laser Imaging

According to the meanings of laser imaging and target detection, the target detection technology based on laser imaging is defined as a method for searching, identifying, and confirming targets. With lasers as the detection media, laser beams are radiated to illuminate targets according to certain spatial distribution law so as to acquire the data including time delay, intensity, waveform, phase and polarization of laser echoes reflected from the targets. After being processed, these data are presented in images. Afterwards, feature extraction and object inversion are conducted based on the images to acquire the information of targets, including distance, position, reflection attribute, overall outline, structure size, and motion feature. In this way, targets are tracked, identified and confirmed. Here, the term-target detection based on a laser imaging system is used, rather than laser imaging detection. This is because that detections are different for different tasks and purposes: one kind is the detection and surveillance, which aims to precisely and real-timely provide information for people so as to help people know the dynamic conditions of targets and conduct related behaviors; the other kind is target acquisition, which provides target information for homing systems so as to make full use of these systems [4]. The target detection using a laser imaging system introduced in this book contains the above two kinds. Adding the word target is to emphasize that the laser imaging detection essentially aims to acquire target information. Moreover, it pays more attention to the characteristics of targets and serves information detection, target surveillance and target indication.

Similar to laser imaging, the target detection technology based on laser imaging can generate the laser images of targets by detecting the modulation action of the objects detected on the characteristics of the incident laser. Afterwards, the target information extracted from the acquired data including time delay feature, waveform feature, and signal intensity and distribution characteristics of echoes is processed. Finally, we obtain the laser images of targets. Unlike laser imaging, the method of applying a laser imaging system in target detection is not ultimately to acquire the laser images of targets. Instead, it also needs to process, interpret, classify and evaluate the generated laser images to obtain intelligence information so as to find, identify and confirm targets. In addition, according to the requirement of detection, various laser images of targets can be generated in the processing, such as waveform images, range images, gray images and hierarchical images. By observing these images from a single point or a comprehensive view, we can analyze and understand targets more accurately so as to master target information in detail.

By using pulse lasers with high brightness, directivity and coherence as the detection media, the proposed detection technology utilizes the laser imaging technology to obtain various laser images of targets. In this way, more target characteristic parameters and information are provided so that targets can be more precisely found, identified and confirmed at faster detecting and processing speeds. Therefore, this technology is especially suitable for precisely and fast detecting ground, space and air targets. Figure 1.2 shows a laser detection image for a target in a complex scene acquired by using the target detection technology based on a laser imaging system.

1.2.3 Characteristics of Target Detection Based on a Laser Imaging System

Compared with visible light imaging, infrared imaging and radar imaging, target detection based on a laser imaging system presents the following features in terms of detection ability and accuracy in addition to the similar characteristics to laser imaging.

(i) It has high detection precision and is slightly influenced by external environment. On the one hand, by using lasers with high frequency, short wavelength and high directivity as the media for imaging detection, the obtained images show higher spatial resolution so that targets can be observed more clearly. On the other hand, irradiated by the narrow beams of

Fig. 1.2 Laser detection image for a target in a complex scene

lasers, targets are not influenced by other lights and therefore slightly disturbed. That is, this technology can be used to effectively detect targets in a continuous and uninterrupted way.

(ii) It can work in multiple modes and the obtained target information is more abundant. According to users' demands, the target detection technology based on laser imaging can obtain not only one of the range images, gray images, waveform images or hierarchical images for targets, but also multiple above mentioned images so as to more adequately acquire characteristic information of targets from various views. Moreover, laser detection images can clearly reflect the position and 3D characteristic information of targets with abundant image information. As a consequence, by performing target detection using a laser imaging system, we can sufficiently master the target information and more precisely identify targets.

(iii) It can detect and identify camouflaged targets. By radiating lasers to illuminate targets, the target detection based on a laser imaging system can receive laser echoes reflected and scattered from targets. In this way, the influence of changed surrounding environment on targets is reduced. Accordingly, camouflage on target surfaces can be recognized. In addition, such detection system can favorably detect and identify targets hidden in reticular objects like vegetations or camouflage mats by processing the fission of waveform according to the waveform characteristics of laser echoes. Moreover, laser imaging detection can identify hidden targets using various working modes. Firstly, the range images or gray images of targets are acquired in a single range imaging mode or gray imaging mode. Then, the hierarchical images are obtained in a level imaging mode. Finally, according to the character that level images can reflect the structural characteristics of targets, hidden targets are identified by combining hierarchical images with range images or gray images.

Though being developed from laser imaging, the target detection technology based on a laser imaging system is different with the laser imaging. Compared with laser imaging, the target detection technology presents the following characteristics.

(i) The system is more complex. Although the target detection based on a laser imaging system and laser imaging both can acquire target images, the former has higher requirements for the spatial resolution and precision of the generated images, which makes the system of the technology more complex. For instance, when laser imaging is performed to an airport, the obtained images merely require to be able to clearly reflect the number and size of airplanes at the airport; by conducting the target detection based on a laser imaging system, the number and size of airplanes, and the subtle features of airplanes including attributes and shapes are shown in the acquired images. Therefore, compared with laser imaging, the target detection technology proposed needs to be equipped with adaptive variable-speed scanning devices and devices for precisely analyzing images. In this way, target information can be obtained from images with high spatial resolution and quality.
(ii) Data are processed in a more complex way. Apart from conventional processing of laser imaging including laser echo data, GPS data and attitude data, the target detection based on a laser imaging system calls for other necessary data processing methods. They include waveform decomposition for laser echoes, laser echoes processing for more than one time, and the extraction, detection and identification for target features. Therefore, the target detection based on a laser imaging system is more complex in both technical realization and processing procedure.

1.3 Principle of the Target Detection Based on a Laser Imaging System

Starting with the range equation of laser detection under direct detection, this section analyzes the basic principles of laser imaging for targets. It mainly contains the range equation of laser detection for different types of targets, target sampling for laser imaging, and information acquisition and processing of laser imaging detection.

1.3.1 Range Equation of Laser Detection

Range equation of laser detection describes the relationship between echo power and a series of factors including range, target characteristics and detection efficiency. It serves as the basis of the target detection based on a laser imaging system, and also an important basis for designing the detection systems. On this basis, we can determine the operating range of target detection based on a laser imaging system. When processing the signals of laser echoes, with the help of the range equation of laser detection, we can deduce the backward scattering rate of targets based on the echo intensity and target range. When designing the devices for target detection based on a laser imaging system, we can use this scattering rate to estimate the echo power reflected by the targets a certain distance away and received by a photodetector, thus helping to select detectors.

1. **General form of range equation**

The general form of the range equation of laser detection is as follows [5].

$$P_r = \frac{4KP_S T_{A1}\eta_t}{\pi\psi^2 R_1^2} \cdot \Gamma \cdot \frac{T_{A2}}{4\pi R_2^2} \cdot \frac{\pi D^2 \eta_r}{4} \tag{1.1}$$

where P_r and P_S show the signal power of laser echoes received (W) and the transmitting power of lasers (W), respectively; K shows the distribution function of beams; T_{A1} is the atmospheric transmittance from the optical transmitting system to the target; η_t presents the efficiency of the optical transmitting system; ψ denotes the width or divergence (rad) of laser beams; R_1 shows the distance from the transmitted lasers to the target (m), while R_2 presents the distance from the target to the optical receiving system (m); Γ is the laser scattering cross section of the target with unit of m^2; T_{A2} denotes the atmospheric transmittance from the target to the optical receiving system; D shows the effective receiving aperture of the optical receiving system with unit of m and η_r is the efficiency of the optical receiving system.

It can be found from Formula (1.1) that echo power is mainly influenced by four parts. The first one is the optical power density of laser beams radiated to the unit area of targets; the second one shows the sectional area of targets irradiated by lasers; thirdly, the echo power is uniformly scattered by Lambert surface into the whole space; while the fourth one presents the efficiency of optical receiving systems. Therefore, the range equation of laser detection can be considered as the product of four factors: transmission of laser to targets, reflection of light by targets, transmission of scattered light to detectors and collection of scattered light by receivers after transmitting lasers with certain power, as shown in Fig. 1.3.

Given that the launchers and receivers of laser beams in most laser detection systems are at the same position and the laser beams are emitted and received coaxially, $R_1 = R_2 = R$ and $T_{A1} = T_{A2} = T_A$. Then, Formula (1.1) can be simplified as follows.

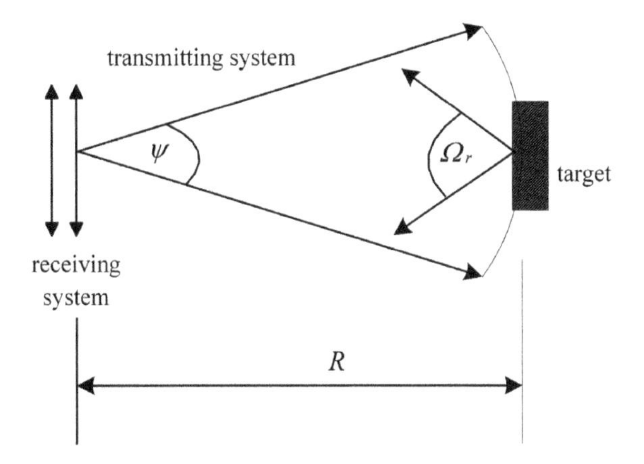

Fig. 1.3 Range equation of laser detection

$$P_r = \frac{KP_S T_A^2 \Gamma D^2 \eta_t \eta_r}{4\pi \psi^2 R^4} \tag{1.2}$$

Based on Formula (1.2), the range equation of laser detection can be obtained:

$$R = \left(\frac{KP_S T_A^2 \Gamma D^2 \eta_t \eta_r}{4\pi \psi^2 P_r} \right)^{\frac{1}{4}} \tag{1.3}$$

If a target shows wider field angle than laser beams in the directions of position and depression angle, that is, the commonly known large target, its laser scattering cross section is positively related to its irradiated area. That is, Γ is proportional to ψ^2 and R^2, and laser beams are approximately in uniform Lambert intensity distribution. Then, Formula (1.2) can be simplified as follows.

$$P_r = \frac{KP_S T_A^2 \rho D^2 \eta_t \eta_r}{4R^2} \tag{1.4}$$

Then, the range equation of laser detection is converted as follows:

$$R = \left(\frac{KP_S T_A^2 \rho D^2 \eta_t \eta_r}{4P_r} \right)^{\frac{1}{2}} \tag{1.5}$$

The laser scattering cross section Γ of targets in Formula (1.2) is calculated using the following formula.

$$\Gamma = \frac{4\pi}{\Omega} \rho A_s \tag{1.6}$$

where Ω shows the solid angle of the target scattering; ρ denotes the average reflection coefficient of the target, while A_s presents the target area.

2. **Special forms of the range equation**

According to the general form of the range equation of laser detection, the power of lasers received is positively related to that of lasers transmitted, while it is inversely proportional to the certain power of the distance between the transmit-receive system of lasers and targets. The larger the power of lasers transmitted or the closer the distance between laser source and targets, the larger the power of lasers received is. In fact, since laser beams are narrower than microwave beams, the relative size of laser spots and targets can influence the signal power of laser echoes reflected from targets, which then affects the power of lasers received. Therefore, different forms of the range equation of laser detection are derived for different types of targets in the target detection system based on laser imaging. Here shows the range equations of laser detection for several typical targets.

(1) Point targets

If the energy detected by a photoelectric detector contains all the energy reflected from the illuminated spots on one target, the whole illuminated region on the target needs to be considered for calculating the range equation. Thus, the standard equation for computing the scattering cross section of targets illuminated by laser beams is supposed to be changed.

For the point target subjecting to Lambert scattering, the laser scattering cross section Γ_{PT} of this target is as follows.

$$\Gamma_{PT} = 4\rho A_s \tag{1.7}$$

where the parameters show the same meaning with those in Formula (1.6). By substituting Formula (1.7) into Formula (1.3), we can obtain the range equation of laser detection for point targets.

$$R = \left(\frac{KP_S T_A^2 \rho A_s D^2 \eta_t \eta_r}{\pi \psi^2 P_r}\right)^{\frac{1}{4}} \tag{1.8}$$

(2) Linear objects

Like a general electric wire, if a target is longer than the length of an illuminated region but its width is less than that of the illuminated region, the laser scattering cross section Γ_w of this diffused linear target is shown as Formula (1.9). Here, d and $r\theta_T$ show the wire diameter and length of the linear target separately and θ_T presents the limited diffraction angle of the transmitted laser.

$$\Gamma_w = \frac{4\pi}{\Omega} \cdot \rho \cdot d \cdot r\theta_T \tag{1.9}$$

By substituting Formula (1.9) into Formula (1.3), the range equation of laser detection for linear objects can be acquired as follows.

$$R = \left(\frac{KP_S T_A^2 D^2 \eta_t \eta_r \rho d r \theta_T}{\Omega \psi^2 P_r} \right)^{\frac{1}{4}} \tag{1.10}$$

(3) Extended targets

When imaging in a short range or the target is large, such as ground remote imaging, the target can be considered as an extended one which is related to the size of the target. All radiations nearby the spots illuminated can be reflected. At this time, all echo beams of the target are received.

The area of a target irradiated by circular spots is presented as follows.

$$A_s = \frac{\pi R^2 \theta_T^2}{4} \tag{1.11}$$

where R shows the distance between laser source and the target while θ_T presents the limited diffraction angle of lasers transmitted. For extended targets, their limited diffraction angle θ_T equals to the width θ of laser beams.

By substituting Formula (1.11) into Formula (1.6), the scattering cross area of extended Lambert targets is expressed as follows.

$$\Gamma_{Ext} = \rho \cdot \frac{\pi^2 R^2 \theta_T^2}{\Omega} \tag{1.12}$$

By substituting Formula (1.12) into Formula (1.3), the range equation of laser detection for extended targets is as follows.

$$R = \left(\frac{KP_S T_A^2 D^2 \eta_t \eta_r \rho \pi \theta_T^2}{4 \psi^2 P \Omega_r} \right)^{\frac{1}{2}} \tag{1.13}$$

It is notable that in a short range, the influence of atmosphere can be considered as that in one-way transmission.

1.3.2 Target Sampling of Laser Imaging

Laser imaging is conducted based on the laser echoes reflected from targets. Therefore, obtaining the reflected laser echoes is the key for laser imaging. However, if a laser beam is radiated to illuminate targets, we can merely obtain the target information in a certain direction from the reflected laser echoes. Thus, in order to completely and clearly image targets, targets need to be sampled according to certain rules, that is, laser scanning. Therefore, when laser imaging is performed to targets, laser beams have to be able to illuminate all positions of targets so as to obtain the laser echoes of various positions of targets. According to different modes of target sampling, laser imaging can be divided into scanning imaging and staring imaging.

1. Scanning imaging

Scanning imaging refers to scan object planes through regularly deflecting laser beams transmitted by the laser source using a beam scanning system. Meanwhile, targets are scanned synchronously by the receiving field so as to acquire the laser echo data from different positions of targets. According to different scanning systems used, scanning imaging can be separated into mechanical scanning imaging, non-mechanical scanning imaging and scanning imaging using binary optics technique.

Mechanical scanning imaging is the earliest emerged detecting method used in the laser imaging and has been most widely used. By periodically rotating reflectors using motors, the directions of radiated lasers and reflected lasers are changed accordingly so that targets are scanned with all laser echoes collected from laser points. This method is characterized by mature technology and large scanning angle; however, it also shows the disadvantages of low imaging rate and large power consumption per volume, which influence the effect of airborne and space-borne laser imaging. Typical mechanical scanning modes include scanning using oscillating mirror, rotating prism, regular polyhedron prism, etc., as shown in Fig. 1.4.

Non-mechanical scanning imaging which applies electro-optic devices or acousto-optic devices as deflectors to substitute the motors in mechanical scanning is performed by using modes including electro-optic scanning, acousto-optic scanning, etc. According to the sensitivity of devices to changed voltage or frequency, driving signals with different voltages or frequencies are loaded. Afterwards, the positions of transmitted beams are changed so that all the positions of targets can be irradiated. This method is predominately characterized by no mechanical motion, fast scanning and that beams can irradiate all points in the scanning range. However, there are also some shortcomings: the circuit is complex and costs high; in addition, scanning can merely be performed in a small angle which is generally several tenths of milliradian. Moreover, as the light transmittance is low and beams are poor, large power is consumed. As a consequence, the optical system must be cooled.

Binary optic imaging technique applies a binary optical scanner as the scanning system. The scanner is composed of a pair of positive and negative microlens arrays distributed in confocal manner. When one of the arrays is moved horizontally, the laser beams scan the objects flexibly by penetrating these arrays.

2. Staring imaging

The progress of microelectronic technology, micro-optical-electro-mechanical system (MOEMS) technology and computer image processing technology promote the development of multi-element detector arrays and flat-panel detectors. Therefore, in addition to scanning imaging, there appears staring imaging, which is

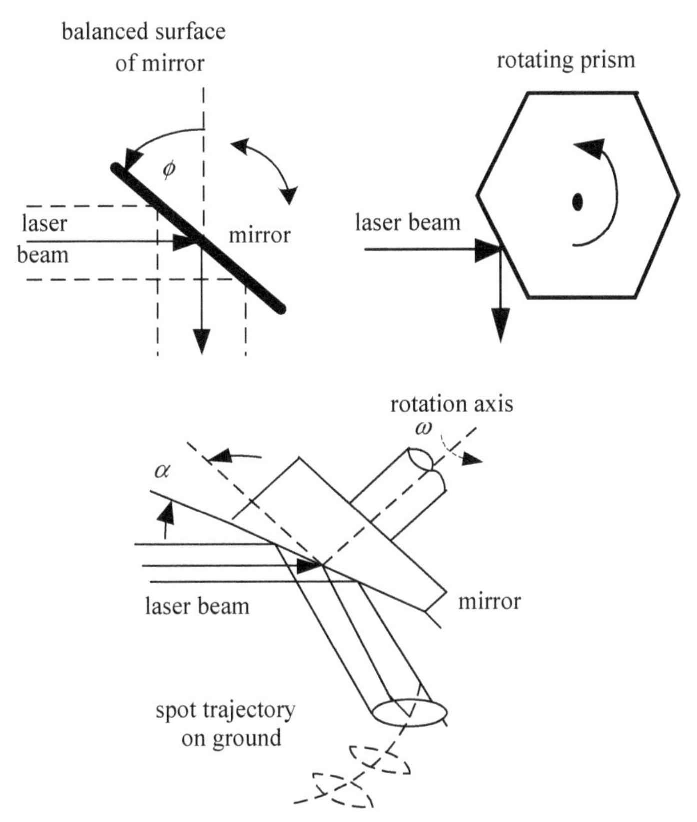

(a) scanning using oscillating mirror; (b) scanning using rotating prism;
(c) scanning using regular polyhedron prism

Fig. 1.4 Typical scanning modes of mechanical scanning imaging

predominately characterized that it can change the directions of radiated beams and receiving fields without the help of scanning mechanism. Moreover, the sensitive elements of detectors can maintain a long time, thus increasing the radiation response time of targets. In this way, targets are seemed to be stared, thus calling it as staring imaging. According to different mechanisms, staring imaging can be divided into area array staring imaging and linear array push-broom staring imaging.

(1) Area array staring imaging

Area array staring imaging is conducted by using focal plane array detectors to parallel transmit multiple laser beams to illuminate targets in beam splitting mode or floodlight targets. Each array element of the area array detectors receives the echo signals of corresponding pixel points and meanwhile obtains the distance and intensity data of all pixel points, as shown in Fig. 1.5. As area array staring imaging

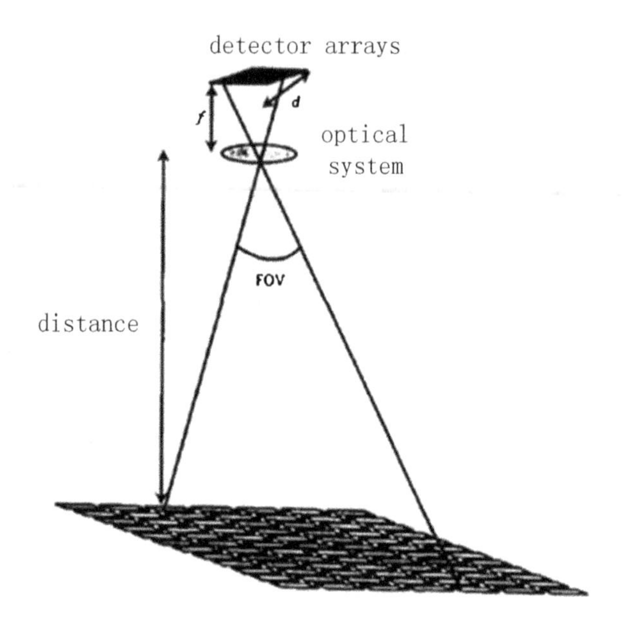

Fig. 1.5 Principle of area array staring imaging

does not need to use scanning mechanism, it is characterized by high sensitivity, fast imaging, and compact structure in small size. However, limited by the size of array elements and readout circuit, it is difficult to apply area array detectors to acquire images with high resolution or images for large fields. Moreover, limited by the production technologies of area array detectors at present, there are little array elements available and the detectors show low yield. As a result, area array detectors cost high, thus restricting the popularization and application of area array staring imaging.

(2) Linear array pushbroom staring imaging

Linear array push-broom staring imaging adopts linear array detectors instead of area array detectors. By using the laser transmitting technology utilized in the area array staring imaging, range linear array detectors can obtain the distance of pixels and echo feature data in one line or column per time; meanwhile, by moving the detection platform along the predetermined direction, laser beams irradiate the object planes along the straight line which is perpendicular to the track. In this way, targets are scanned by laser beams along a series of parallel lines, thus obtaining the distance of pixel and intensity data of echoes in each line or column. Finally, two-dimensional scanning for targets is realized and all the pixel data are acquired. This technology is suited to moving platform and large field imaging and its imaging speed is far higher than that of scanning imaging. In addition, as there are various linear array detectors with numerous array elements, we can obtain images with high resolution. The push-broom mode can be considered as extending the array detectors by another dimension and can increase the number of array elements of array detectors so that its imaging field is larger than that of area array staring imaging.

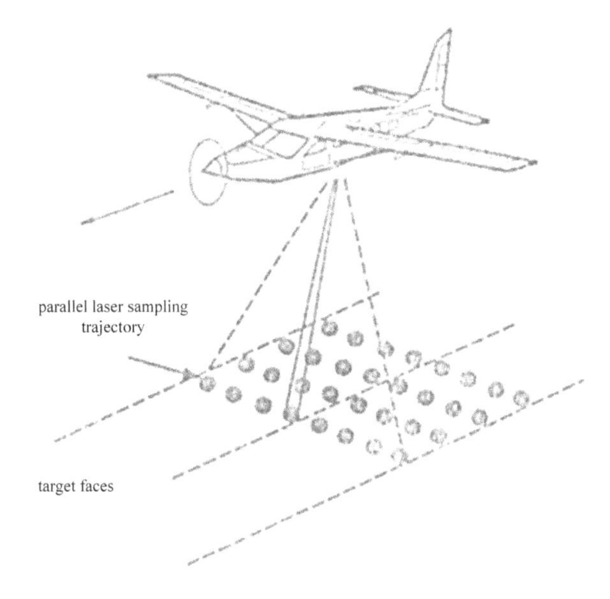

Fig. 1.6 Push-broom laser scanning imaging

Supported by National High-tech Research & Development Program (863 Program) of China, the author developed an airborne push-broom 3D imaging system. In the system, the linear array push-broom staring imaging technology based on split laser beams and linear array avalanche photodiode detectors (APD) is first adopted. The imaging mode of the system is shown in Fig. 1.6. By using multiple linear array APDs, the launcher of laser beams is carried on an airplane to one-dimensionally scan targets along the flight direction. Through splitting laser beams and array detection, parallel detection is performed to multi-point in one line along cross-track direction, thus realizing laser 3D imaging detection.

No matter area array staring imaging or linear array push-broom staring imaging, they present the following advantages: (i) imaging resolutions are determined by the pixels of detector arrays; (ii) laser beams are narrow which not only improve the spatial resolution, but also can cover large space; (iii) high repetition frequency is not needed for the lasers, which reduces the requirement for the laser devices; and (iv) images can be obtained instantly using single pulse so that no high speed scanning devices are needed, for which reason the imaging system is small and light, and therefore is suitable for airborne, missile-borne and space-borne laser imaging.

1.3.3 Information Acquisition and Processing of Target Detection Based on a Laser Imaging System

To acquire and process the target information detected using laser imaging, target echo pulses with high fidelity needed to be obtained by receiving, amplifying and digitizing the laser echoes reflected from targets. This process is realized on the basis that target sampling is finished through traversal scanning using a single laser beam, parallel irradiation or floodlighting with multi-laser beams. Based on the acquired target echo pulses with high fidelity, data including time delay and echo intensity of targets are collected by using the technology for measuring time delay and intensity with high resolution. Then, laser images of targets are generated after comprehensively processing the position and attitude data of the system. By processing the generated laser images, we can extract the image features and target features, and therefore design suitable classifiers to identify targets. The key of the information acquisition and processing of target detection based on a laser imaging system lies in the receiving of echo signals in laser detection with high sensitivity, the accurate measurement and real-time imaging for target parameters, and imaging processing and target information extraction.

1. **Receiving laser echo signals with high sensitivity**

For laser imaging detection, to break the limits of range and meteorological conditions and therefore realize imaging detection for long-distance or weak targets, it is important to receive and detect the weak signals of laser echoes. This mainly depends on the sensitivity of the laser imaging detection system. The factors influencing the sensitivity of laser imaging detection systems include the detection methods used, types of detectors, and effective optical receiving aperture. The sensitivity of detectors whose maximum value is single photon is related to the materials used and their photoelectric conversion efficiency. Therefore, sometimes people hope that each pixel of imaging devices is with single photon sensitivity [6, 7]. APDs are a kind of linear detectors with high sensitivity and bandwidth, low noise, large dynamic range, current gain and good reliability, thus being one of the most commonly used detectors for laser imaging. APDs are arranged in linear array or area array, while the pixel of the latter has reached to 128 × 128 when using a Geiger mode. Currently, the Lincoln Laboratory in the Massachusetts Institute of Technology in USA has developed a laser imaging system based on 32 × 32 APD array.

2. **Accurate parameters measurement and real-time imaging of a target**

Accurate parameters measurement and real-time imaging for a target is the key of laser imaging detection. Whether parameters including range, energy and waveform of targets can be precisely measured directly determines the resolution of images obtained and the subsequent detection results. As for the real-time imaging, its realization influences the efficiency of subsequent image processing and target detection, and is directly related to the real time and availability of laser imaging detection. There are various measurement principles and methods for precisely measuring target parameters. For instance, the delay measurement for the pulses of laser echoes contains analog-digital conversion, analog interpolation, and delay line conversion. In practical application, the measurement principles, precision, and real time of implement methods as well as the complexity of the technology are supposed to be taken into consideration. In addition, to achieve real-time imaging, methods such as non-scanning imaging and multi-element parallel detection can be utilized. After obtaining the parameter data of each point of targets, methods such as coordinate transformation, distortion correction and digital quantization are expected to be applied to convert these data into those in image formats so as to improve the accuracy of the obtained laser images.

The images obtained using laser imaging mainly contain the range images, gray images and waveform images of targets composed of time delay, energy, waveform, and distribution of laser echoes. Time delay and distribution law of echoes constitute the range images of targets. Range images show the 3D spatial relationship of targets and surrounding objects with the laser imaging detection systems. They are obtained based on the time delay information of laser echo pulses at each laser point. Time delay of echoes presents the time interval from the emission of pulse laser beams to the receiving of the laser echoes reflected from targets. The spatial distribution of targets and the spatial relationship between these targets and surrounding objects can directly influence the time delay of echoes. Therefore, range images reflect the appearance characters of targets and characteristics of the surrounding environment and directly mirror the target characteristics including structure, shape and size. By further analyzing range images, we can also obtain the deep relation between target characteristics and imaging positions, and therefore can deduce more characteristics of targets.

The combination of energy and distribution law of echoes forms the gray images of targets which are acquired by comprehensively analyzing the distribution law of the attitude and position data of the detection system based on the energy extraction of laser echoes. Gray images intuitively demonstrate the laser back-reflection in the illuminated regions of targets. For different objects, the reflectivity is different. According to this difference, we can distinguish the targets from the background by using the gray images so as to obtain target characteristics including approximate contour, shape, and spatial distribution.

The waveform of echoes and their fission and distribution law compose the waveform images of targets. Target detection based on the waveform images of targets can obtain the in-depth structure information of targets using a single pulse signal, thus being suitable for the rough detection of linear targets with structure fluctuation. In addition, this technology can be used to distinguish target attribute information including types under simple condition so as to provide reference for the threat assessment and attribute discrimination of a target. Ni-Meister et al. once established a model regarding the relation between the waveform of laser remote sensing echoes and vegetation with various types. By using this model, they deduced the information of trees including height, vertical structure and distribution of sub-crowns. Moreover, the target detection based on the waveform images of targets is also suitable for imaging targets moving at high speed. This is because that, on the one hand, we cannot obtain the 3D imaging data of moving targets with high speed through multiple scans, but can merely find and identify moving targets according to the waveform information of echoes; on the other hand, though Doppler profiles can be obtained for moving targets, they merely reflect the motion characteristics of targets but cannot attribute the information of targets, such as their types. While waveform images can satisfy the requirement.

3. **Imaging processing and target information extraction**

Imaging processing and information extraction of targets is the core of laser imaging detection. Imaging processing generally includes laser imaging pre-processing (e.g. data smoothing, contour reinforcement, and geometric correction) and image segmentation. For laser imaging detection, imaging processing is supposed to be real-time and realizable. At present, the image processing algorithm based on digital morphology and wavelet transform is the research focus.

Target information extraction includes target feature extraction and target classification identification. The key of the former lies in extracting stable and typical features of targets, including geometrical characteristics, shape features, moment features, etc. [8]. Thereinto, geometrical characteristics contain length, width, height, perimeter, area, etc. of targets; shape features include factors describing rectangularity, circularity, shape, etc.; while central moment, Hu invariant moment, line moment, Zernike moment, and spectral moment are contained in moment features. For the waveform images of targets, typical characteristics include broadening, energy, distribution and statistics characteristics of the waveform of echo pulses. Laser imaging detection finally aims to classify and identify targets. To realize the classification and identification of targets with high reliability, the theories and methods of modern pattern recognition need to be used. The key of target classification and identification lies in the design of classifiers with both favorable efficiency and high precision in classification and identification. At present, the commonly used classification and identification methods include those based on neutral network, support vector machine, models and knowledge database.

References

1. Li L (2008) Reconnaissance and surveillance: useful tool in space operation. National Defense Industry Press, Beijing
2. The 210 Institute of China North Industries Group Corporation (2008) Development of laser guidance technology. Beijing
3. Space Countermeasure (2005) American air force file 2-2.1. Military Science Publishing House, Beijing
4. Reconnaissance and surveillance with high technology (2002) Chinese People's Liberation Army Publishing House, Beijing
5. Tan X (2001) Research on ranging equation for laser radar. Electronics Optics & Control 1:12–18
6. Ni-Meister W, Jupp DLB, Dubayah R (2001) Modeling Lidar waveforms in heterogeneous and discrete canopies. IEEE Trans Geosci Remote Sens 39(9)
7. Wei G (2007) G-APD arrays—a three-dimensional imaging detector with single photon sensitivity. Laser Technol 31(5):452–455
8. Chen X, Ma J, Zhao H et al (2009) Survey of automatic target recognition technology of imaging laser radar. Mod Defense Technol 37(6):114–117

9. Krawczyr R, Goretta O, Kassighian A (1993) Temporal pulse spreading of a return lidar signal. Appl Opt 32(33):6784–6788
10. Zhao N, Hu Y (2006) Relation of laser remote imaging signal and object detail identities. Infrared Laser Eng 35(2):226–229
11. Hu Y, Shu R (2008) Airborne and space-borne laser sounding technology and application. Infrared Laser Eng 38(Supplement)
12. Min H, Hu Y, Nanxiang Z et al (2008) Application of airborne three-dimensional laser imaging. Laser Optoelectron Prog 45(3):43–49
13. Hu Y (2003) The Laser positioning technology and its precision in remote sensing imaging. The international geoscience and remote sensing symposium (IGARSS03), Toulouse, France, 2003
14. Tao X, Hu Y, Huang Y (2005) Passive location by optical imaging and error analysis. Optoeletron Technol Inform 18(1):57–60
15. Lai X (2010) Principe and application of airborne laser radar. Publishing House of Electronics Industry, Beijing

Chapter 2
The Atmospheric and Target Modulations of the Lasers

During the receiving and emitting of lasers in laser imaging, the interactions among atmosphere, targets, and the lasers, namely, the atmospheric and target modulations of the lasers, directly influence the waveform characteristics of the targeted laser echoes. As a result, various subsequent processes such as information and data acquisition, data processing, image generation, feature extraction, classification and recognition are directly or indirectly influenced. By taking the mechanism of the target detection based on a laser imaging system as the object, this chapter explores the atmospheric and target modulations of lasers. This chapter mainly studies the concepts and characteristics of the atmospheric and target modulations of lasers, the characteristics of the energy and signal modulations, and the influences of the atmospheric and target modulations.

2.1 The Concepts of the Atmospheric and Target Modulations of Lasers

The link of the target detection based on a laser imaging system is usually composed of a system platform, a laser source, the atmosphere, a laser imaging sensor, a GPS, an attitude measurement unit, a data acquisition and processing system, an image generation system, a feature extraction system, a classification and recognition system, a detection and location system, etc. The block diagram of the signal and data link for the target detection based on a laser imaging system constituted by the aforementioned parts is demonstrated in Fig. 2.1.

As demonstrated in the link shown in Fig. 2.1, the laser signals radiated by the laser source are expected to experience complicated interaction with the matters in the atmosphere while passing through the atmosphere. Meanwhile, the incident lasers on the target are expected to interact with the target, while the reflected lasers cannot reach the laser imaging sensor unless passing through the atmosphere.

Y. Hu, *Theory and Technology of Laser Imaging Based Target Detection*,
DOI 10.1007/978-981-10-3497-8_2

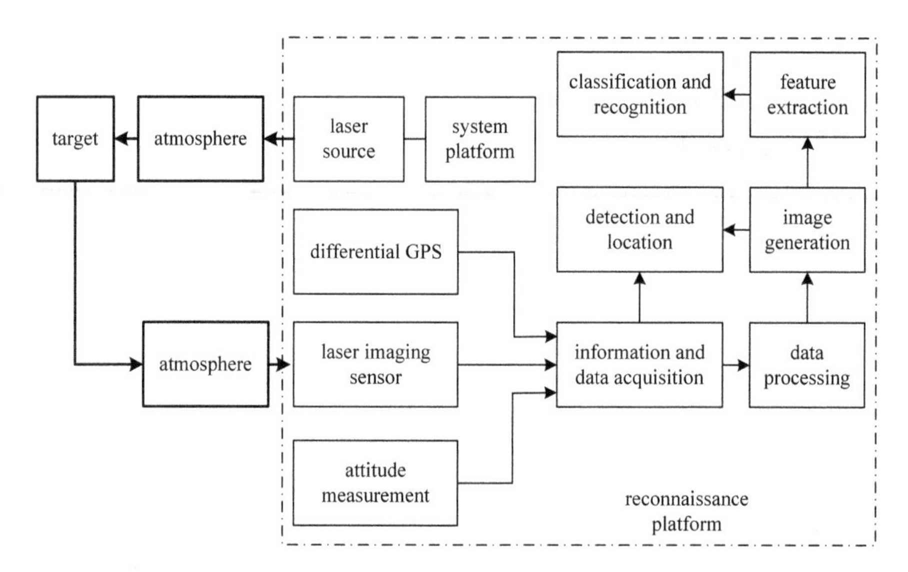

Fig. 2.1 The signal and data link for the target detection based on a laser imaging system

Therefore, while carrying out the target detection using laser imaging, laser signals require to contact and interact with the atmosphere and target respectively, so that they can be modulated.

The atmosphere is comprised of various gases and suspended impurity particles of various sizes, which show different components and chemical properties. When lasers propagate from the laser source to the targets, they are expected to interact with the gas molecules and particles in the atmosphere, which gives rise to the absorption, scattering, reflection, diffusion, refraction and diffraction phenomena. As to the main influences:

(i) the selective absorption of several gas molecules in the atmosphere for the lasers leads to the attenuation of the radiation intensity and the variation of the propagation direction of the lasers;
(ii) the scattering of the lasers by the suspended particles in the atmosphere results in the attenuated intensity and changed propagation direction of the lasers;
(iii) the violent change of the physical properties of the atmosphere leads to the flickers of the lasers and thus results in the modulation and variation of the lasers illuminance;
(iv) the variation of the physical properties of the atmospheric molecules and suspended particles results in the variation of the laser radiation;
(v) atmospheric turbulence gives rise to the random variation of the optical refractive index, resulting in the wave-front distortion, and thus changes the intensity and coherence of the lasers. All these indicate that the existence of the atmosphere changes the intensity, phase, polarization, and propagation

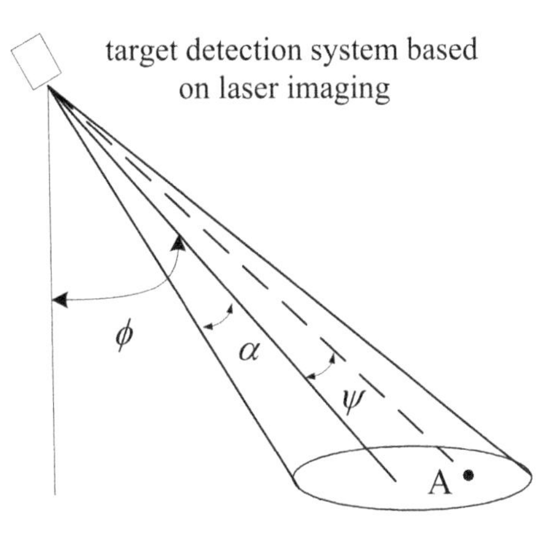

Fig. 2.2 The relationship between the laser spot and the target

direction of the laser signals, which is generally represented by the attenuation of the laser signals. Besides, the attenuation of the laser signals changes with the atmospheric conditions.

When it comes to the target modulation of lasers, the relationship between the laser echo signal reflected from any a point on the target and the target can be obtained according to that between the target and the incident beams shown in Fig. 2.2 [1]:

$$f_i(t) = a_i \exp[-k(t_i - t)^2/\tau^2] \cdot g(\psi) \cdot \cos(\zeta) \cdot h/s_i^2 \tag{2.1}$$

where, a_i represents the surface reflectivity, characterizing the attribute of the targets, while t_i indicates the echo time-delay of the corresponding target sampling point, representing the spatial distribution of the targets. τ denotes the width of the laser pulses and $g(\psi)$ is the energy distribution function of the laser spots, which is considered to be complied with the Gaussian distribution. ζ signifies the angle between the incident angle of the lasers and the normal line on the surface of the targets. h is a coefficient related to the response of the detector, effective receiving area and atmospheric attenuation. s_i represents the distance between the target sampling point and the target detection system based on laser imaging.

As shown in Fig. 2.2, ϕ indicates the angle between the central axis of the laser beams and the nadir, while α represents the half divergence angle of the lasers. ψ signifies the angle between any a beam in the lasers and the central axis.

The echo signal of the laser reflected by the target is the composition of the echo signals from various points on the target, and can be expressed as:

$$p(t) = \int_{\psi 1}^{\psi 2} f_i(t) d\psi \tag{2.2}$$

It can be seen from formulas (2.1) and (2.2) that echo signals of the laser are modulated by the various sampling points on the target form both the time and the space. The waveform characteristic of the target echo is a combined modulation function related to the time (echo time-delay), space (the position in light spots) and attributes (back-reflection and structural fluctuation) of the various parts in the targets [2]. Meanwhile, different target types lead to the generation of the target echoes with different waveform characteristics such as width, amplitude, phase and so on.

When it comes to the lexical meaning, modulation refers to changing several parameters (such as amplitude, frequency, etc.) of a signal (such as light, electrical oscillation and so on) according to the variation of the other signal (modulation, acoustic and TV signals). The interactions of the atmosphere, target and the lasers lead to the direction, energy, shape, amplitude and width of the reflected laser signals different from those of the incident pulsed lasers, which coincides with the implicit meaning of the modulation. Therefore, it is considered that the target echo is generated by the atmospheric and target modulations of the pulsed lasers, namely, the product of the atmospheric and target modulations of the lasers. Hence, the atmospheric and target modulations of the lasers refer to the phenomenon that the interactions of the lasers with the atmosphere and the target lead to the variation of the parameters of the laser signals. Meanwhile, the parameters of the laser signals are changed with the atmospheric components and the characteristics of the target.

The atmospheric modulation of the lasers is represented by the absorption, scattering and turbulence phenomena, all of which attenuate the signal intensity. In contrast, the target modulation of the lasers is slightly complicated and represented by the reflection and scattering phenomena. However, target modulation gives rise to the variations of several or all of the parameters including the intensity of the echo signals, and the width, shape and number of the pulses. Besides, the variations of these parameters are inevitably related to the characteristics of the target, which provides conditions for obtaining the characteristics of the targets based on the echo signals of the laser.

2.2 The Atmospheric Modulation of the Lasers

2.2.1 The Absorption Characteristics of the Atmosphere for the Lasers

The atmospheric absorption of the lasers refers to the fact that the gases such as H_2O, CO_2, O_3, N_2O, CO and CH_4 in atmosphere experience energy state transition under the influence of the lasers. In this way, light energy is converted into other forms such as power and heat energy or distributed in other types of spectra, which thus results in the energy attenuation of the light beams. The energy attenuation caused by the absorption of lasers by the atmosphere can be described by the

Lambert-Bougner law. Assume that the intensity of the laser (with the intensity being $I(v)$) is $I(v) + dI(v)$ after passing through the atmosphere layer that is dh thick, we obtain:

$$dI(v) = -I(v)\beta_{ab}(v, h)dh \tag{2.3}$$

where, $\beta_{ab}(v, h)$ is called the absorption coefficient. By solving formula (2.3), we obtain the following formula:

$$I(v) = I_0(v)\exp[-\int_{h_0}^{h_1} \beta_{ab}(v, h)dh] \tag{2.4}$$

If the atmosphere between h_0 and h_1 is homogeneous, and $\beta_{ab}(v, h)$ is unrelated with the thickness h of the atmosphere layer, formula (2.4) can be simplified as:

$$I(v) = I_0(v)\exp[-\beta_{ab}(v)L] \tag{2.5}$$

where, L represents the distance between h_0 and h_1.

By transforming formula (2.5), the parameter τ that can quantitatively describe the attenuation degree caused by the atmospheric absorption is obtained:

$$\tau = \frac{I(v)}{I_0(v)} = \exp[-\beta_{ab}(v)L] \tag{2.6}$$

where, the parameter τ is called the transmittance, namely, the ratio of the intensity of the laser after propagating for the distance of L to that of the output laser. The dimension of the absorption coefficient β_{ab} is L^{-1} with the commonly used unit being km^{-1}. For the convenience of calculation, the two sides of formula (2.6) are taken the logarithm and then enlarged for 10 times. Meanwhile, decibel dB is utilized to represent I/I_0, and the unit of β_{ab} is changed as $dB\ km^{-1}$.

The absorption of light energy by these gases makes the absorption spectrum of the atmosphere become several separated absorption bands, which are composed of considerable amounts of the spectral lines with different overlapping degrees and intensities. The overlapping degree of these spectral lines is related to the half width of the position of the spectra lines and the type of the absorption molecules (Fig. 2.3). Table 2.1 illustrates the wavelengths of the absorption line and band of the gas molecules.

Theoretically speaking, the absorption coefficient can basically be obtained by determining the intensity and position of the absorption line or band of the molecules. Then, the attenuation of the light intensity induced by the atmospheric absorption can be calculated according to the Lambert-Bougner law. However, as a matter of fact, absorption coefficient is not only related to the type of the gas

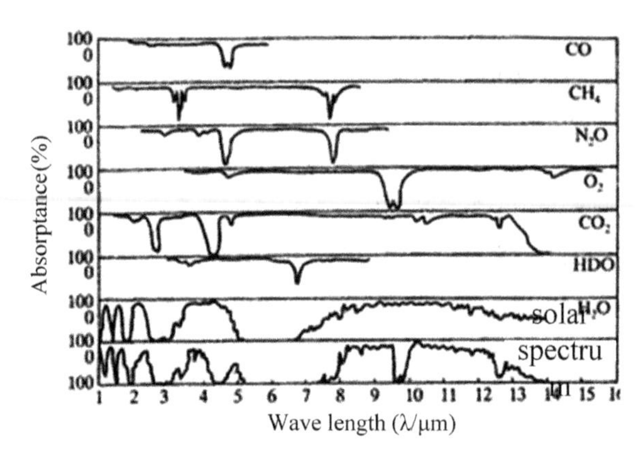

Fig. 2.3 The absorption spectra of the various absorption components in the atmosphere

Table 2.1 The absorption lines of the main gas molecules absorbing the light energy (μm)

Molecule	Type	Ultraviolet region	Visible region	Infrared region	
				Strong absorption line	Weak absorption line
CO	Linear	–	–	4.67	1.19, 1.57, 2.34
CH_4	Symmetric gyroscope	–	–	3.31, 3.8, 7.6	6.5
N_2O	Linear	–	–	4.5	3.9, 4.05, 7.7, 8.6, 17.1
O_2	Linear	0.175–0.2026 0.242–0.26	0.63, 0.69, 0.76		1.0674, 1.2683, 3.23, 10.33
CO_2	Linear	–	–	2.7, 4.3, 14.7	0.4–1.24, 1.4, 1.6, 2.0, 4.8, 5.2, 9.4, 10.4, 11.4–20
H_2O	Asymmetric gyroscope	–	0.72	1.87, 2.70, 2.67	–
HDO	–	–	–	3.7	–

molecules, but also the following factors: (i) absorption coefficient is changed with the wavelength; (ii) temperature and pressure can influence the shape of the absorption lines, and then affects the absorption coefficient; (iii) absorption coefficient is a function of the concentration of the absorption molecules; and (iv) the concentration, temperature and total pressure of the absorption molecules change with the geographical position, altitude, season and weather. The existence of the aforementioned factors makes it difficult to determine the absorption coefficient. In order to satisfy the requirements of the engineering practice, the following methods including the absorption coefficient method (line-by-line method), band model, LOWTRAN based semi-empirical method are primarily adopted to determine the atmospheric transmittance [3].

2.2.2 The Characteristics of Atmospheric Scattering for the Lasers

When lasers propagate in the atmosphere, apart from the energy attenuation caused by the selective absorption of the atmospheric molecules, there also exists the energy attenuation induced by atmospheric scattering. The so called atmospheric scattering refers to the fact that the atmospheric molecules and aerosol particles are polarized under the irradiation of the lasers, and thus produce the vibrating electromagnetic multipoles. Afterwards, electromagnetic vibration is generated by the vibrating multipoles, and irradiates secondary electromagnetic waves to various directions to form the light scattering. The energy distribution of the scattered light is related to the wavelength and intensity of the incident light, and the size and shape of the particles, as well as the refractive index. Pure scattering not consumes the total energy of the laser beams, but is expected to change the original spatial distribution of the energy of the laser beams. Therefore, the scattering process will attenuate the energy of the laser beams in the original propagation direction.

After the one-wavelength laser with the wavelength and power being λ and $P_\lambda(0)$ respectively propagates in the inhomogeneous media for x, the power of the laser beams is attenuated as $P_\lambda(x)$ after the scattering process. Then, the relationship between the distance x and the power $P_\lambda(x)$ is expressed as follows:

$$P_\lambda(x) = P_\lambda(0)\exp[-\gamma(\lambda)x] \tag{2.7}$$

where, $\gamma(\lambda)$ represents the scattering coefficient and is the function of the wavelength λ of the incident light.

A parameter $\tau_S(x, \lambda)$ that can characterize the attenuation capacity of the atmospheric scattering is obtained by transforming formula (2.7). This parameter is the function of the wavelength λ and distance x, and is defined as the atmospheric transmittance:

$$\tau_S(x, \lambda) = \frac{P_\lambda(x)}{P_\lambda(0)} = \exp[-\gamma(\lambda)x] \tag{2.8}$$

In general, the atmospheric scattering of the lasers is composed of two parts, namely, the scattering of the atmospheric molecules and suspended particles (aerosols in general) in the atmosphere. Therefore, scattering coefficient can be divided into two coefficients, namely:

$$\gamma(\lambda) = \gamma_m(\lambda) + \gamma_a(\lambda) \tag{2.9}$$

where, $\gamma_m(\lambda)$ and $\gamma_a(\lambda)$ represent the scattering coefficients of the atmospheric molecules and aerosols respectively.

Table 2.2 The size and concentration of the various particles in the atmosphere

Particle type	Radius (μm)	Concentration (cm^{-3})
Air molecules	10^{-4}	10^{19}
Aitken nucleus	10^{-3}–10^{-2}	10^{2}–10^{4}
Haze particles	10^{-2}–100	10^{1}–10^{3}
Fog droplet	10^{0}–10^{1}	10^{1}–10^{2}
Cloud droplet	10^{0}–10^{2}	10^{1}–10^{2}
Raindrop	10^{2}–10^{3}	10^{-5}–10^{-3}
Ice crystals and snowflakes	10^{3}–10^{4}	–
Dust	10^{2}–10^{3}	10^{0}–10^{1}

It can be seen from formula (2.9) that if the scattering coefficient $\gamma(\lambda)$ is known, the atmospheric transmittance $\tau_S(x, \lambda)$ of the laser after propagating for the distance of x can be obtained according to formula (2.8).

The sizes of the particles in atmosphere vary greatly, as shown in Table 2.2. The particles with different sizes are endowed with different scattering properties. The scattering properties of the particles are basically determined by the ratio of the particle size to the incident wavelength. To be specific, different particle sizes indicate different ratios of the particle size to the incident wavelength, which call for different theoretical models. When the particle size is much smaller than the wavelength, the simple Rayleigh scattering model is applicable here. In case the particle size is nearly equal to the wavelength, it is appropriate to adopt the complicated Mie scattering model. When the particle size is much greater than the wavelength, the scattering is considered to be non-selective.

1. **Rayleigh scattering**

Rayleigh scattering occurs when the wavelength of the lasers is much greater than the radius of the particles, and mainly takes place in gas molecules. Therefore, Rayleigh scattering is also called the molecular scattering. The scattering cross section of the lasers is utilized to describe the scattering ability of the target to the incident lasers. Similarly, the scattering cross section σ_m can be used to represent the scattering ability of the gas molecules. According to this parameter, it is defined that the energy of the incident lasers that pass through this cross section σ_m is equal to the total power scattered in various directions of this molecule, namely:

$$\sigma_m E_0 = I \tag{2.10}$$

where, E_0 represents the irradiance of the incident lasers on the cross section, and I indicates the scattering intensity of the molecules.

Another expression of the scattering cross section can be obtained by transforming formula (2.10), namely:

$$\sigma_m = \frac{E_0}{I} \tag{2.11}$$

That is to say, the scattering cross section σ_m is the ratio of the scattering intensity to the irradiance of the incident lasers of the molecule. It is worth noting is that the scattering cross section is different from the geometrical cross section in terms of the meaning and value.

According to the Rayleigh molecular scattering formula, the total scattering cross section of a single molecule is demonstrated as [4]:

$$\sigma_m = \frac{8\pi^3 (n^2 - 1)^2}{3N^2 \lambda^4} \tag{2.12}$$

where, n represents the refractive index of air, and N indicates the number of the molecules per unit volume. λ signifies the wavelength of the incident light.

It can be seen from formula (2.12) that the scattering cross section of the molecules is inversely proportional to the fourth power of the wavelength. Therefore, it is obtained that molecular scattering has a particularly significant influence on the light wavebands with small wavelength in Rayleigh scattering. For example, if $\lambda = 0.55$ μm, then $\sigma_m \approx 4.6 \times 10^{-27}$ cm^2; while when $\lambda = 5.5$ μm, we obtain $\sigma_m \approx 4.6 \times 10^{-31}$ cm^2 under the standard state.

When the scattering process happens on the small solid particles, if the radius of the particles satisfies the condition of $2\pi\lambda/r \leq 1$, it is still appropriate to adopt the Rayleigh scattering model. Accordingly, the scattering cross section is changed as:

$$\sigma_p = \frac{128\pi^3 r^6}{3\lambda^4} \left(\frac{n^2 - 1}{n^2 + 2}\right)^2 \tag{2.13}$$

where, n and r represent the refractive index and the radius of the particles separately.

According to formula (2.13), the scattering cross section of the condensation nucleus with the wavelength, radius and refractive index being 1 μm, 0.032 μm and 1.60 respectively is 1.17×10^{-14} cm^2. It can be seen from the calculation result of the scattering cross section of single a molecule or particle that this value is very small. However, if the scattering cross sections of all the atmospheric molecules or particles per unit volume are added together, the obtained scattering cross section will be great in value. The volume scattering coefficient β_m is the ratio of the total energy scattered in various directions of the gas molecules per unit volume to the irradiance of the incident light. It represents the sum of the scattering cross sections of each molecule per unit volume, namely $\beta_m = N\sigma_m$:

$$\beta_m = N\sigma_m = \frac{8\pi^3 (n^2 - 1)^2}{3N\lambda^4} \tag{2.14}$$

where, N indicates the number of the gas molecules per unit volume, with the dimension being L^{-3}. Therefore, the dimension of β_m is L^{-1}. If there are N particles of same size per unit volume, similarly, the volume scattering coefficient β_p of the particles can be obtained:

$$\beta_p = N\sigma_p = \frac{128\pi^3 r^6 N}{3\lambda^4}\left(\frac{n^2 - 1}{n^2 + 2}\right)^2 \tag{2.15}$$

Unlike gas molecules, particles generally are not consistent. As a result, it is difficult to calculate the volume scattering coefficient of the particles using the linear superposition. Instead, the distribution function requires to be utilized to solve the integration.

2. **Mie scattering and Non-selective scattering**

Mie scattering occurs when the particle size is nearly equal to the wavelength of the laser beams, and primarily happens on the fallout and aerosol particles. The refractive indexes of these particles are significantly different from that of the gas molecules. Compared with the Rayleigh scattering, Mie scattering is endowed with the following characteristics:

(i) The distribution of the intensity of the scattered light becomes increasingly complicated with the angle: the greater the particles compared with the wavelength, the more complicated the distribution is;
(ii) With the increase of the particle size, the ratio of the forward scattering to the backscattering is increased, leading to the increased lobe of the forward scattering;
(iii) When the ratio of the particle diameter to the wavelength increases, the scatting process presents increasingly insignificant dependence on the wavelength.

The particles where Mie scattering occur are composed of considerable amounts of closely arranged molecules, which form a multipole subarray. The vibrating multipoles are expected to be produced by the multipole subarray under the motivation of the incident lasers, and then irradiate outwardly the secondary electromagnetic waves, namely, the partial waves. Later, the partial waves are overlapped in the far field to form the scattered waves.

The scattering interface σ_p of the particles with Mie scattering is endowed with the same definition with that of the particles with Rayleigh scattering. That is to say, the power of the incident waves on σ_p is equal to the total power scattered in various directions of this particle. The Mie scattering cross section of single a particle can be obtained using the distribution function of the intensity of the Mie scattering [5]:

$$\sigma_p = \frac{\lambda^2}{2\pi}\sum_{n=1}^{\infty}(2n+1)(|a_n|^2 + |b_n|^2) \tag{2.16}$$

The scattered waves generated in the Mie scattering are composed of multiple partial waves. Therefore, a_n and b_n in formula (2.16) represent the complex amplitudes of the electric wave and magnetic wave of the nth partial wave separately.

The scattering efficiency factor Q_{sc} can be obtained by dividing the two sides of formula (2.16) using the geometric cross section πr^2, namely:

$$Q_{sc} = \frac{2}{v^2}\sum_{n=1}^{\infty}(2n+1)(|a_n|^2 + |b_n|^2) \tag{2.17}$$

where, v is called the scale parameter, and $v = 2\pi r/\lambda$. The scattering efficiency factor Q_{sc} reflects the ability of the geometric cross section per unit in scattering light-wave. Figure 2.4 illustrates the curve describing the variation of the scattering efficiency factor Q_{sc} of the spherical water drops with the scale parameter when the refractive index is $n = 1.33$. It can be discovered from the figure that when the value of the scale parameter is small, the value of the scattering efficiency factor Q_{sc} is far smaller than 1. That is to say, the scattered energy is far smaller than the incident energy of the particles. With the increase of the scale parameter v, the value of the scattering efficiency factor increases rapidly with the maximum value being close to 4, accompanying with the appearance of the damped oscillation. Finally, the value is converged to 2.

Since the scattering process is incoherent, the luminous flux of the volume scattering is equal to the sum of the luminous fluxes of each particle in the scatterer. According to the definition of the scattering interface, volume scattering interface is equal to the sum of the scattering cross sections of each particle. Assume that there are N particles per unit volume, the volume scattering coefficient is expressed as:

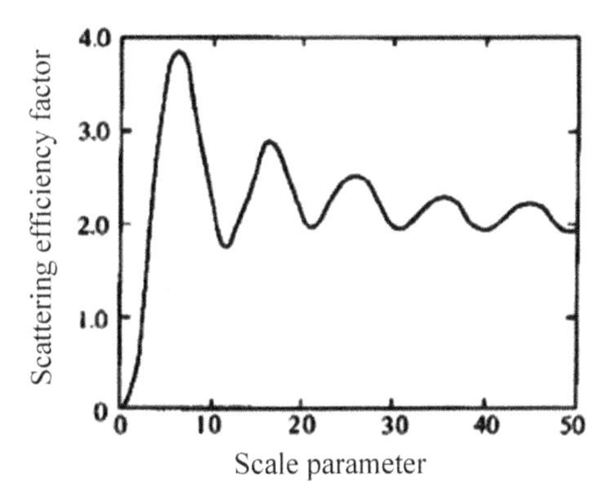

Fig. 2.4 The variation curve of the scattering efficiency factor of the spherical water drops with the scale parameter [6]

$$\beta_p = N\sigma_p = \frac{N\lambda^2}{2\pi}\sum_{n=1}^{\infty}(2n+1)(|a_n|^2 + |b_n|^2) \tag{2.18}$$

When the particle size is far greater than the wavelength of the lasers, non-selective scattering is expected to take place. Under such circumstance, the scattering intensity has nothing to do with the wavelength, and lasers with different wavelengths show the same scattering intensity.

3. **Multiple scattering**

The Rayleigh scattering, Mie scattering, and non-selective scattering are all analyzed based on the following conditions: the scattering process of each particle is independent of other particles, and particles merely scatter the direct light rays such as the incident light of the laser source or the lasers reflected by the target. This is the so called single scattering. As a matter of fact, a scatterer usually contains multiple scattering particles. As a result, apart from the direct lights, each scattering particle is likely to be exposed to the scattered lights of other particles, and therefore scatters these lights. Therefore, before being emitted from the scatterer, several lights that have been scattered once are likely to be scattered once again or for multiple times, which is the so called multiple scattering. Although multiple scattering has a little influence on the total light energy eliminated from the direct lights, it greatly changes the patterns of the compound distribution of the scattering intensities for all particles [5].

2.2.3 *The Atmospheric Turbulence of Lasers*

Atmospheric turbulence is an irregular motion of each atmospheric molecule group and is different from the mean motion of the overall atmosphere. The micro and random variation of the atmospheric temperature caused by human activities and solar irradiation is expected to randomly change the wind speed of the atmosphere, and thus forms the atmospheric turbulence. Meanwhile, the random variation of the atmospheric temperature is supposed to lead to the random variation of the atmospheric density and then that of the atmospheric refractive index. Furthermore, the cumulative effect of the random variation of the refractive index gives rise to the obvious inhomogeneity of the profile of the atmospheric refractive index. As a result, the wave-front of the light beams turbulently propagating in the atmosphere randomly fluctuates. In this way, a series of atmospheric turbulence effects including intensity scintillation, beam extension, phase fluctuation, jitter of source images and beam drift generated during the propagation processes of the lasers are triggered.

The degree and form of the influence of the atmospheric turbulence on the beam characteristics are related to the relative sizes of the diameter d of the beams and the turbulent scale l [6]. In case $d \gg 1$, that is, the diameter of the laser beams is far greater than the turbulent scale, the section of the laser beams is expected to contain numerous vortexes, each of which diffracts the small part of the light beams irradiated on them. As a result, the intensity and phase of the light beams present random variation in terms of the time and space. Meanwhile, the area of the light beams expands increasingly, which results in the generation of the effects such as intensity scintillation, beam extension and phase fluctuation. When $d \cong l$, turbulence gives rise to the random deflection of the section of the light beams, leading to the fluctuated angle of arrival. As a result, the source image is supposed to dither on the sensitive surface of the detector. While, as $d \ll 1$, that is, the diameter of the laser beams is far smaller than the turbulent scale, light beams are randomly deviated under the influence of the turbulence, causing the generation of the beam drift. What is worth noting is that turbulence induced effects are mutually related, and the turbulent scale is within certain a range. The vortexes of different sizes play their own roles respectively.

1. **Intensity scintillation**

Intensity scintillation is the most common and obvious effect caused by the atmospheric turbulence. Intensity scintillation refers to the phenomenon that lasers experience intensity fluctuation while propagating in the atmosphere. It is a kind of power modulation of the received signals induced by the atmospheric turbulence. The fundamental reason leading to the intensity scintillation is the inhomogeneous optical properties in certain an area of the atmosphere nearby the ground. When laser beams propagate the areas with varying temperature, the density and refractive index of the air are changed, which makes the laser beams deviate from the original propagation direction. As a result, the section of the light beams is varied, accompanying with the variation of the angular distribution of the laser energy within the section of the light beams. Since the inhomogeneity of the optical properties of the atmosphere is unstable and varies randomly, the influence of the intensity scintillation is therefore unstable and random. The main influence of the intensity scintillation on the receiving part of the target detection system based on laser imaging is the scintillation modulation. That is to say, intensity scintillation leads to the random variation of the energy of the received lasers reflected by the targets in a certain form, and the frequency generally varies in the range of 2.5–450 Hz.

Since there exists scintillation in the intensity of the received lasers reflected by the targets, it is necessary to measure this scintillation using a parameter. Therefore, the variance $\sigma_{\ln I}^2$ of the intensity logarithm $\ln I$ is commonly utilized to quantitatively describe the scintillation degree of the intensity. Both the measured data and theoretical studies demonstrate that in weak fluctuation regions, the variance of the intensity logarithm satisfies the following condition [4]:

$$\sigma_{\ln I}^2 = C_o C_n^2 k_o^{7/6} z^{11/6} \tag{2.19}$$

where, C_o is a constant, and is taken as 0.496 and 1.23 respectively for the spherical and plane waves. C_n represents the structure constant of the atmospheric refractive index ($m^{-2/3}$), while k_o indicates the wavenumber and $k_o = 2\pi/\lambda$. z signifies the propagation distance of lasers (m).

It can be seen from formula (2.19) that the variance $\sigma_{\ln I}^2$ of the intensity fluctuation is in direct proportion to the structure parameter C_n^2 of the atmospheric refractive index and $k_o^{7/6}$. C_n^2 reflects the degree of the turbulence, which indicates that the intensity fluctuation is directly proportional to the turbulence intensity. The stronger the turbulence is, the more obvious the fluctuation of the light intensity in the receiving part. k_o is inversely proportional to the wavelength: the shorter the wavelength, the more obvious the intensity fluctuation. Besides, the variance of the intensity fluctuation is in direct proportion to the 11/6 power of the propagation distance z, which demonstrates that the farther the propagation distance, the more obvious the intensity fluctuation is.

While as the light beams undergo the slope-way transmission from the ground to the air or from the air to the ground, the aforementioned calculation formula of the variance of the intensity logarithm can be rewritten as [4]:

$$\sigma_{\ln I}^2 = C_o k_o^{7/6} \int_{h_o}^{h_T} C_n^2(h)(\sec \acute{\omega})^{11/6} h^{5/6} dh \tag{2.20}$$

where, $\acute{\omega}$ represents the zenith angle in the transmission path of the light beams, while h_o indicates the launching height of the light source, and refers to the target height. h_T denotes the height of the receiving part in the system.

In the transmission from the surface to the air with $\psi < 60°$, the typical value of $\sigma_{\ln I}^2$ is 20%, which can be reduced to 4% in the case of favorable visibility. The upper limit of $\sigma_{\ln I}^2$ can be estimated using the following formula:

$$\sigma_{\ln I}^2 < 0.436 \left(\frac{\lambda}{2.1 \rho_0 \vartheta_0} \right)^{5/6} \tag{2.21}$$

where, $\vartheta_0 = 0.508 \lambda^{6/5} (\cos \acute{\omega})^{8/5} \left[\int_{h_o}^{h_T} C_n^2(h) h^{5/3} dh \right]^{-3/5}$ represents the isoplanatic angle, with the typical value being 10 μrad in the visible light bands and usually varying from 3 to 10 μrad. $\rho_0 \sim \left(C_n^2 k_o^2 R \right)^{-3/5}$ indicates the coherent length of the turbulence, and is related not only to the turbulence intensity, but also to the wavelength and propagation distance. Its typical value is 10 cm in the visible light bands, and usually in the range of 5–20 cm.

According to the test on the theoretical formula (2.20), when $\sigma_{\ln I}^2$ is small, it is in direct proportion to C_n^2. However, as C_n^2 increases to some extent, $\sigma_{\ln I}^2$ basically maintains certain a fixed value and even reduces sometimes, which is called the saturated scintillation. Due to the existence of the saturated scintillation, the result of formula (2.20) is not coincident with the observed intensity fluctuation. Therefore, another expression is required to be utilized to describe the intensity fluctuation. The intensity variance $\sigma_I^2 = (\overline{I^2} - \bar{I}^2)/\overline{I^2}$ of the plane waves is expressed as [6]:

$$\sigma_I^2 = \begin{cases} 1 + \frac{2.03}{\left(\sigma_o^2\right)^{1/6}} & \sigma_o^2 = \left(\frac{2\pi}{l_o}\right)^{7/3} \quad C_n^2 z^3 \to \infty \\ 1 + \frac{0.86}{\left(\sigma_0^2\right)^{2/5}} & \varsigma = \frac{\lambda z}{l_0^2} \to \infty \end{cases} \tag{2.22}$$

The intensity variance $\sigma_I^2 = (\overline{I^2} - \bar{I}^2)/\overline{I^2}$ of the spherical waves is demonstrated as [6]:

$$\sigma_I^2 = 1 + \frac{0.86}{\left(\sigma_0^2\right)^{2/5}} \tag{2.23}$$

2. **Beam expansions**

Beam expansions refer to the fact that the spot area on the receiving system is larger than that generated after lasers propagating in the free space. The effective diameter of the expanded light spots is expressed as:

$$d_c = d_o \left(\frac{I_o}{\bar{I}_r}\right)^{1/2} \tag{2.24}$$

where, d_o represents the diffraction limited diameter of the light spots, and I_o indicates the limited intensity of the emitted light beams. $\bar{I}_r$ denotes the received average intensity.

The root mean square is utilized to represent the size of the beam extension. Then, the root mean square of the expanded light beams induced by the atmospheric turbulence is expressed as [7]:

$$\sigma_0 = 1.71 C_n^2 R^{1/2} D^{1/6} \tag{2.25}$$

where, C_n represents the structure constant of the atmospheric refractive index, and R indicates the distance between the target and the receiving part of the target detection system based on laser imaging. D denotes the transmitting aperture of lasers.

Figure 2.5 illustrates the values of the beam extension under different intensities of atmospheric turbulence when the transmitting aperture is 10 cm. As observed in the figure, the extension of the laser beams caused by the turbulence is generally small when the imaging distance is small. In such case, the influence on the echoes can be ignored. However, in the case of long-distance imaging, the extension of the

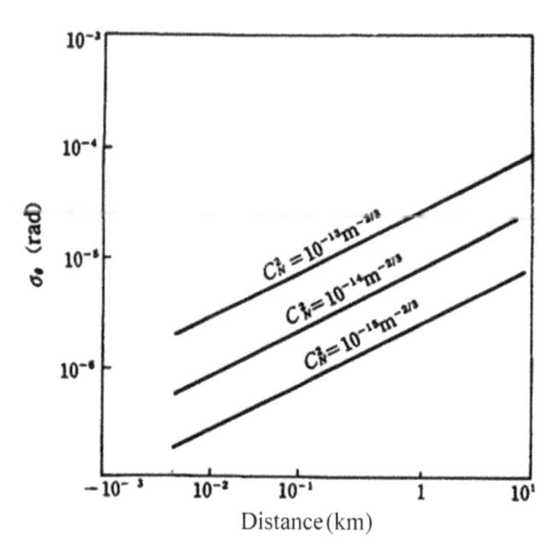

Fig. 2.5 The broadening of the beam angle of the root mean square induced by the atmospheric turbulence

laser beams induced by the turbulence is great. Under such circumstance, the influence on the echoes cannot be ignored.

3. **Phase fluctuation**

The fluctuation and inhomogeneity of the atmospheric refractive index caused by the atmospheric turbulence is expected to distort the wave-front in the transmission of the laser beams. With the increases of the propagation distance and time, the phase of the laser beams that reach the receiving optical aperture of the laser imaging system is expected to present significant irregular variation, which is called the phase fluctuation.

The intensity scintillation in the propagation direction is the function of the 3D space and time. Therefore, the inhomogeneity and random fluctuation of the refractive index within the section of the light beams are supposed to lead to the unequal fluctuations of each optical path, and thus produces the phase jitter. In the coherent laser imaging system, the coherence can be obviously destroyed by the phase jitter under the influence of the atmospheric turbulence. Consequently, beams are diverged with the shift of the frequency, leading to the blurred laser images.

4. **Source images Jitter**

When the diameter *d* of the light beams is close to the turbulent scale *l*, the influence of the atmospheric turbulence is primarily demonstrated as follows: the fluctuated atmospheric refractive indexes in different parts within the interface of the light beams result in the different phase shifts of the different parts in the wave-front of the light beams. In this way, the randomly fluctuated equiphase surfaces are generated. Later, such phase distortion leads to the fluctuation of the wave-front angle of the light beams, resulting in the jitter of the light spots in the receiving part, which gives rise to the jitter of the image points in the laser source, namely, source images jitter. During the dithering of the source images, the propagation direction of the laser beams randomly changes with the statistical average direction as the center.

The dithering degree of the source images is represented by the deflection angle between the actual direction of the laser beams and the statistical average direction, and the variation range of this angle is demonstrated as [7]:

$$(\Delta\theta)^2 = 1.048(A_C)^{-\frac{1}{3}}RC_n^2 \tag{2.26}$$

where, A_C represents the area of the receiving aperture of the target detection system based on laser imaging, and R indicates the distance between the target and the target detection system.

The jitter of the image points is unrelated to the wavelength, and the dithering frequency varies from several to dozens of Hertz. The dithering angle is associated with the geographic position and weather, with the maximum value being less than or equal to 50 μrad.

5. **Beam drift**

Beam drift refers to the random movement of the central position of the light spots under the disturbance of the atmospheric turbulence. After light beams propagate in the air for a period of time, the inhomogeneity of the refractive index of the atmospheric vortexes larger than that of the light beams is likely to result in the random deviation of the propagation direction of the light beams. Meanwhile, the central position of the light beams in the plane that is perpendicular to the propagation direction is supposed to vary randomly.

The drift effect of the light beams can be represented by the statistical variance $\langle\rho_c^2\rangle$ of the displacement of the light beams. Under the approximation in geometrical optics, the statistical variances of the displacements of the collimated and focused beams can be theoretically calculated using [8]:

$$\langle\rho_c^2\rangle_{collimation} = 0.97C_n^2(2a_0)^{-1/3}z^3 \tag{2.27}$$

$$\langle\rho_c^2\rangle_{focus} = 1.10C_n^2(2a_0)^{-1/3}z^3 \tag{2.28}$$

where, a_0 represents the outlet diameter of the transmitting end.

Figure 2.6 demonstrates the relationships of the ratio of the drift of the uniaxial beams to that of the collimated beams with the distance z and focal length F. As discovered in the figure, the beam drifts of the three kinds of light beams are arranged in an ascending order, namely, divergent beam < collimated beam < convergent beam. Particularly, when the focused beams are focused on 35% of the transmission path, the beam drift reaches the maximum value. Since the approximation in geometrical optics is valid merely in the weak fluctuation regions where the intensity variance is $\sigma_I^2 < 1$, the results of formulas (2.27) and (2.28) coincide well with the measured beam drift. While for the strong fluctuation region

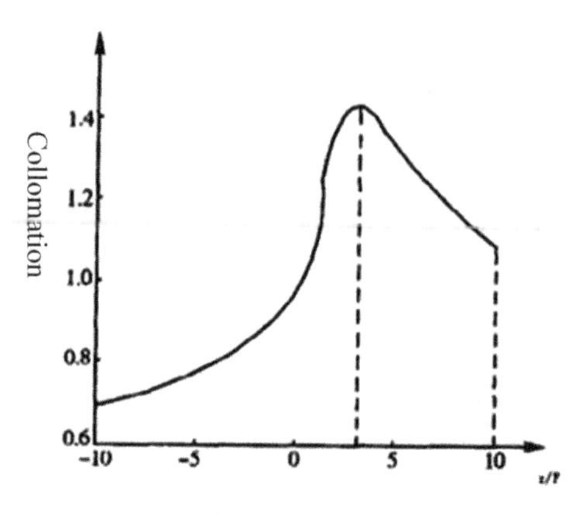

Fig. 2.6 The relationship between the beam drift and the ratio *Z*/*F* of the distance to the focal length

with the intensity variance being $\sigma_I^2 > 1$, it calls for a new theoretical model. The research results demonstrate that the beam drift is expected to present the saturation effect: it is not supposed to increase with the increase of the distance z and C_n^2, under certain a condition, and the range of the beam drift can reach 8 cm.

2.3 Target Modulation of Lasers

The target modulation of lasers is the basis for conducting the target detection based on the laser imaging. The analysis of the mechanism and model of the target modulation of the laser signals is expected to provide a theoretical support for the recognition of the target detection.

2.3.1 The Characteristics and Analytical Method of the Target Modulation of Lasers

1. The characteristics of the target modulation of lasers

   (i) Various action modes: the interaction between the targets and the incident laser gives rise to the phenomena such as reflection, scattering, absorption and transmission, and thus changes the energy and direction of the laser. What's more, the characteristics such as the material, shape, hierarchy and fluctuation of the targets are expected to change the intensity, width and shape of the waveform of the laser echoes. In this way, the characteristics of the waveform of the laser echoes are changed, so as to endow the laser echoes with the aforementioned characteristics of the targets.

   (ii) Complicated modulation mechanism: the parameters such as intensity, width and shape of the pulse of the laser echoes are influenced by the characteristics including the material, shape, hierarchy and fluctuation of the targets.

Meanwhile, modulation factors are mutually coupled, and the strict relationship and accurate influence mechanism for the modulation factors to modulate the laser signals are complicated.

(iii) The difficulty in precise description: in order to understand the characteristics of the targets, it requires to establish the mathematical model of the mapping relation between the echo parameters and the target characteristics based on the characteristics such as the waveform of the laser echoes. However, it is difficult to establish the mathematical model due to the complicated mechanisms and various action models of the modulation of the laser signals using the targets with different characteristics. Thereby, precisely describing the modulation relationship between the characteristics of the targets and the laser signals cannot be easily realized.

2. Methods for analyzing the characteristics of the target modulation of lasers

The theoretical analysis, laboratory measurement of the target materials and scaled models, and static and dynamic measurement of the full-scale target sites can be adopted to study the characteristics of the target modulation of lasers [3].

Theoretical analysis is primarily utilized to reveal the interaction mechanism between the targets and the incident lasers. As an important approach to study the reflection and scattering characteristics of the targets for the lasers, theoretical analysis plays an equal role with the other two methods. Without being mathematically modeled, the data measured in the experiment cannot be favorably utilized to design the lidar, laser imaging and the target detection system based on laser imaging.

The static and dynamic measurement of the full-scale target sites is a basic approach used for quantitatively analyzing the reflection and scattering characteristics of the lasers. As to the principle, it firstly measures relevant data, which are then verified using the mathematical model. By doing so, the accurate data used for describing the parameters are obtained and then employed to design and evaluate the system proposed.

The laboratory measurement of the target materials and scaled models is another important method to study the reflection and scattering characteristics of the targets for the lasers. Compared with the static and dynamic measurement of the full-scale target sites, laboratory measurement is endowed with the following characteristics including simple equipments, readily controlled environmental conditions, convenient repeated measurement and low measurement cost. Laboratory measurement is divided into two types, namely, the measurement of the bidirectional reflectance distribution function (BRDF) of the materials and the measurement of the scaled models. The BRDF of the surface materials of the targets is an important factor determining the characteristics of the target modulation of lasers. According to the BRDF of the surface materials and the geometric parameters of the targets, the laser radar cross section (LRCS) of the targets can be estimated. Therefore, the BRDF measurement of the materials is of great importance for the theoretical analysis and modeling. The measurement of the scaled models is important for it can reveal the characteristics of the modulation of the lasers using the targets with different structures. Meanwhile, it can obtain the numerical relationship between the modulation results and the structure of the targets. However, it presents great measuring difficulty. This is because it is difficult to establish the scaled model and analyze the scale factors in the optical frequency regions. Meanwhile, it is not easy to obtain the characteristics of the modulation of lasers by actual targets through processing and reversing the data measured using the scaled models.

2.3.2 The Modulation Characteristics of the Laser Echoes Energy

According to the detection range equation of lasers, the power of the laser echo is not only related to the atmosphere and the reception performance of the optical system, but also the characteristics of the target itself. These characteristics mainly include the reflection and scattering abilities of the targets for lasers, which are relevant to the type of the surface materials, surface roughness and geometric structure of the targets. Therefore, the reflection and scattering abilities of the targets for lasers are actually the reflection and scattering abilities of the target surface for the lasers. The parameters that can specifically characterize the reflection and scattering properties of the targets for lasers include the reflectivity, BRDF and LRCS.

1. **BRDF**

The reflection on the surface of the targets is divided into the specular and diffuse reflections according to different surface roughnesses. However, as a matter of fact, there exists no absolute specular and diffuse reflections, and the two kinds of reflections are generally interwoven at different degrees. As a result, the distribution curve of the reflection obtained through actual measurement is distorted, and cannot be simply regarded as the specular reflection or diffuse reflection. It is also inappropriate to linearly superpose the components of the two kinds of reflections. Therefore, a parameter that can describe different reflection conditions is required here, so as to accurately reflect the reflection performance of the targets for lasers. Hence, BRDF is adopted as one of the characteristic parameters to describe the reflection performance of the surface of the targets.

According to the geometric distribution shown in Fig. 2.7, it is assumed that the light source with the radiance being L is radiated and uniformly illuminates the isotropous surface A_i. Then, the luminous flux on the bin dA_i nearby the point (x_i, y_i) of the light being emitted from the solid diagonal entry $d\omega_S$ along the direction $s(\theta_S, \varphi_S)$ is expressed as:

Fig. 2.7 The geometrical relationship of BRDF

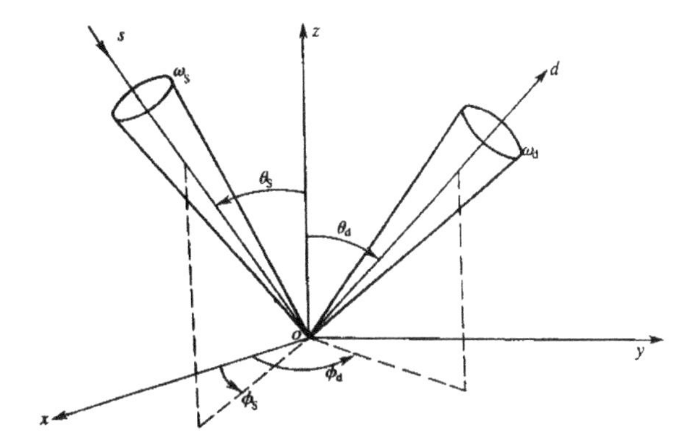

$$d^2\Phi_i = L_i \cos\theta_S d\omega_S dA_i = dE_i dA_i \tag{2.29}$$

where, dE_i indicates the irradiance of the light source on dA_i, and L_i represents the radiance on dA_i.

$$dE_i = L_i \cos\theta_S d\omega_S \tag{2.30}$$

As a general rule, the radiance L_r of the reflected light is in direct proportion to the luminous flux on the plane A_i, namely:

$$d^2L_r = Sd^2\Phi = SdE_i dA_i \tag{2.31}$$

where, the proportional coefficient S is related to the irradiation position, and the incident and detection direction d of the incident light:

$$S = S(\theta_S, \varphi_S, x_i, y_i; \theta_d, \varphi_d) \tag{2.32}$$

The contribution of the whole plane A_i to the radiance of the reflected light can be obtained through solving the integral of formula (2.32):

$$f_r(\theta_S, \varphi_S; \theta_d, \varphi_d) = \int_{A_i} S(\theta_S, \varphi_S, x_i, y_i; \theta_d, \varphi_d) dA_i \tag{2.33}$$

It is defined that:

$$dL_r = dE_i \int_{A_i} S(\theta_S, \varphi_S, x_i, y_i; \theta_d, \varphi_d) dA_i \tag{2.34}$$

According to formula (2.34), we obtain:

$$f_r(\theta_S, \varphi_S; \theta_d, \varphi_d) = \frac{dL_r(\theta_S, \varphi_S; \theta_d, \varphi_d; E_i)}{dE_i(\theta_S, \varphi_S)} \tag{2.35}$$

where, $f_r(\theta_S, \phi_S; \theta_d, \phi_d)$ is called the BRDF of the targets, and is the ratio of the radiance of the reflected light in the direction (θ_d, φ_d) to the irradiance dE_i of the incident light in the solid diagonal entry $d\omega_S$ in the direction (θ_S, φ_S). The value range is from 0 to infinity, with the unit being sr^{-1}. As a differential component, f_r cannot be measured directly, but its average value can be acquired by being measured within the scope of non-zero parameters.

2. **Reflectivity**

Reflectivity $\rho(\omega_S, \omega_d, L_i)$ is an important characteristic parameter describing the reflection characteristics of the surface of the targets for the lasers, and is defined as the ratio of the reflected radiation flux to the incident radiation flux. The incident radiation flux on the bin dA_i within the solid angle ω_S is demonstrated as follows:

$$d\Phi_i = dA_i \int_{\omega_S} L_i(\theta_S, \phi_S) \cos\theta_S d\omega_S \tag{2.36}$$

The reflected radiation flux $d\Phi_r$ measured within the solid angle ω_S is expressed as:

$$d\Phi_r = dA_i \iint_{\omega_S \omega_d} f_r(\theta_S, \phi_S; \theta_d, \phi_d) L_i(\theta_S, \phi_S) \cos\theta_S \cos\theta_d d\omega_S d\omega_d \tag{2.37}$$

Thus, the reflectivity $\rho(\omega_S, \omega_d, L_i)$ is formulated as:

$$\rho(\omega_S, \omega_d, L_i) = \frac{dA_i \iint_{\omega_S \omega_d} f_r(\theta_S, \phi_S; \theta_d, \phi_d) L_i(\theta_S, \phi_S) \cos\theta_S \cos\theta_d d\omega_S d\omega_d}{\int_{\omega_S} L_i(\theta_S, \phi_S) \cos\theta_S d\omega_S} \tag{2.38}$$

If the incident radiation within the incident light beams is isotropous and uniform, then the L_i in formula (2.38) is a constant. Thereby, we obtain:

$$\rho(\omega_S, \omega_d) = \frac{\iint_{\omega_S \omega_d} f_r(\theta_S, \phi_S; \theta_d, \phi_d) \cos\theta_S \cos\theta_d d\omega_S d\omega_d}{\int_{\omega_S} \cos\theta_S d\omega_S} \tag{2.39}$$

Since ω_S and ω_d individually contain three conditions, namely, orientation ($\omega = 0$), cone ($\omega < 2\pi$) and hemisphere ($\omega = 2\pi$), there exit 9 different reflectivities. Formula (2.39) represents the reflectivity of the double-cones. f_r can be considered as a constant within ω_S if the value of ω_S is very small. Then, the reflectivity $\rho(\theta_S, \phi_S; \omega_d)$ of the directional-cone can be obtained, namely:

$$\rho(\theta_S, \phi_S; \omega_d) = \int\limits_{\omega_d} f_r(\theta_S, \phi_S; \theta_d, \phi_d) \cos\theta_d d\omega_d \tag{2.40}$$

Similarly, the reflectivity $\rho(\theta_S, \varphi_S; 2\pi)$ of the directional-hemisphere can be obtained:

$$\rho(\theta_S, \phi_S; 2\pi) = \int\limits_{\omega_d} f_r(\theta_S, \phi_S; \theta_d, \phi_d) \cos\theta_d d\omega_d \tag{2.41}$$

In this way, other reflectivities under different geometric conditions of the incidence and detection can be obtained. The 9 reflectivities describe the 9 possible combinations of the geometric conditions of the incident and reflected beams. Therefore, when referring to the reflectivity, the geometric condition of the incident and reflected beams has to be provided.

Although there exists no absolute diffuse and specular reflections, many conditions are approximated as the diffuse reflection or specular reflection in practice for the sake of simplicity.

The surface with complete diffuse reflection is also called the Lambertian surface, the reflected radiance of which is isotropous within the hemispheric solid angles, that is, the value of the reflected radiance is same in all the direction (θ_d, φ_d). According to the definition of BRDF, we obtain:

$$\begin{aligned} L_r(\theta_d, \phi_d) &= \int f_r(\theta_S, \phi_S; \theta_d, \phi_d) dE_i \\ &= \int f_r(\theta_S, \phi_S; \theta_d, \phi_d) L_i(\theta_S, \phi_S) \cos\theta_S d\omega_S \end{aligned} \tag{2.42}$$

Therefore, it can be seen that in order to make the reflected radiance unrelated to (θ_S, φ_S), f_r has to be a constant. Thus, for a Lambertian surface, formula (2.42) can be simplified as:

$$f_{r,d} = \frac{L_{r,d}}{E_i} \tag{2.43}$$

where, the subscript d indicates there exists a relationship with the Lambertian surface.

The cone-hemisphere reflectivity of the Lambertian surface can be obtained according to the definition of reflectivity:

$$\rho(\omega_S, 2\pi) = \frac{\iint_{\varpi_S \varpi_d} \cos\theta_S \cos\theta_d d\omega_S d\omega_d}{\int\limits_{\varpi_S} \cos\theta_S d\omega_S} = f_{r,d} \cdot \pi \tag{2.44}$$

Thereby, the BRDF of the Lambertian surface is formulated as:

$$f_{r,d} = \frac{\rho(\omega_S, 2\pi)}{\pi} \tag{2.45}$$

Under the condition of complete specular reflection, reflected beams comply with the law of reflection, namely, $\theta_d = \theta_S, \varphi_d = \varphi_S \pm \pi$. Meanwhile, the reflected radiance $L_r(\theta_d, \varphi_d)$ is directly proportional to the incident radiance $L_i(\theta_d, \varphi_{d\pm\pi})$, that is:

$$L_r(\theta_d, \varphi_d) = \rho_{sp} L_i(\theta_d, \varphi_{d\pm\pi}) \tag{2.46}$$

where, ρ_{sp} represents the specular reflectivity.

The f_r of the complete specular reflectivity is an impulse function δ:

$$f_{r,sp}(\theta_S, \varphi_S; \theta_d, \varphi_d) = 2\rho_{sp}(\sin^2\theta_d - \sin^2\theta_S)\delta(\varphi_d - \varphi_{d\pm\pi}) \tag{2.47}$$

3. **LRCS**

LRCS is a physical quantity describing the scattering ability of the targets to the radiated laser. It is defined that when the light intensity generated on the receiver is equal to that generated by a sphere with total reflection, the cross sectional area of the sphere is the LRCS of the targets.

The LRCS of the sphere target is expressed as:

$$\sigma_d = \rho_d \pi r^2 \tag{2.48}$$

where, ρ_d and r indicate the reflectivity and radius of the sphere respectively.

Generally speaking, the more approximate the target to a sphere, the closer the relationship between the LRCS and the size of the target is, while the less the correlation with the orientation and azimuth.

The commonly used methods to establish the calculation model of the laser scattering on the targets include the Kirchkoff's approximation (physical optics method), geometrical optics method, perturbation method, dual-structure method and so on. However, due to the random surface structure of the targets, the aforementioned methods present drawbacks including great workload and time consumption while being used for calculating the distribution of the lasers scattered by the targets. From the perspectives of the feasibility and practicability of the projects, measuring the BRDF of the surface materials is simpler. Then, the effective cross section of the targets for reflecting lasers is calculated according to the test on the geometric structure of the given target.

(1) Estimation of the LRCS

As to the full-scale target characterized by the backscattering with a fixed incident angle, its surface can be divided into many small bins dA_i. Then, the LRCS

$\sigma(\xi, \eta)$ of the whole target can be obtained by performing the area integral on the backscattering coefficient $f_r(\theta_d)$:

$$\sigma(\xi, \eta) = \int_{A_i} \pi f_r(\theta_d) \cos^2 \theta_d dA_i \tag{2.49}$$

where, ξ and η represent the angle parameters between the direction of the incident light and the proprio-coordinate of the target. θ_d indicates the angle between the incident light and the normal line of the target surface, while A_i represents the surface area of the projection of the lasers on the target.

Thus, the relationship between the BRDF and the LRCS is established. The LRCS of the targets can be obtained as long as the analytical expressions of the BRDF and geometric shape of the targets are known.

There are simpler methods to obtain the LRCS of the targets with simple geometric shapes. The backscattering cross section of the targets is generally determined by the product of the reflectivity and irradiated sectional area of the targets. Then, this characteristic quantity can be favorably described by measuring the width of the reflected beams on the targets. The microwave scattering based method can be employed to define LRCS:

$$\sigma = \rho_0 G A_0 \tag{2.50}$$

where, ρ_0 represents the reflectivity of the target surface, while A_0 and G indicate the actual projection area and the optical gain of the targets respectively.

$$G = \frac{4\pi A_c}{\lambda^2} = \frac{4\pi}{\Omega_r} \tag{2.51}$$

where, Ω_r and A_c denote the backscattering solid angle and the coherence area separately, and:

$$A_c = \frac{\lambda^2}{\Omega_r} \tag{2.52}$$

It can be seen from formula (2.50) that ρ_0, A_0 and G are all variables. ρ_0 has been discussed above, and A_0 is the function of the direction of the target type. G is determined by the materials, surface condition, and direction of the targets.

According to Lambertian's law, the cross section areas of several targets with simple geometric shapes are deduced as follows:

(1) For spheres

$$A_0 = \frac{8}{3} a \pi r^2 \tag{2.53}$$

(2) As to planes and plates

$$A_0 = 4a\pi \cos^2 \beta_x \tag{2.54}$$

(3) As for cylinders

$$A_0 = 2a\pi rl \cos^2 \beta_x \tag{2.55}$$

(4) Regarding hemispheres

$$A_0 = \frac{8}{3}a\pi r^2(1 + \frac{\sin 2\beta_x + 2\beta_x}{2\pi}) \tag{2.56}$$

(5) For cones

$$A_0 = \frac{2a\pi r^2}{\sin \varphi}(\sin \beta_x \cos^2 \varphi + 2\cos^2 \beta_x \sin \varphi) \quad 0 < \beta_x < \varphi, \varphi \le \pi \tag{2.57}$$

where, a and r represent the coefficient and radius respectively. l and ϕ indicate the length of the cylinder and the half-angle of the cone respectively, while β_x denotes the incident angle of lasers.

By calculating the cross section areas of these targets with different shapes, the LRCSs of high resolution of these targets and the complex-shaped targets constituted by these shapes can further be obtained.

(2) The LRCS of typical targets

The laser scattering ability of the targets is related to the surface toughness and geometric structure of the targets. Therefore, different surface toughnesses and geometric structures of the targets lead to the different LRCSs which characterize the laser scattering ability of the targets. The LRCSs of some typical targets are demonstrated as follows.

(i) Targets with specular reflection: a perfect reflecting sphere is defined as a kind of the targets with specular reflection. Specular echoes can be generated on any spherical objects when the root mean square of the surface toughness is far smaller than the wavelength of the lasers.

    ① Solid corner reflector: since the projection area of the reflector reduces with the increase of the incident angle, the LRCS of the

corner reflector is approximately in direct proportion to the cosine of the incident angle (for small angles).
The LRCS of the solid corner reflector is demonstrated as:

$$\sigma = 4\pi \frac{l^4}{3\lambda^2} \tag{2.58}$$

where, l represents the side length of the solid corner reflector, and λ indicates the wavelength of the incident lasers.
When the curvature of the laser signals reflected by the corner reflector is smaller than $\lambda/4$, formula (2.58) is still valid.

② Reflecting plates: the LRCS of the reflecting plates is related to the coatings and the shape of the plates, and the cross sectional area is 100–1500 times greater than that of the original plates with white coatings and diffuse reflection.

(ii) Targets with diffuse reflection: diffuse reflection takes place when the root mean square of the surface toughness of the targets is equal to or slightly greater than the wavelength of the lasers. The amplitude and distribution of the reflected light can be represented by BRDF. The BRDF of the targets with diffuse reflection is determined by many special factors related to the material of the targets. The BRDF of the targets with special materials requires to be determined according to the experimental measurement.

(iii) Targets with Lambertain surfaces: Targets with Lambertain surfaces refer to the targets on which the intensity of the scattered light complying with the Lambert cosine law. As pointed in the law, the light intensity (the flux of the solid angles per unit) reflected from any a direction on the surface of the materials is in direct proportion to the cosine of the angel between this direction and the normal line on the surface.
The LRCS σ of the Lambertain plate with the cross section smaller than that of the transmitted beams is expressed as:

$$\sigma = 4\pi\rho r^2 \cos\theta \tag{2.59}$$

where, ρ and r represent the hemisphere reflectivity on the Lambertain surface and the radius of the Lambertain plate respectively, while θ signifies the incident angle (rad) of the laser source on the Lambertain surface.
The ideal Lambertain reflection surface is able to completely depolarizing the incident lights. However, as a matter of fact, there exist few ideal Lambertain surfaces, and the ideal Lambertain surface is assumed while analyzing problems. Therefore, the polarized LRCS of actual Lambertain targets is 1/2 of the non-polarized LRCS.

(iv) Point targets: formulas (2.58) and (2.59) can be applied in the point targets which are far smaller than beam diameters while beam distribution function is stochastic.

(v) Extended targets: the target detection system based on laser imaging is endowed with narrow beamwidth, and the widths of some targets (ground vehicles or airplanes) are always greater than the beamwidth. Under such circumstance, the LRCS of the Lambertain surface is demonstrated as:

$$\sigma = \pi\rho\varphi^2 r^2 \tag{2.60}$$

where, the definitions of each parameter are same to those of the aforementioned ones.

(vi) Linear targets: the diameter of the linear targets is usually greater than the light beams in certain a direction, and electric wires are the special example of the linear targets. The LRCS of the electric wires or thin targets with diffuse reflection is exhibited as:

$$\sigma = 4\rho D_w \varphi r \tag{2.61}$$

where, D_w represents the diameter of the cross section of the electric wires (m), and the definitions of other parameters are demonstrated above.
What is worth noting is that formulas (2.60) and (2.61) are established based on the precondition that the surface of the targets is uniformly irradiated, and these formulas need to be used in combination with the distribution functions of the homogeneous beams. Similarly, electric wires and cables contain components of the specular reflection that have strong intensity and are related to the wavelength of the irradiated lasers as well. At present, it has been discovered that the angular position and chord of the components of the specular reflection are associated with the torsional angle of naked electric wires or the wiring harnesses in cables.

(vii) Targets with high resolution: as for the target detection system based on laser imaging, the phenomenon that targets are outside of the resolution cells of the system is likely to occur due to the narrow width of light beams. Under such circumstance, merely part of the targets scatters the laser beams in the resolution cells. In addition, when the short pulses utilized in the detection system cannot illuminate the whole subject simultaneously, but illuminate the target segmentally, the receiver can only receive the light intensity reflected by part of the targets. Under these conditions, the LRCS cannot be represented by a single value. Different irradiation positions on the targets lead to different LRCS.

2.3.3 *The Modulation Characteristics of the Waveform of the Laser Echoes*

After the pulse signals of lasers undergo the space and reflectivity modulation of the targets, the signals received by the detector presents a time-domain modulation waveform. The information contained in this waveform mainly includes the characteristics of the space fluctuation and reflectivity, as well as the distribution law of the targets.

The laser pulses emitted from the laser present Gaussian lineshape in time domain, and the power of the transmitted lasers is demonstrated as [9]:

$$P(t') = P_0 \exp[-\frac{k(t' - \tau/2)^2}{\tau^2}] \tag{2.62}$$

where, P_0 represents the maximum power in the center of the light beams, and $k = 4\ln 2$. τ denotes the half-power width of the laser pulses, while t' indicates the time delay relative to the starting point of the transmitted pulses.

Formula (2.62) represents the waveform that is in symmetrical relationship with $t' = \tau/2$, and the starting point of the transmitted pulses is $t' = 0$.

As for the basic-mode lasers, the spatial distribution of the light intensity also shows Gaussian lineshape, namely:

$$I(r) = I_0 \exp(-\frac{r^2}{w^2}) \tag{2.63}$$

where, represents the central intensity of the laser beams. As demonstrated in Fig. 2.8, $w = w_0\sqrt{1 + (\lambda z/\pi w_0^2)^2}$, and w denotes the radius of the laser beam at the place presenting the distance of z to the beam waist. Thereinto, w_0 and λ represent the radius of the beam waist and the wavelength of the lasers respectively, while r signifies the distance between a certain point in the laser beam presenting a distance of z to the beam waist and the center of the beams. Laser beams are emitted

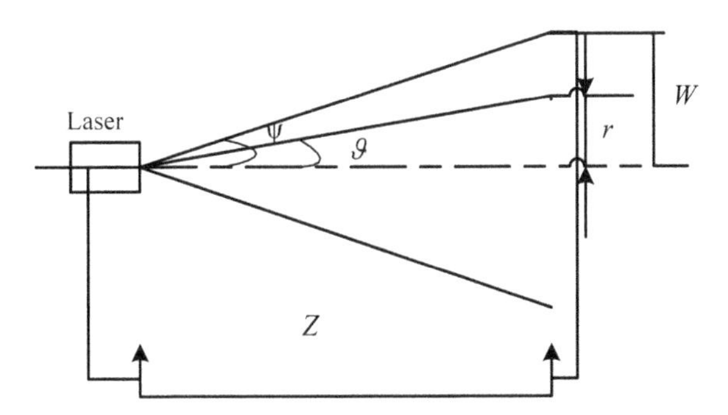

Fig. 2.8 Parameters related to the pulsed laser beams

in the form of wave beams at a divergence angle, and if α represents the half divergence angle of the laser beams, we obtain:

$$\alpha = \lim_{z \to \infty} \frac{dw}{dz} = \frac{\lambda}{\pi w_0} \tag{2.64}$$

As shown in Fig. 2.8, it is assumed that the angel between a line (connected by a point with the radius being r and the laser) and the central axis of the beams is $\psi(-\alpha \leq \psi \leq \alpha)$. Then, formula (2.63) can be converted into the following formula by using the expressions of α and w when the condition $\lambda z / \pi w_0^2 \geq 1$ in the case of far fields is satisfied.

$$I(\psi) = I_0 \exp(-\frac{2tg^2\psi}{\alpha^2}) \tag{2.65}$$

Thereby, the temporal and spatial distribution of the power of the transmitted laser beams can be expressed as [9]:

$$P(t', \psi) = P_0 \exp[-\frac{k(t' - \tau/2)^2}{\tau^2}] \cdot \exp(-\frac{2tg^2\psi}{\alpha^2}) \tag{2.66}$$

where, P_0 signifies the maximum power in the center of the light beams.

As demonstrated in Fig. 2.9, as the emitted laser beams present a divergence angle, when the target is illuminated by the laser beams, all parts of the target contribute to the laser echoes. Under such circumstance, it is considered that the echoes reflected from the target are caused by the joint imaging of the various points on the target for the laser echoes. The echoes reflected from each point are mutually superposed on the receiving photosensitive surface to form the compound echo. However, the adoption of the direct laser detection gives rise to the loss of the

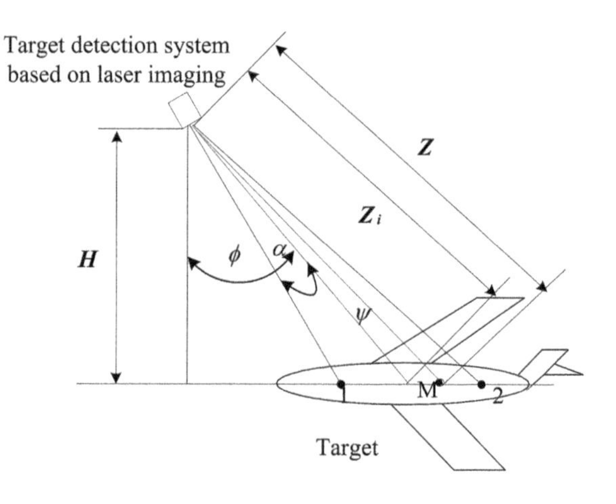

Fig. 2.9 The irradiation of the laser beam on the target

phase information of each sub-echo, and therefore, the compound echo is represented by the superposition of the powers of each echo. As exhibited in Fig. 2.9, α indicates the half divergence angel of lasers, and ψ represents the angel between the line passing through any a point and the central axis of the laser beam. ϕ signifies the angle between the central axis of the laser beam and the nadir of the system.

As demonstrated in Fig. 2.9, H denotes the vertical distance from the laser and detector to the target, and the area between the points 1 and 2 represents the irradiation range of a beam on the target. M is an arbitrary point between 1 and 2. All the points between 1 and 2 are expected to emit the backscattering laser signals with the duration equal to the pulse width. For the convenience of analysis and maintaining the generality, it is assumed that the target being analyzed is in one dimension. The purpose is to find out the time relation of the waveform of the echo pulses. It is obtained that there is always a point M with the distance to the laser being z in the irradiation area at any a moment t within the waveform of the pulses to make $z = ct/2$ valid. Apart from the contribution of the backscattering power of the point M, many points located between the points 1 and 2 contribute to the received power at the moment t. However, according to the assumption of the time domain of the transmitted pulses shown in formula (2.63), as for the transmitted pulses with the width being τ, merely the echoes scattered by the points with the distance z_i to the detector satisfying the condition of $z_i = z - ct'/2$ $(0 \le t' \le \tau)$ can reach the detector simultaneously at the moment t. Therefore, it is concluded that merely the points to the left of the point M have contributions to the echo power at the moment t.

According to the principle of the backscattering of the laser detection, the power received by the detector is the product of the transmit power and a propagation factor $\gamma(z)$. The power at the moment t can be obtained through spatial integral, namely:

$$P_R(t) = \int_{\psi_2}^{\psi_1} P_T(\psi, t')\gamma(z_i)d\psi \tag{2.67}$$

Thereinto:

$$\gamma(z_i) = S_r\rho A(z_i)/z_i^2 \tag{2.68}$$

where, $\gamma(z_i)$ represents the propagation factor with the receiving field angle greater than the divergence angle of lasers of the large targets, and S_r indicates the effective receiving area. ρ denotes the reflectivity of the targets, and is generally considered to be consistent in various directions. $A(z_i)$ is the double-pass total attenuation system of lasers related to the distance, and is assumed as a constant A for the convenience of analysis.

It can be seen from Fig. 2.9 that:

$$z_i = H/\cos(\phi + \psi)\ (-\alpha \leq \psi \leq \alpha) \tag{2.69}$$

$$t' = t - 2H/c \cdot \cos(\phi + \psi) \tag{2.70}$$

By introducing formulas (2.66), (2.68), (2.69) and (2.70) into formula (2.67), we obtain:

$$P_R(t) = \frac{P_T S_r \sigma A}{\pi} \int\limits_{\psi_2}^{\psi_1} \exp\left[-\frac{k}{\tau^2} \cdot \left(t - \frac{2H}{c \cdot \cos(\phi + \psi)} - \frac{\tau}{2}\right)^2 - \frac{2tg^2\psi}{\alpha^2}\right] \cdot \frac{\cos^2(\phi + \psi)}{H^2} d\psi \tag{2.71}$$

It can be seen from formula (2.71) that the power distribution of the pulses of the laser echoes in time domain is related to the directional angle ϕ and divergence angle α of the wavebeam, the vertical distance H to the target, and the width τ of the transmitted pulses. However, in practical laser scanning detection, the values of α, H and τ are basically constants, merely the value of ϕ keeps changing.

2.4 Typical Target and Atmospheric Modulations

2.4.1 Typical Target Modulation

The target modulation of lasers directly determines the characteristics of the pulse of the laser echoes, specifically, changing the energy and waveform of the echo pulses. In the following part, the modulation of the typical factors including the shape, fluctuation and reflectivity of the targets to the pulse of laser echoes are analyzed.

1. **The modulation using the target with a tilted surface**

When the laser detection is performed on the targets with a tilted surface, a common condition is taken into account, that is, beams irradiate the inclined surface at certain an angle. This is a commonly seen working mode of laser imaging at present. In this part, the relationship between the waveform of the echoes and the inclined surface is discussed.

When lasers illuminate the target obliquely, the relationship between the laser beams and the target is demonstrated in Fig. 2.10. Assume that the laser source transmits laser beams to the target obliquely at the point A, and laser beams are received in the emitting axis. Meanwhile, it is considered that the targets within the laser spots are flat, and the half divergence angle of the laser beams is α.

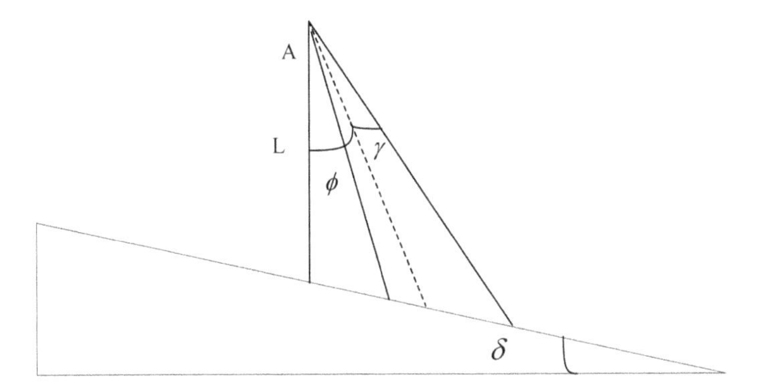

Fig. 2.10 The situation when laser beams irradiate the target obliquely

The height difference within the light spots leads to the enlarged time delay of the echoes on each point, and thus increases the width of the echo pulses. The broadening of the echoes in one dimension is formulated as:

$$\tau = \frac{4L\,\sin(\delta+\phi)\cos(\delta+\phi)\sin\gamma}{c\,\cos(\delta+\phi+\gamma)\cos(\delta+\phi-\gamma)} \tag{2.72}$$

where, L represents the vertical distance from the detector to the inclined plane, and δ denotes the dip angle of the ground. ϕ and γ signify the angle between the central axis and the nadir, and the half width of the light beams respectively.

Figure 2.11 illustrates the relationship between the dip angle and the broadening of the echoes in case L and γ are 300 m and 3 mrad respectively, and ϕ is 0°, 10° and 20°.

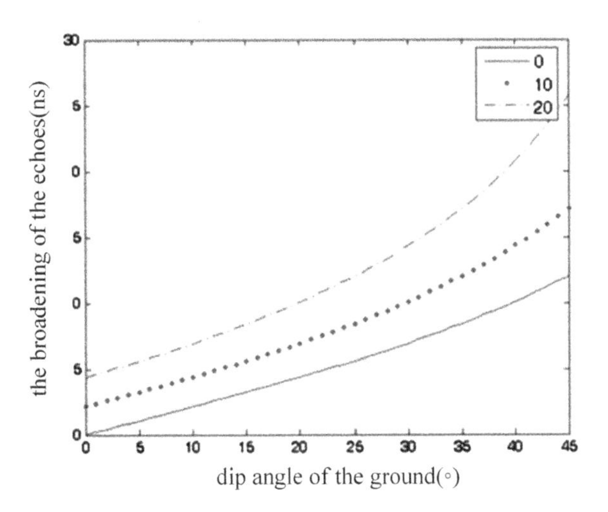

Fig. 2.11 The relationship between the dip angle of the ground and the broadening of the echoes

As discovered in Fig. 2.11, with the increases of the surface slop and dip angle of the irradiation of the laser beams, the broadening of the waveform of the pulses is gradually increased. Then, the increased broadening of the pulses is expected to decrease the receiving power of the echoes, and thus reduce the SNR of the echoes. This finally leads to the increased ranging error and reduced ranging precision. However, the relationship between the broadening of the echo pulses and the slope of the targets and the variation of the frequency components of the echoes caused by the increased broadening of the echo pulses can be obtained. This provides important basis for the extraction of target features in laser detection and the inversion of the attributes of the targets.

2. **The modulation using the targets with a fluctuant surface**

As for the targets with a fluctuant surface, the surface can be divided into the uniformly distributed rough surface and the surface with strong fluctuation. The uniformly distributed rough surfaces such as grasslands and ordinary rocks are endowed with small height variation on the surface. The density function of the height distribution on the surface is generally utilized to describe the roughness. For ordinary targets, it is assumed that they comply with the Gaussian distribution:

$$P_b(H) = \frac{1}{\sqrt{2\pi\sigma_b^2}} \exp\left(-\frac{H^2}{2\sigma_b^2}\right) \tag{2.73}$$

where, σ_b represents the root mean square of the height distribution.

Figure 2.12 illustrates the condition when laser beams illuminate the target with a rough surface. Laser echoes can be regarded as the reconvolution of the transmitted pulses,the response of the planar targets and the roughness function. Therefore, when the height of the target surface shows small fluctuations, it mainly

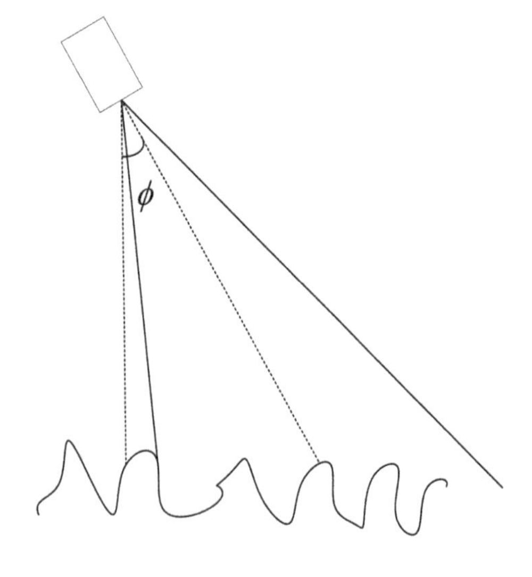

Fig. 2.12 The illumination of a rough surface by the lasers

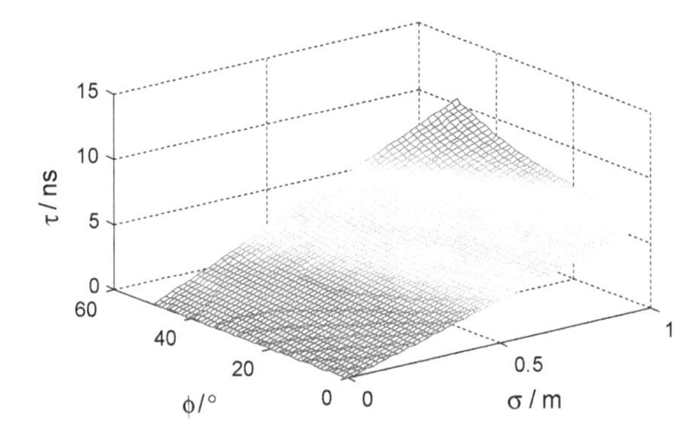

Fig. 2.13 The relationships of the broadening of the echoes with σ_b and ϕ

results in the broadening effect of the echo pulses while having a little influence on the waveform of the echoes.

If the root mean square of the height distribution of the surface is σ_b, the broadening of the echoes caused by the surface fluctuation can be expressed as:

$$\tau = \frac{2\sigma_b}{c \cos \phi} \tag{2.74}$$

where, ϕ indicates the angle between the radiated lasers and the nadir. The broadening, and the variations of the echoes of laser pulses radiated by the nanosecond lasers with the root mean square σ_b with height distribution and the angel ϕ are demonstrated in Fig. 2.13.

When the target surface presents strong fluctuations within the light spots, the violent height variation of the target surface is expected to result in the increased difference in time delay of the echoes reflected from different points. This kind of surface includes trees and buildings on the ground, airplanes in the air, and ships at sea. As a result, the broadening of the compounded echoes is supposed to increase as well. Besides, when the time delay difference caused by the different time-delays of the echoes from each point is increased to some extent, the waveform of the echoes is expected to be changed and even fractured to form multiple sub-echoes. Figure 2.14 illustrates the echo waveform of the small trees in the lawn being obliquely illuminated in the experiment. Under such circumstance, the small trees are regarded as the strong fluctuation in the lawn.

As discovered in Fig. 2.14, a laser echo is broken into two sub-echoes. Thereinto, the first and second sub-echoes represent the echoes from the small trees and the lawn respectively. Meanwhile, the first sub-echo is fractured to form three peaks. The reason for the generation of this waveform is the great height difference between the small trees and the lawn. As a result, part of the laser beams is reflected by the crown or the lawn directly, while some illuminate the lawn after penetrating through the gaps in the crown and then reflected back by the lawn. Finally, two echoes are compounded in the time domain while reaching the detector. As the crown is endowed with certain thickness, and the branches and leaves in the crown

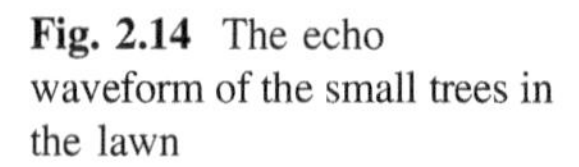

Fig. 2.14 The echo waveform of the small trees in the lawn

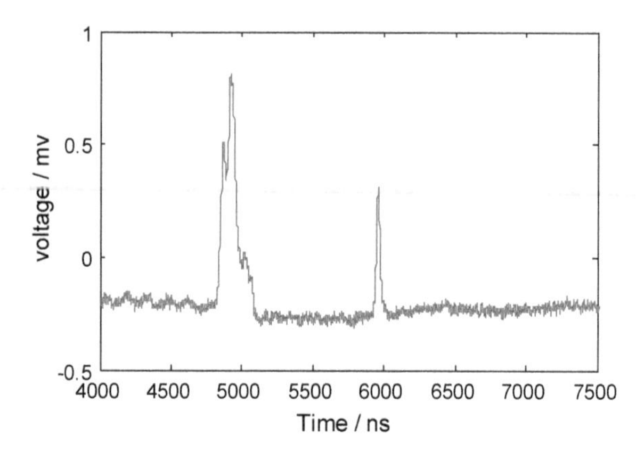

present complicated height distribution, broadening and fracturing occur to the first echo.

3. **The modulation based on the variation of surface reflectivity of the targets**

According to the analysis above, the echo signals of lasers are the superposition of the echoes reflected from each point in the target within the light spots. When the reflectiveity of the surface in the laser spots changes, the waveform characteristics of the echoes are expected to change accordingly. As for the actual targets, the surface reflectivity varies in most situations. For example, in earth observations, different surface reflectivities can be found in the junctions of the vegetations, water surfaces and snowfields with the bare lands, and the boundaries between the cement ground and the vegetations. The phenomenon can also be discovered for the targets in the air and at sea with different surfaces, as well as the boundaries between the paints and other materials. Due to the different surface reflectivities, the energy of the laser echoes reflected from each point does not coincide with Gaussian distribution any more in airspace. The waveform of the compounded echoes is therefore changed. Figure 2.15 illustrates the one-dimensional simulation results of the echoes reflected from the terrains with same height distribution but different distributions of the reflectivity.

Given the target detection system based on laser imaging uses the pulse laser, the broadening of the laser echoes reflected from the targets has a great influence on the imaging of the targets. Firstly, in reduces the vertical resolution of the target detection based on a laser imaging system. The vertical resolution is inversely proportional to the width of the laser pulses: the narrower the laser pulses, the higher the resolution. Therefore, the existence of the broadening increases the pulse width to some extent, and thus virtually reduces the range resolution. Secondly, it affects the ranging precision. The ranging precision is closely related to the measurement precision of the time delay of the echoes. However, the broadening of the laser echoes reflected by the targets damages symmetry and uniformity of the front and back edges of the echo pulses present, causing the rising and falling edges

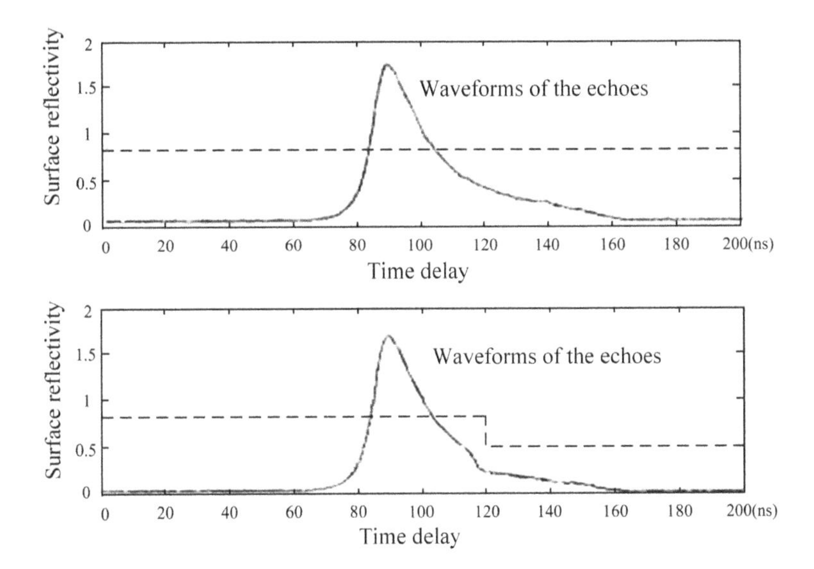

Fig. 2.15 Comparison of the waveforms of the echoes reflected from the terrains with different distributions of the reflectivity

losing the symmetric relation. As a result, the time distribution of the energy of the echoes fluctuates, which reduces the measurement precision of the time based on the energy. Furthermore, it affects the ranging precision and then the precision of the range profile. In addition, the broadening of the echo waveforms is expected to reduce the echo power and thus affect the operating range of the imaging detection. Additionally, the broadening of the laser echoes is supposed to change the waveform of the laser echoes. For the waveform based imaging detection mode for identifying the attributes of the targets, there exists a mapping relation between the waveform of the laser echoes and the targets. However, while performing the target identification using the matching method, the irregular variation of the waveform of the echoes is expected to give inaccurate detection results.

2.4.2 *Typical Atmospheric Modulation*

Due to the modulation of the typical factors including the atmospheric absorption, scattering and turbulence of the lasers, the phenomena such as wavefront distortion, intensity scintillation and beam extension generally take place on the laser echoes. As a result, these phenomena give rise to the reduced SNR and resolution of the imaging system, which thus affects the receiving, processing, image quality and detection results of the target detection system based on laser imaging.

1. Atmospheric absorption and scattering reduce the SNR of the imaging system, the operating range, and the image quality

Both the atmospheric absorption and scattering can reduce the flux of the given light beams [8]. Atmospheric scattering does not change the internal energy of the molecules, while atmospheric absorption gives rise to the variation of the energy state.

As a laser with a determined wavelength propagates in the atmosphere, according to the Lambert-Bougner law, we obtain:

$$P_{\lambda}(x) = P_{\lambda_0} \exp(-\mu(\lambda)x) \tag{2.75}$$

where, P_{λ_0} represents the laser power at $z = 0$, and $P_{\lambda}(x)$ indicates the power of the lasers after propagating in the atmosphere for a distance of x.

Thereby, the attenuation degree of the lasers after propagating in the atmosphere for the distance of x is obtained:

$$\begin{aligned} \tau(\lambda) &= \frac{P_{\lambda}(x)}{P_{\lambda_0}} = \exp(-\mu(\lambda)x) \\ &= \exp(-\alpha(\lambda)x) \cdot \exp(-\gamma(\lambda)x) \\ &= \tau_{\alpha}(\lambda) \cdot \tau_{\gamma}(\lambda) \end{aligned} \tag{2.76}$$

It can be seen that the atmospheric absorption and scattering make the echo power of the lasers received by the system become $\tau_{\alpha}(\lambda) \cdot \tau_{\gamma}(\lambda)$ when the lasers propagate in free spaces. In case the noise is constant, the SNR of the imaging system is expected to become the normal $\tau_{\alpha}(\lambda) \cdot \tau_{\gamma}(\lambda)$.

For the target detection system based on laser imaging, the ranging precision is inversely proportional to the square root of the SNR of the echo signals. Therefore, the higher the SNR, the higher the ranging precision is. The greater the distance precision is, the more accurate the distance profile of the targets obtained in the range imaging; and vice versa. The declined SNR caused by the atmospheric absorption and scattering inevitably reduces of the ranging precision and thus the image quality.

2. Atmospheric turbulence changes the broadening of the pulse width and the echo waveform, and thus affects the precision of the detection

One of the influences of the atmospheric turbulence is the beam extension. The existence of the bean extension leads to the subtle variation of the dips of the radiated light beams on the various sampling points of the targets. Meanwhile, the paths between the various points in the light spots and the optical receiving system are expected to be changed as well, which gives rise to the variation of the time delay of the echoes reflected from the various sampling points on the targets. Afterwards, the superposition of the laser echoes reflected from different points broadens the pulse width of the laser echoes, and thus damages the symmetry between the front and back edges of the pulses.

3. Atmospheric turbulence reduces the azimuth resolution and the image resolution

One of the advantages of the target detection system based on laser imaging is the narrow beam width, which contributes to the high azimuth resolution. However, one of the influences of the atmospheric turbulence on the laser propagation is to diverge the lasers. In this way, the width of the laser beams is increased. Since the azimuth resolution is generally 1/2 of the beam width, the azimuth resolution is therefore reduced.

There exists the following relationship between the pixel of the laser images and the field of view:

$$P = \frac{FOV}{M} \tag{2.77}$$

where, P is the size of the pixel, and determines the image resolution. FOV and M represent the field angle and the image matrix of the lasers on the targets respectively. Atmospheric turbulence is expected to diverge the lasers, which actually increases the field angle and spot area of the lasers. The increase of the spot area leads to the increased blur degree of the sampled targets, and thus reduces the size of the image matrix. According to formula (2.77), the increased field angle and reduced matrix inevitably lead to the reduction in the image resolution.

References

1. Krawczyr R, Goretta O, Kassighian A (1993) Temporal pulse extension of a return lidar signal. Appl Opt 32(33):6784–6788
2. Zhao N, Hu Y (2006) Relation of laser remote imaging signal and object detail identities. Infrared Laser Eng 35(2):226–229
3. Dai Y (2010) LiDAR (volume 1). Publishing House of Electronics Industry, Beijing
4. Chen X (2006) Atmospheric environment and microwave laser weapons. Chinese People's Liberation Army Publishing House, Bejing
5. McCartney EJ (1988) Atmospheric optical molecules and particle scattering (trans: Pan N, Mao J et al.). Science Press, Beijing
6. Song Z (1990) Elements of applied atmospheric optics. China Meteorological Press, Bejing
7. Xiong F (1994) LIDAR. Astronavigation Press, Beijing
8. Zhang Y, Chi Q (eds) (1997) The propagation and imaging of the light waves in the atmosphere. National Defense Industry Press, Bejing
9. Hu Y, Wei Q, Liu J et al (1997) An analysis of the characteristics of the return pulse in airborne laser beam scanning. Appl Laser 17(3):109–111

Chapter 3
The Target Detection System Based on Laser Imaging

To more comprehensively understand the target detection technology based on laser imaging and therefore learn the realization means, constitutional features and action effects of this technology on the whole, it is necessary to analyze the detection system before studying each technique. The target detection technology based on a laser imaging system is the equipments using for target detection using a laser imaging system integrating many techniques. They include laser receiving and transmitting, dynamic measurement of carrier attitude, dynamic differential global positioning system (GPS) with high precision, data processing, target detection and location, target classification and identification, and photoelectric transceiver and calibration. This chapter mainly introduces the components, functions, and working principles of the target detection using laser imaging systems.

3.1 Components of the System

The target detection system based on laser imaging generally includes a laser transmission system, an optical receiving and detection system, a system for delay measurement and echo acquisition, and a positioning and orientation system (POS). In addition, a system for data processing and image generation, a target detection and location system, a target classification and identification system, units for controlling, monitoring and recording, and a photoelectric transceiving and calibration system are also contained. Figure 3.1 presents the basic structure of a target detection system based on laser imaging. In practice, the structures and components of the system are varied according to the detection modes applied. If scanning imaging is used, the system needs to be equipped with corresponding scanning mechanisms.

Y. Hu, *Theory and Technology of Laser Imaging Based Target Detection*,
DOI 10.1007/978-981-10-3497-8_3

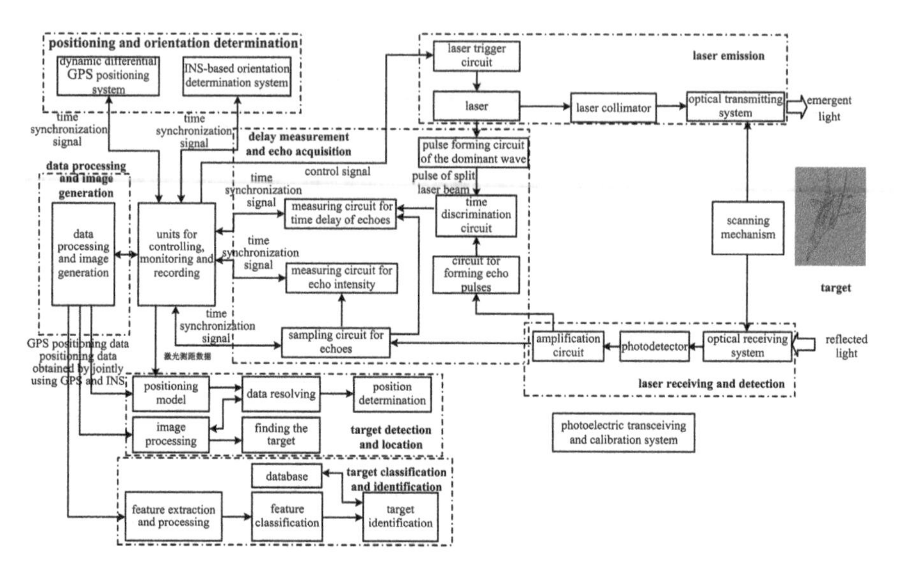

Fig. 3.1 Basic structure of a target detection system based on laser imaging

3.1.1 Laser Transmitting System

The laser transmitting system consists of a laser trigger circuit, a laser, a laser collimator and an optical transmitting system. After being collimated, the generated pulse laser beams are radiated to illuminate targets.

(i) The laser trigger circuit is used to generate current pulses so as to drive the laser to output pulse laser and control the parameters of the generated pulse laser beams including pulse amplitude, pulse width and repetition frequency.

(ii) Under the effect of the driving circuit, the laser generates the needed laser beams. The laser, as the key device in the detection system, its performance and the quality of the generated laser beams directly influence the quality of the laser images of targets and thereby affect the accuracy of the target detection system based on laser imaging. For this reason, the lasers used in target detection based on laser imaging systems generally require to present high peak power and repetition frequency, narrow pulse width, and controllable parameters.

(iii) The laser collimator is applied to expand the collimated laser beams and compress the divergence angles of laser beams output by the laser so as to gather laser energy. In this way, laser beams are focused on the detected targets in a strong directivity, thus satisfying the requirements of detection view field and angular resolution.

(iv) The optical transmitting system aims to transmit the collimated laser beams through a rotating mirror or scanning mirror according to a certain angle.

Meanwhile, a spectroscope is needed to sample a small part of the transmitted laser beams, which then are used as the dominant wave pulse data for laser ranging.

3.1.2 *Optical Receiving and Detection System*

The optical receiving and detection system is comprised of an optical receiving system, a photodetector and an amplification circuit. This system receives the signals of laser echoes reflected by targets and amplify these signals by changing the optical signals to electric ones.

(i) The optical receiving system is used to gather the energy of laser signals reflected from targets and calibrate the wavefront so as to concentrate laser echoes on the photosensitive surface of the photodetector.

(ii) As the core of the optical receiving system, the photodetector is to convert the received signals of laser echoes to electric signals. The photodetectors that can be used in imaging detection include single-detectors, linear array detectors and area array detectors. The types and materials of the photodetectors need to be selected by comprehensively considering various requirements in the design of the target detection system based on laser imaging.

(iii) The amplification circuit is utilized to amplify and process the electric signals output by the detector. It aims to make the amplitude and waveform of echo signals meet the requirements in the subsequent acquisition and ranging. If merely for laser ranging, the bandwidth of the amplifier is supposed to be designed according to the maximal SNR; while if the waveform of echoes is acquired for the target detection system, the bandwidth is expected to be designed based on the criterion that the maximal waveform is not distorted so as to guarantee the best fidelity for the waveform of the output laser echoes.

3.1.3 *System for Delay Measurement and Echo Acquisition*

This system is mainly composed of pulse forming circuits of the dominant wave and the echoes, a time discrimination circuit, and measuring circuits for echo time delay and echo intensity, and an echo acquisition circuit. By digitizing laser echoes, this system can obtain echo pulses with high fidelity and the amplitude and width information of pulses so as to measure the time delay and intensity of echoes.

(i) By sampling the laser beams in the laser transmitting channel, the pulse forming circuit of the dominant wave obtains the laser transmitting pulses which are synchronous with the emitted laser beams. Afterwards, these pulses are amplified to be served as the initial pulses for laser ranging.

(ii) The circuit for forming echo pulses is employed to detect the super-thresholds of the signals of laser echoes sent by the optical receiving and detection system according to a certain threshold which is generally determined based on the constant false-alarm rate. In this way, the terminate pulses of laser ranging can be formed. This circuit is generally designed by using automatic control and constant fraction timing technologies so as to ensure that the false-alarm rate satisfies the requirement and the front of echo pulses is minimally influenced by the signal intensity.

(iii) The time discrimination circuit is utilized to identify and synthesize the front time for the pulses of the dominant wave and the echoes so as to form pulse pairs for time measurement.

(iv) With the help of other processing measures, the measuring circuit for echo time delay is generally realized by using a pulse counter. By controlling pulse pairs, the time delay of echoes and the distance to targets are measured so as to prepare data for range imaging.

(v) The measuring circuit for echo intensity generally measures the pulse intensity of laser echoes by using a peak detector circuit.

(vi) The echo acquisition circuit is used to sample and digitize pulses of laser echoes so as to obtain echo pulses with high fidelity. Therefore, the sampling rate and gray scale quantization bits of this circuit have to meet the need of waveform imaging. Meanwhile, the initial time and terminate time for sampling echoes are expected to be adaptively controlled so as to reduce the demand for data storage space. Waveform digitization is the basis for processing laser imaging data, and its technical performance determines the level of laser 3D imaging and classification.

3.1.4 Positioning and Orientation System

The positioning and orientation system (POS) is composed of a dynamic differential GPS positioning system and an inertial navigation system (INS) based orientation determination system. It is used to measure the position and orientation parameters of the projection center of laser imaging sensors. These parameters are processed with the measured data of laser echoes to generate laser images for detecting and identifying targets.

(i) For the dynamic differential GPS positioning system, it works in the following aspects: ① it supplies the geographic coordinate (X_G, Y_G, Z_G) of the projection center of the optical system at the moment of pixel imaging; ② it provides relevant data for the INS-based orientation determination system to constitute an IMU/GPS compound orientation determination system, so as to improve the precision of attitude angle $(\varphi, \omega, \kappa)$ measured; and ③ it offers the navigation and control data to ensure the stability of the platform of the laser imaging system.

(ii) The INS-based orientation measurement system is applied to obtain the instantaneous orientation parameters of the projection center. These parameters mainly include three external azimuths $(\varphi, \omega, \kappa)$, namely, pitch angle, roll angle and course angle of the main optical axis.

3.1.5 Data Processing and Image Generation System

The data processing and image generation system is the key for performing the target detection based on a laser imaging system. The precision of data processing and the quality of images obtained directly influence the accuracy of the detection results. This system is mainly composed of the embedded software and post-processing software on the detection platform and shows the following functions: ① pre-processing various data; ② providing the distance information of the sampling points in targets after data processing; ③ obtaining the backward scattering rate of targets according to the pulse intensity and distance information of laser echoes; ④ processing the stored GPS data and orientation data; ⑤ acquiring the information of echo pulses including waveform, and distributions of pulse width and echo amplitude so as to prepare data for waveform imaging; and ⑥ generating laser images for targets, such as range images, gray images and waveform images by comprehensively using information including location, orientation, range, gray, waveform data and distribution law of sampling points based on data processing.

3.1.6 Target Detection and Location System

Composed of various kinds of embedded processing software, the target detection and location system is used to process the data of the obtained laser images so as to find and locate targets. It can ① detect and find targets from the background, and ② precisely locate targets by comprehensively analyzing GPS positioning data, orientation data and laser detection data based on the geometrical principle of space positioning.

3.1.7 Target Classification and Identification System

The target classification and identification system is utilized to extract the stable and typical characteristics of targets by using processing software. Moreover, under the support of databases, targets are classified and identified based on the target features so as to obtain the types of targets and further identify the attributes of targets and their platforms.

3.1.8 Units for Controlling, Monitoring and Recording

These units are the controlling and data acquisition center of the target detection system based on laser imaging. Equipped with multiple data interfaces and control interfaces, these units are realized by a high-speed computer. They are mainly used to ① generate the control signals of the circuit driven by laser pulses; ② produce the synchronous pulses of the whole system so as to synchronously realize the measurements of time delay, echo intensity, and orientation, the acquisition of echoes, and GPS positioning; and ③ record and store the measured distance and intensity, echo acquisition data, dynamic GPS positioning data, and orientation data of the platform so as to make preparations for image generation for targets.

3.1.9 Photoelectric Transceiving and Calibration System

This system generally adjusts optical paths and circuits through parameter contrast and parameter conversion based on measurable and controllable observation results so as to realize the on-line and laboratorial calibration of optical paths and circuits.

Among the above compositions, the laser transmission system, the optical receiving and detection system, the calibration system of receiving and transmitting optical paths, the POS, and the system for time delay measurement and echo acquisition are hardware systems. While units for data processing and image generation, target detection and location, and target classification and identification are composed of various kinds of software. The subsequent sections in this chapter introduce the hardware units of the detection system, including transmitting, receiving, calibration, and POS systems. While the systems for time delay measurement and echo acquisition, data processing and image generation, target detection and location, and target classification and identification are elaborated in the following chapters by combining the demands of the target detection system based on laser imaging.

3.2 Laser Transmitting System

The laser transmitting system is the main sub-system of the target detection system based on laser imaging and its performance parameters directly influence the performance of the whole detection system. Therefore, the performance indexes of the laser transmitting system require being determined according to the functional

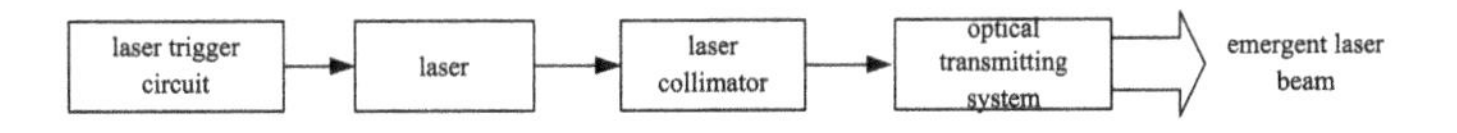

Fig. 3.2 Structure of an optical transmitting system

requirements of target detection system based on laser imaging. The performance indexes mainly include the peak power of laser beams emitted, the pulse width, the repetition frequency, and the divergence angle.

The laser transmitting system is mainly composed of a laser source, a laser trigger circuit, a laser collimator and an optical transmitting system, as shown in Fig. 3.2.

3.2.1 *Lasers and Laser Trigger Circuits*

Lasers and their required trigger circuits, as the key components of the optical transmitting systems, can transmit pulse laser beams with needed wavelength, power and repetition frequency. In principle, any lasers can be served as the laser source of the target detection system based on laser imaging. However, various factors have to be taken into consideration in the practical application of lasers. The lasers used in the system need to satisfy the following requirements: ① the generated laser pulses are expected to show high peak power and repetition frequency, and the laser waves present favorable transmission characteristics in the atmosphere. In addition, photoelectric detectors with favorable performance are available; ② the laser beams show small divergence angles; ③ lasers are small and light with little energy consumption; and ④ lasers show high reliability and mature technology. At present, the commonly used lasers contain diode pump solid state lasers (DPSSL), CO_2 lasers, semiconductor lasers and microchip lasers.

(1) **DPSSL**

The research on the application of DPSSLs in laser imaging systems dates from the late 1980s and has been rapidly developed in the late 1990s. DPSSLs are characterized by high pulse peak power and pulse repetition frequency. Moreover, they are small and light with low cost. Currently, the wavelength of DPSSLs that are suitable for laser imaging is mostly in ranges of 1 and 2–3 μm. Mainly applying Nd:YAG, Nd:YLF, Nd:YVQ4, etc., this kind of lasers shows good atmospheric transmission. Furthermore, lasers doped with Tm and Ho, the wavelengths of which are 2.0 and 2.1 μm separately, also have been developed [1].

Characterized by strong concealment, high efficiency and repetition frequency, Nd:YAG lasers can output infrared waves with a wavelength of 1.06 μm. There have appeared Si avalanche detectors which show favorable performance for laser waves with wavelength of 1.06 μm. The diode pump Cr:Nd:YAG lasers are adopted in the laser detector for spacecraft use developed by the Johns Hopkins

University Applied Physics Laboratory of America in 2000. With repetition frequency varying from 1/8 to 8 Hz and single pulse energy of 15.3 mJ, this detector can detect laser beams with wavelength of 1.064 μm and pulse width of 15 ns. In addition, it can be used in long ranging distance of 327 km with an error less than 6 m and a resolution of 0.312 m [2]. The airborne 3D imaging system developed by the research group also adopts a Nd:YLF solid laser for detecting laser beams with wavelength of 1.064 μm. With repetition frequency reaching to 10 kHz and energy of 500 μJ, this laser can generate laser beams with pulse width of 7 ns. Meanwhile, the first set of Chinese space laser detection system–Chang'E-1 satellite laser altimeter was developed by the team including the author with the application of the DPSSLs with pulse width of 7 ns and repetition frequency of 1 Hz. The altimeter can generate laser beams with single pulse energy of 150 mJ and wavelength of 1.064 μm.

(2) **CO_2 lasers**

CO_2 lasers are characterized by large output power, high efficiency and good atmospheric transmission characteristic under bad weather. Their continuous output power of energy conversion has reached hundreds of thousands of watts, and their pulse power density has grown to tens of billions scale with an energy conversion efficiency (ECE) as high as 20–25%. Therefore, they have been a kind of lasers with highest energy efficiency. Meanwhile, CO_2 lasers can output a wide band with wavelengths mostly in a range of 9–11 μm so that they can easily realize heterodyne detection and 3D imaging with high sensitivity.

The Defense Evaluation and Research Agency (DERA) of British and the Defense Procurement Agency (DGA) of France jointly developed an air-based CO_2 laser imaging radar–CLARA, which employs coherent detection. The adopted lasers are wave guided CO_2 lasers with continuous wave output. They can also be used for pulse operation through "Q switch" with the repetition frequency of pulse reaching 100 kHz. France has tested this system on the HS748 and "Jaguar" helicopters [3]. In 1990, the Navy and Air force of the United States formulated the ATLAS plan, which applies AGM-130 which is compatible with infrared detection as the object with the utilization of wave guided CO_2 lasers. This plan obtained imaging results in the airborne laser imaging tests conducted in 1992 and 1993, respectively.

However, CO_2 lasers show disadvantages including short storage lifetime, large size and the necessity for refrigeration. In recent years, with the development of the technology, some technologies have been developed to solve these shortcomings. The storage lifetime of CO_2 lasers developed in China have been lengthened to 10 years, and the size of the lasers with the radio frequency (RF) excited power being 150 W can be reduced to 100 mm × 70 mm × 20 mm.

(3) **Semiconductor lasers**

Compared with gas lasers and solid lasers, semiconductor lasers are small, light with compact structure, and show long service life and high electro-optical efficiency. In addition, they can be easily pumped and modulated. Developed from homojunction, heterojunction and quantum-well ones, semiconductor lasers have developed to be wide band-gap GaN-based ones [4]. In addition to improving the materials and manufacturing technology, semiconductor lasers are developed to be with longer service life, improved reliability and increased output power. To obtain high output power, the units of semiconductor lasers can be integrated to be one-dimensional linear array ones and two-dimensional area array ones. With the continuous improvement for the output power and output characteristics of semiconductor laser, semiconductor lasers can be used in various fields. In addition, as a kind of novel laser sources, semiconductor lasers with lager power present high electro-optical efficiency and reliable working stability, thus have being widely applied [5]. For example, semiconductor lasers or semiconductor laser arrays have been used in semiconductor laser imaging systems. Since the detectors are arranged in APD/PIN or APD array, semiconductor laser imaging generally adopts direct detection and presents functions such as ranging and 3D outline imaging of targets [6].

(4) **Microchip lasers**

The Lincoln Laboratory of America has developed sophisticated lasers which are highly compatible with 3D laser imaging radars. Among these lasers, microchip ones which are mainly Nd:YAG-pumped and Cr^{4+}-pumped are the most extensively used. As the length of the resonant cavity is about 1 mm, merely the lasers in longitudinal mode less than gain bandwidth (GB) are supported. Microchip lasers are developed with the aim to increase the output energy of a single pulse. At present, lasers show the highest pulse repetition frequency of 1 kHz and output energy of 250 μJ. While, lasers with pulse width less than nanosecond scale (380 ps) are still under development [7]. According to the empirical relation between the repetition frequency and output power of a single pulse of microchip lasers, lasers with the highest photon counting efficiency show a very high repetition frequency. In this way, the energy of a single pulse is very small and the magnification of driving oscillators is supposed to be correspondingly adjusted. Under such situation, the driving oscillation of the system comes from microchip lasers [8].

3.2.2 Laser Collimators and Optical Transmitting Systems

Laser collimators are used to collimate the laser beams with larger divergence angles and asymmetric shapes output by lasers so as to make them with small divergence angle and required shapes for detection. For a single-mode Gaussian beam with waist radius of w_0, the Gaussian divergence angle θ_0 is expressed as follows.

$$\theta_0 = \frac{2\lambda}{\pi w_0} \tag{3.1}$$

The product of the far divergence angle and waist radius of the Gaussian beam is a fixed value and is presented as follows.

$$\theta_0 w_0 = \frac{2\lambda}{\pi} \tag{3.2}$$

Therefore, to decrease the divergence angle, we can increase the waist radius.

After being transformed by a single lens with focal length of F, the divergence angle is changed to $\theta_1 = \frac{2\lambda}{\pi w_1}$, then:

$$\theta_1 = \frac{2\lambda}{\pi}\sqrt{\frac{1}{w_0^2}\left(1 - \frac{l}{F}\right)^2 + \frac{1}{F^2}\left(\frac{\pi w_0}{\lambda}\right)^2} \tag{3.3}$$

Thus it can be found that for a Gaussian beam with a finite waist radius w_0, its divergence angle cannot be 0. However, when $l = \Gamma$, $\theta_1 = \frac{2\lambda}{\pi w_1} = \frac{2w_0}{F} = \frac{4\lambda}{\pi F \theta_0}$. That is, for a Gaussian beam with a determined waist radius w_0 or a fixed Gaussian divergence angle θ_0, to obtain a small far-field divergence angle θ_1, we have to increase the focal length F of the lens. Aspherical single lenses present many advantages in terms of laser collimation: with a large numerical aperture, they can calibrate wave aberration and improve the light energy coupling efficiency. Therefore, they are the first choice for collimating mirrors.

The collimation of lasers can compress the divergence angles of Gaussian beams output by lasers to a certain extent. However, the divergence angle needs to be further reduced in practice. According to the characteristics of Gaussian beams, to decrease the divergence angles, we can increase the beam waist. However, as beam waist cannot be at infinity, a beam expanding system needs to be equipped behind the collimator. Devices used for expanding laser beams include telescopes, prisms and cuneiform prisms. Generally, telescopes are the most commonly used for expanding beams as they can increase spot diameters and reduce divergence angles. For example, Galileo telescopes are adopted as beam expanders.

In general, laser expanding systems can be divided into refractive expanding systems and reflective ones. The former is characterized by the simplicity in design, processing and alignment; however, the expanded beams are small. When the output beams require to present large diameters, reflective expanding systems are supposed to be used. Figure 3.3 shows the aspherical reflective beam expanding system composed of two aspherical mirrors.

When a target detection system based on laser imaging adopts parallel array detection, a parallel laser launcher is expected to be equipped on the optical transmitting system so as to split laser beams with required patterns. The commonly used launchers for parallel transmitting split beams include diffractive grating lenses, fiber splitters, reflectors and prism splitters.

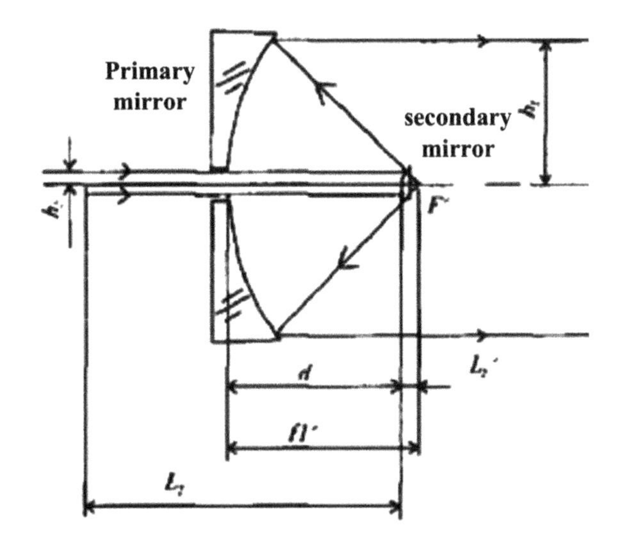

Fig. 3.3 Structure of the aspherical reflective beam expanding system

The expanded laser beams are transmitted towards a certain direction through a rotating mirror or scanning mirror. When the lasers are transmitted and received coaxially, the laser launcher is generally located in the center of the receiving system. However, at this time, it can shade part of the receiving field, thus reducing the effective receiving area. When the operating range is far, parallel axial laser transmitting systems can be utilized. In these systems, the receiving and transmitting axes are close to each other and parallel, and synchronous scanning can be guaranteed through follow-up design. The laser collimating mirrors and optical transmitting systems generally require to be coated so as to meet the requirements for transmission and reflection with strong laser power.

3.3 Optical Receiving and Detection System

The optical receiving and detection system is an important component of the target detection system based on laser imaging, and also one of the key factors determining the performance of the systems. The optical receiving and detection system aims to receive the signals of laser echoes reflected from targets, and then covert optical signals to electric signals. After being amplified, these electric signals are transmitted to the delay measurement and echo acquisition system for waveform sampling, intensity measurement, time discrimination and time interval measurement. Therefore, the system requires to collect more laser energy reflected by targets and inhibit the interference of stray light as far as possible, so as to ensure certain imaging quality and meet the requirement of detection field.

The optical receiving and detection system is designed with the consideration of various perspectives including effective receiving aperture, total field of view, instantaneous receiving field, types of detectors, detecting sensitivity, and SNR of detection. Therefore, when we design an optical receiving and detection system and select optical systems and detectors, we need to select those with suitable receiving field and SNR. Moreover, the inhibiting ability of them to stray light is expected to be taken into account. Besides, the optical receiving and detection system needs to show simple structure, small size and low cost as much as possible.

3.3.1 Optical Receiving Systems

A telescope is adopted in the optical receiving system of the target detection system based on laser imaging. Its mechanism and corresponding angular magnification are shown in Fig. 3.4. Optical gain equals to the square of the angular magnification. In the system, a universal objective lens is used to generate the scattering images of targets. Then, the energy scattered by targets is collected and transmitted to the detector by a component for collecting light. The detector is generally positioned at the exit pupil of the telescope to receive the laser energy reflected by targets. This optical design aims to increase the angular density radiated by laser through the objective lens so as to make laser energy distributed on the whole detector.

Currently, there are mainly two optical receiving systems, namely, the transmission-type and the reflective ones. They are selected depending on the whole performance requirement, types of photodetectors, limitation for the size of optical receiving systems, etc. For the detection systems for remote and ultra-remote laser imaging, both the aperture and the focal length of objective lenses of the optical receiving systems are large. To make the structures of the systems more compact, reflective or catadioptric optical receiving systems are generally used, such as Newton, Gregory, Cassegrainian, Schmid, Masksutov (Masksutov-Cassegrain, abbreviated as Maca), and coude optical systems. For the optical receiving systems in medium and small laser imaging systems, transmission-type optical systems are

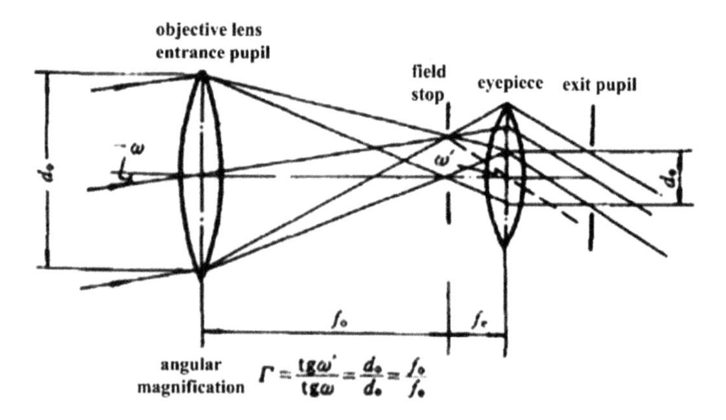

Fig. 3.4 A universal telescope system

generally utilized, such as transmission-type telescopes, single lenses, and combined lenses as the receiving mirrors with a large field of view.

The target detection system based on laser imaging receives the laser signals which are transmitted nearby the telescope and then reflected by targets. To make full of the energy, it is important to consider various factors so as to design telescopes that can most effectively collect the signals of laser echoes. Here, we introduce several typical telescope systems. To ensure the image quality and the field of view for laser imaging, transmission-type telescopes are generally used in staring imaging, while reflective ones are applied in scanning imaging. The typical transmission-type optical receiving systems include Kepler (Fig. 3.5) and Galileo ones (Fig. 3.6), while the typical reflective optical receiving systems contain Newton (Fig. 3.7), Cassegrain (Fig. 3.8) and Gregory (Fig. 3.9) ones.

As shown in the figures, the Kepler optical receiving system is simple and composed of an objective lens and an eyepiece. Both of which are positive lenses and can observe inverted images. The objective lens can be a lens or a reflector. The Galileo optical receiving system uses an eyepiece lens with negative optical power to substitute the positive eyepiece in the Kepler optical receiving system. The rear focus of the objective lens is overlapped with the object focal length of the eyepiece. The Galileo optical receiving system is small with a short cylinder. The images projected on the objective lens of Galilean telescope are virtual ones. For this reason, measuring marks such as cross wires and reticules cannot be placed on the images. Moreover, Galilean telescopes can be used as beam expanders.

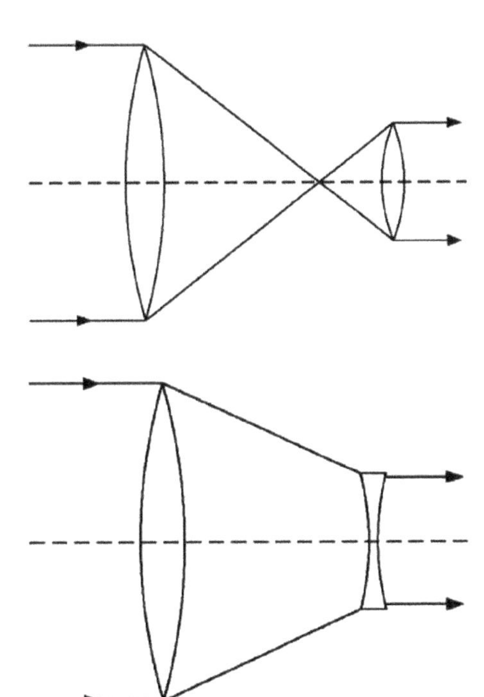

Fig. 3.5 Kepler optical system

Fig. 3.6 Galileo optical system

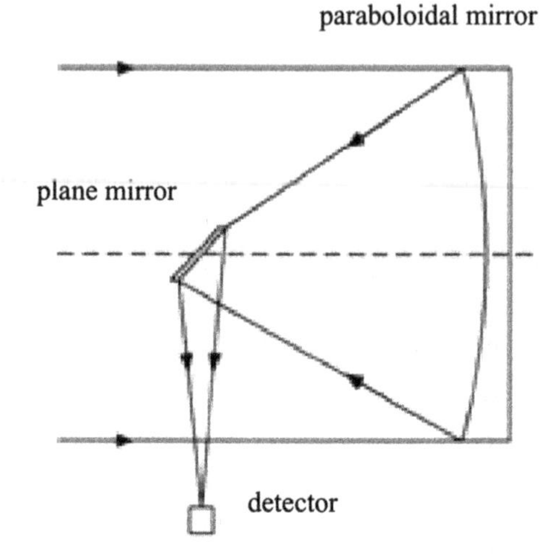

Fig. 3.7 Newton optical system

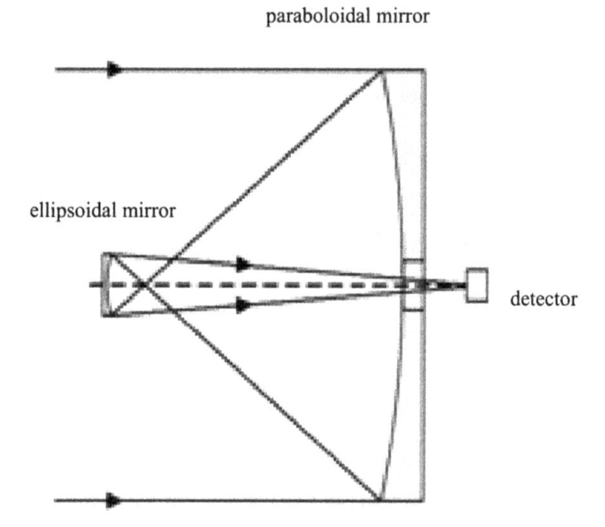

Fig. 3.8 Cassegrain optical system

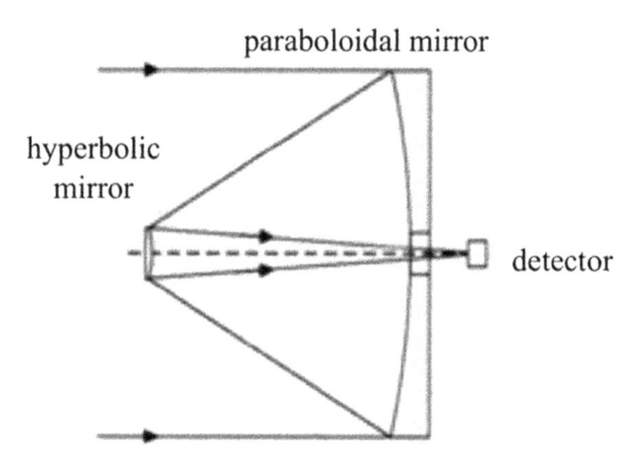

Fig. 3.9 Gregory optical system

The Newton optical system is originally comprised of a primary paraboloidal mirror and a plane mirror which forms an angle of 45° with the optical axis. The plane mirror is used to reflect the images formed by the primary mirror outside the cylinder of the primary mirror so as to make it convenient for detection and observation. Since the secondary mirror of the system is a plane mirror, it is also called a single reflector system.

The Cassegrain optical system is most widely applied. It is originally composed of a primary paraboloidal mirror and a secondary hyperbolic mirror. The focus of the primary mirror is overlapped with the virtual focus of the secondary mirror. The light transmitted from the targets on the axes at infinity is converged at the real focus of the hyperboloidal reflector, thus forming limited diffraction images. As the primary and secondary mirrors of the system show different field curvatures, they can correct the whole field curvature to some extent. However, there exist serious coma aberrations and the field curvature limits the field of view. In addition, there are no intermediate images between the primary mirror and the secondary mirror. Therefore, the system presents compact structure, small size, short cylinder, and high resolution for the points on axes.

The Gregory optical system is composed of a primary paraboloidal mirror and a secondary ellipsoidal mirror. The focus of the primary mirror is overlapped with a focus of the secondary mirror. Light reflected by point targets on the axis at infinity passing through the primary mirror and is converged on a focus of the ellipsoid, and then intersected at another focus of the ellipsoid after being reflected by the ellipsoid. Similar to the Cassegrain optical system, the points on the axis of the system show no aberration. However, points beyond the axis present obvious coma aberrations and astigmatism. Since there are intermediate images between the primary mirror and secondary mirror, the cylinder is long, causing a heavy structure. Gregory optical systems can also be designed in an aplanatic structure with an ellipsoidal primary mirror so as to correct the spherical and coma aberrations.

Transmission-type optical receiving systems are applied in the optical receiving systems with concentrated light energy and small size. With a large field of view, they are suitable for the target detection system based on laser imaging for close range imaging. While, reflective optical receiving systems are mostly used in those systems for long-distance imaging. Be causing of running in a long ranging distance, the target detection system based on laser imaging requires to increase the aperture of the optical receiving systems. The optical receiving systems with a large aperture generally use reflective optical receiving systems with small receiving view angles.

3.3.2 Main Technical Parameters

Since there are many performance parameters for the optical receiving and detection systems, this section mainly analyzes the key parameters including the receiving field angle, types of detectors, and SNR.

3.3.2.1 Receiving View Angle

The receiving view angle is one of the important parameters of optical receiving systems. They are determined by considering whether the intensity of the received signals reach the requirement of detectors and whether the spatial resolution of the imaging pixel is realized. Generally, the larger the view angle, the larger the covering range is. Under such circumstance, though the target signals received are increased, the relative amplitude of target signals reduces as the received background light power increases correspondingly. When the view angle is too small, though the received background light power reduces, the relative amplitude of target signals still reduces owing to the significantly reduced intensity of target signals. In addition, the larger the view angle, the smaller the focal length is. As a result, the relative aperture and the aberration increase, thus making laser imaging more difficult. On the contrary, reducing the view angle can increase the focal length of the receiving system, thus reducing the relative aperture. Accordingly, the system can be more easily designed.

The background light power received by detectors in daytime is expressed as follows [9].

$$P_b = \rho_T E_{sun} \Delta\lambda A_r \tau_a \eta_r \cos\theta_i \sin^2\frac{\theta_r}{2} \tag{3.4}$$

where, ρ_T is the reflectivity of targets with the assumption that targets are Lambertain targets with large areas or direct reflection ones; E_{sun} presents the spectral irradiance of sunlight at the receiver; $\Delta\lambda$ shows the spectral bandwidth of the optical receiving system (if a narrow-band filter with the width being 10 nm is used, then $\Delta\lambda$ is 10 nm); A_r denotes the effective receiving area of the optical receiving system, τ_a and η_r are the single-pass transmittance of the atmosphere and the efficiency of the optical receiving system, respectively; θ_i show the angle between the incidence direction of sunlight and the target surface and θ_r presents the receiving view angle of the optical receiving system separately.

Similarly, the background light power received by detectors at night with a full moon is obtained as follows [10].

$$P_b = \rho_T E_{mn} \Delta\lambda A_r \tau_a \eta_r \cos\theta_i \sin^2\frac{\theta_r}{2} \tag{3.5}$$

where, E_{mn} is the spectral irradiance of moonlight on the ground at night with a full moon and other parameters are defined as those in the above formula.

Thus it can be seen that the background light power received by detectors has nothing to do with the detection range R, but is positively related to the receiving view angle $\sin^2\frac{\theta_r}{2}$. Obviously, the receiving view angle of the optical receiving system directly influences the background light power received by detectors. Therefore, while determining the receiving view angle, it is necessary to comprehensively consider the background light power of the optical receiving system, the requirement for imaging resolution and the divergence angle of the laser output by the optical transmitting system.

3.3.2.2 Types of Detectors

Detectors are mainly selected on the basis of the combination of optimal performances at the operating wavelength of lasers. These performances primarily include responsivity, noise characteristics, response speed (bandwidth), and response linearity. On this basis, we can obtain the high receiving sensitivity, echo detection probability and echo acquisition with high fidelity, and low false alarm probability of the target detection system based on laser imaging. Therefore, it is very important to select suitable detectors. Currently, the detectors used in laser detection can be divided into photomultipliers based on external photoelectric effect and photodiode and avalanche diodes based on photoconductivity [11]. For the target detection systems calling for remote and ultra-remote laser imaging, as the detectors used in the optical receiving systems require to have high sensitivity and gain, photomultipliers and single-photon detectors are generally applied; while PIN detectors (silicon photodiode) or APD detectors are commonly used in optical receiving systems of the small- and medium-scale target detection systems based on laser imaging. In the optical receiving systems with APD detectors as the receiving components, since the amplifier noise is generally larger than the background noise, the systems depend slightly on the field diaphragm and filters, and therefore can be simplified.

3.3.2.3 SNR

There are two kinds of noises in the optical receiving system: optical noise and electrical noise. They mainly come from the background radiation of the sun or the moon and detectors. Thereinto, the detector noises primarily include thermal noise, leakage (dark) current noise, shot (quantum) noise, flicker ($1/f$) noise, temperature noise, etc. The $1/f$ noise is inversely proportional to light frequency. As it mainly occurs in the region with low frequencies (less than 1 kHz), it can be neglected. When the target detection systems based on laser imaging perform direct detection, the output noise of APD detectors mainly contain shot noises of signal light and background light, dark current noise, and thermal noise.

The shot noise of signal light refers to the shot noise current caused by the signal light. The mean square of the shot noise current is presented as follows.

$$i_n^2 = 2eBP_rR_iM\Delta F \tag{3.6}$$

where, e is the electronic charge with value of 1.602×10^{-19} C; B shows the spectral bandwidth of the noise, P_r denotes the signal power received by the APD (W); R_i presents the responsivity of the unit multiplication factor of the APD

(A/W); M is the multiplication factor of the APD; ΔF shows the excess noise coefficient of the APD, and can be approximately expressed as follows.

$$\Delta F = M\left[1 - (1-k)\left\langle\frac{M-1}{M}\right\rangle^2\right] \tag{3.7}$$

where, k is related to the multiplication factor M and the ionization rate ratio of hole carriers to electronic carriers. For silicone tubes, $k \approx 0.02$.

The shot noise of background light denotes the shot noise current resulted from the background light. The mean square of the shot noise current is expressed as

$$i_n^2 = 2eBP_bR_iM\Delta F \tag{3.8}$$

where, P_b presents the background light power received by the APD.

Dark current noise is the shot noise current caused by the dark current produced by the APD. The dark current output by the APD can be expressed as follows.

$$i_n^2 = 2e(i_{ds} + i_{db}M^2\Delta F)B \tag{3.9}$$

where, i_{ds} shows the surface leakage current of the APD, which does not participate in the multiplication process; i_{db} presents the body leakage current of the APD and involves in the multiplication process.

Thermal noise is the noise current caused by the thermal motion of electrons in resistances. The mean square of the thermal noise current is shown as follows.

$$i_t^2 = \frac{4kTB}{R_L} \tag{3.10}$$

where, R_L is the load resistance of the APD; k shows the Boltzmann constant and T denotes the thermodynamic temperature.

Since the shot noises of signal light and background light, dark current noise and thermal noise are mutually independent, the mean square (effective value) of the total noise current output by the APD in the presence of laser echoes can be obtained as follows.

$$\begin{aligned} I_n &= \left(i_b^2 + i_n^2 + i_d^2 + i_t^2\right)^{\frac{1}{2}} \\ &= \left[2eB(P_r + P_b)R_iM\Delta F + 2e(i_{ds} + i_{db}M^2\Delta F)B + \frac{4kTB}{R_L}\right]^{\frac{1}{2}} \end{aligned} \tag{3.11}$$

The signal photocurrent output by the APD is presented as follows.

$$I_s = R_i P_r \tag{3.12}$$

Based on Formulas (3.11) and (3.12), the SNR of the current output by the APD can be acquired as follows.

$$SNR = \frac{I_s}{I_n} = \frac{R_i P_r}{\left[2eB(P_r + P_b)R_i M \Delta F + 2e(i_{ds} + i_{db} M^2 \Delta F)B + \frac{4kTB}{R_L}\right]^{\frac{1}{2}}} \tag{3.13}$$

When $\frac{d(SNR)}{dM} = 0$, the optimal multiplication factor of the APD can be acquired. At this time, the SNR has a maximum value.

For Si-APD detectors, the optimal multiplication factor is shown as follows.

$$M_{opt} = \frac{\frac{4kTB}{R_L}}{\left[eB(i_d + P_b R_i \Delta F)\right]^{\frac{1}{2}}} \tag{3.14}$$

3.3.3 Background Light Suppression

It is especially important to filter background light for the target detection systems based on laser imaging when applying in strong light backgrounds such as space. For the detection systems that adopt APD detectors and direct detection, on the premise that the out-of-band background light is sufficiently suppressed, the in-band background light entering in detectors through the reflection of solar irradiation by the detected space targets is the main source for the background light in detectors.

Figures 3.10 and 3.11 show the typical responsivity of APDs and solar spectral irradiance separately [12].

Figure 3.12 presents the responsivity R_λ of an APD to the solar spectral irradiance at each wavelength [12]. Suppose that the central wavelength and bandwidth of the narrow-band filter used in the front of the detector in the satellite-borne target detection systems based on laser imaging are 1.064 and 0.01 μm, respectively. Therefore, the response of the APD shown in Fig. 3.12 can be separated into two shade areas: the area A_1 with transverse lines demonstrates the response of the APD to the solar spectral irradiance in a wavelength of 0.4–1.1 μm, while the area A_2 with oblique lines displays the response of the APD to the solar spectral irradiance in the pass-band of a filter with wavelength of 1.064 ± 0.005 μm.

By solving the integrals of the two shadow areas in Fig. 3.12 separately, it can be found that the ratio of the response of the APD to solar irradiance in the response section of 0.4–1.1 μm to that of the filter in the pass-band of 1.064 ± 0.005 μm is A_1:A_2 = 300: 1. Since laser beams cannot be completely filtered by the filter, certain solar irradiation can be transmitted in the waveband out of the pass-band.

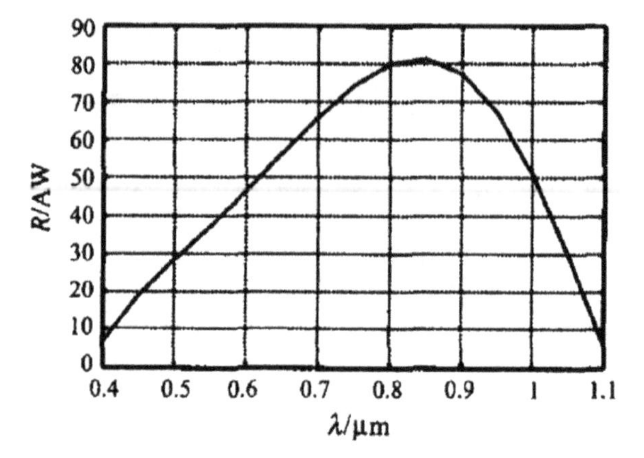

Fig. 3.10 Typical responsivity of APDs (A/W)

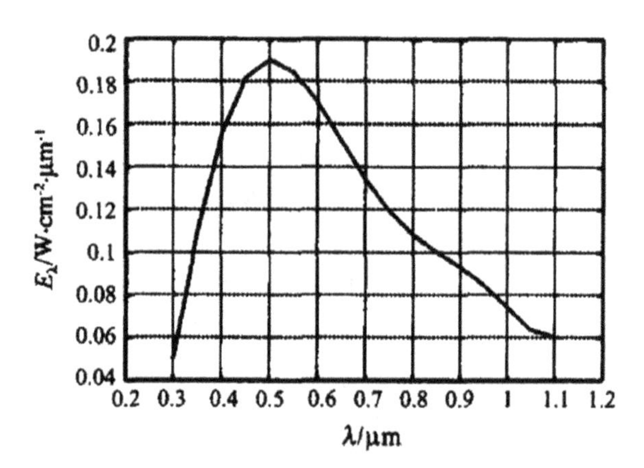

Fig. 3.11 The curve of solar spectral irradiance

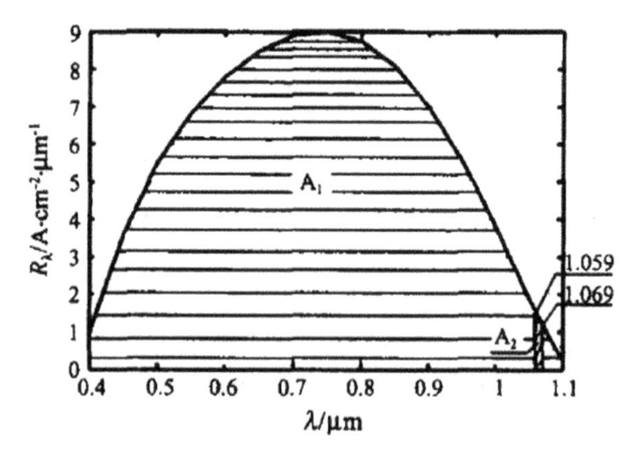

Fig. 3.12 Response of the APD to the solar spectral irradiance in a wavelength range of 0.4–1.1 μm

Moreover, the transmittance value varies in the whole waveband. Suppose that 10% of energy outside the channel enters the detector and the transmittance in the pass-band is larger than 90%, the power of solar radiation entering into the detector is approximately four times of the original one. Therefore, on the premise that the transmittance in the pass-band of the filter is ensured the same, properly reducing the bandwidth of the filter can decrease the power of solar radiation that enters into the detector, thus reducing background noise.

On the premise that the out-of-band background light is fully inhibited, the background noise entering into the APD is mainly the in-band background light noise, that is, the reflection of the target surface for the solar light. It can be expressed as follows at a solar elevation angle of 90°.

$$P_s = E_\lambda \cdot \Delta\lambda \cdot \eta_r \cdot \pi \cdot \left[\frac{\theta_r \cdot R}{2}\right]^2 \cdot \frac{r_{diff}}{\pi} \cdot \frac{\pi}{4} \cdot \frac{\phi_{tel}^2}{R^2} \tag{3.15}$$

where, P_s is the in-band background light power received (W); E_λ shows the spectral irradiance generated by the solar light on the target surface around the central wavelength of the narrow-band filter (W/m^2/nm); $\Delta\lambda$ presents the bandwidth of the narrow-band filter (nm); η_r and θ_r are the optical receiving efficiency and receiving field (rad) separately; R denotes the distance between the target and the detection systems based on laser imaging(m); r_{diff} is the reflectivity of the space target around the central wavelength of the narrow-band filter, while ϕ_{tel} shows the effective aperture of the receiving telescope (m).

The spectral irradiance on the surface of the space target generated by vertically incident solar light can be expressed as follows.

$$E_\lambda = E_\lambda \times \frac{r_{solar}^2}{d^2} \tag{3.16}$$

where, E_λ is the spectral irradiance emitted by the sun at λ (W/m^2/nm); r_{solar} shows the radius of the sun with unit of m, and $r_{solar} = 6.69 \times 10^8$ m; d presents the distance between the sun and the space target with unit of m.

For the satellite-borne target detection systems based on laser imaging, suppose that the distance from the sun to the detected space target is $d = 1.5 \times 10^{11}$ m, the spectral irradiance of the sun around the laser wavelength of 1.064 μm is 2.94×10^4 W/m^2/nm [13]. With receiving optical efficiency of $\eta_r = 0.8$ and a receiving field angle of $\theta_r = 2.5$ mrad, the optical receiving system shows a distance to the space target of $R = 200$ Km. The bandwidth of the filter used is $\Delta\lambda = 5$nm. The reflectivity of the target in the wavelength around 1064 nm is about $r_{diff} \approx 0.15$. In addition, the effective aperture of the receiving telescope is $\phi_{tel} = 128$ mm. By substituting these parameters into Formulas (3.15) and (3.16), the background light power received by the system is $P_b = 6.94$ nW.

According to Formula (3.13), when the background light power entering the receiving system changes, suppose that the body leakage current of the APD keeps

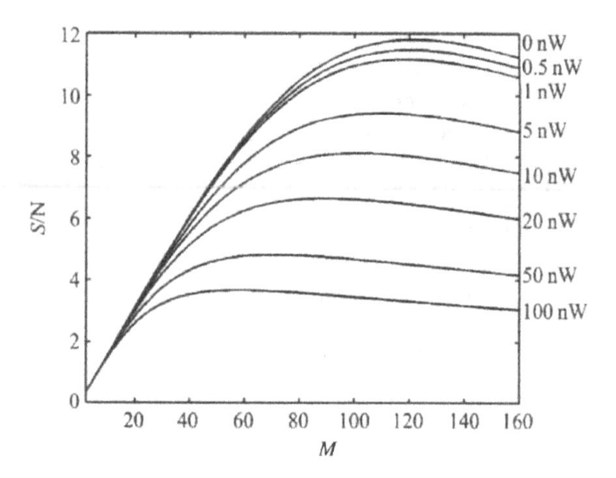

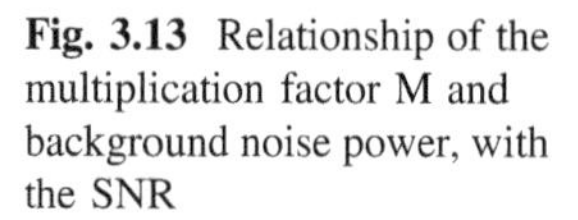

Fig. 3.13 Relationship of the multiplication factor M and background noise power, with the SNR

unchanged, the signal power received is $P_s = 20\,\text{nW}$. The variation of the SNR output by the system with the change of multiplication factor is shown in Fig. 3.13.

Thus it can be seen that the influence of the power of background light noise on the SNR of the system is different with the variation of the power. In the range of P_s = 0–1 nW, the current noise of the optical receiving system caused by the power of background light noise is less than the thermal noise and amplifier noise, thus showing a small influence on the SNR of the system. While, as it increases, the power then becomes the main noise source of the system and seriously affects the SNR of the system. Meanwhile, the multiplication factor *M* also greatly influences the optimal SNR of the system. Based on Formula (3.13), when the received background light power and the signal power are fixed, the signal current increases with the multiplexed factor *M*. However, as the excess noise coefficient and body leakage current i_{db} also grow as *M* increases, the increasing rates of the excess noise power in the first and second items of the denominator surpass that of the signal power, thus aggravating the SNR when *M* is large. When *M* is small as 1, the excess noise is much smaller than the thermal noise and amplifier noise. At this time, the total noise basically equals to the combination of thermal noise and amplifier noise and not increases with the increment of the multiplication factor [13]. Therefore, the optimal multiplication factor varies under different background light powers. Moreover, with the increase of background noise power that enters into the receiving system, the value of the multiplication factor for obtaining the optimal SNR reduces correspondingly.

Therefore, when designing the optical receiving systems of satellite-borne target detection systems based on laser imaging for long-distance targets or targets with complex background irradiation power, we can inhibit the background light. This can be realized by reducing the receiving field and lowering the bandwidth of the narrow-band filters. Apart from this, it is necessary to control the multiplication factors of APD detectors according to the background irradiation power so as to ensure the optimal SNR of the receiving systems.

3.4 Photoelectric Transceiving and Calibration System

As an essential component of the target detection system based on laser imaging, the photoelectric transceiving and calibration system is mainly used to calibrate the consistency of the transceiving path, ranging accuracy and channels. They are more complex and more necessary than general optical systems for the following reasons: (1) since the detection systems can receive and transmit laser beams, the transmitting of the laser and the receiving of the echoes need to be realized coaxially; (2) multi-point scanning or transmitting and receiving arrays are required for performing laser imaging; (3) apart from optical calibration, the consistency of circuits also needs to be calibrated; (4) in addition to laboratorial calibration, on-line calibration is necessary. This section discusses the photoelectric transceiving and calibration system applied in array emission and detection based laser imaging systems. The discussion is performed based on the optical calibration methods for general optical receiving systems and by considering the characteristics of the detection system. The basic principles are also applicable to scanning laser imaging systems based on single-channel transmitting and receiving.

3.4.1 Calibration for Array Transmitting and Receiving Light Path

For the target detection systems based on laser imaging, to focus laser energy on targets and prevent stray light from entering the systems as much as possible, the transmitting angle and the receiving angle need to be small. This requirement is especially predominant for airborne or satellite-borne long-distance detection using laser imaging. Therefore, laser imaging systems require a high parallelism between the transmitting and receiving axes. For this reason, during and after the machine assembly of the systems, it is necessary to detect and correct the registration of the transmitting axes and receiving axes. During the assembly processes, the collimator method is generally used to ensure the registration of the transmitting and receiving axes. After the systems are well assembled, new calibration methods for receiving and transmitting paths are needed to detect and correct the registration of the transmitting and receiving axes of the systems as the output signals are electronic ones.

Calibration for array transmitting and receiving light paths is developed to meet the demand of array laser imaging based on the calibration for a single transmitting or receiving light path. The calibration for a single transmitting or receiving light path is performed by using reference conversion and parameter comparison through the laser imaging calibration system. This system is composed of a collector for spot images (comprised of a charge-coupled device (CCD) and so on), a coupling fiber, a generator for simulating echoes, and an angle adjuster. Based on the obtained measurement results of optical path and circuit deviations with high

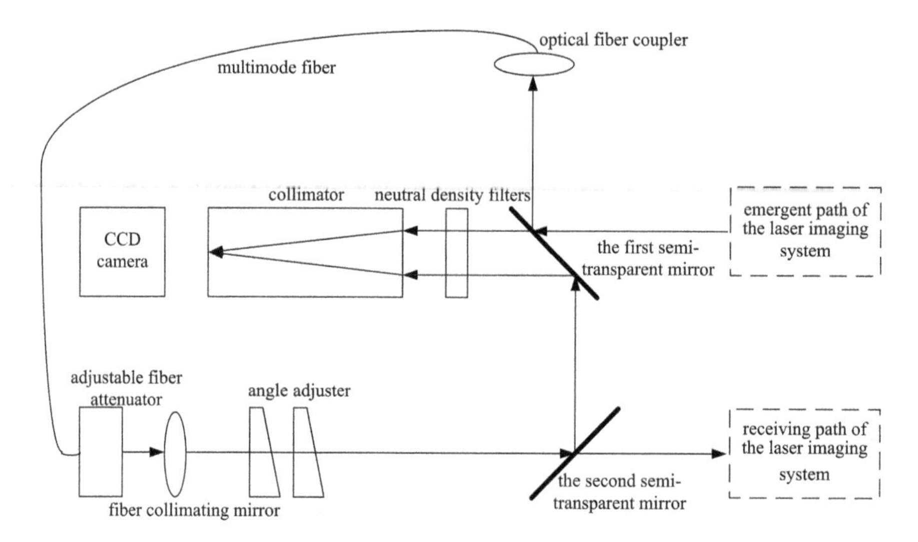

Fig. 3.14 Transmitting and receiving light path calibration for measuring the deviation of the coaxiality

precision, precise calibration parameters are given, thus realizing the calibration of the coaxiality and isolation of the optical path with high precision [14–16]. While, transmitting and receiving optical calibration refers to measuring the coaxiality and isolation deviations using the method similar to the calibration for a single transmitting or receiving light path based on the calibration device of array transmitting and receiving light path. This device is composed of an integrated beam deflector, a collimator, a beam angle adjuster and a generator for simulating echoes that consists of an optical fiber coupler, a multimode fiber, a fiber collimator, and an adjustable fiber attenuator. The specific operating procedure is shown as follows: the reference beam and the simulated laser echoes are adjusted to be parallel using the collimator and the beam angle adjuster so as to precisely calibrate the initial position of the simulated echoes; by adjusting the angle adjuster, the variation of signals output by the receiver in the array transmitting and receiving light path is observed so as to obtain the values of angle adjusted. Afterwards, by comparing these values with those of the initial positions, we can acquire the deviations of the transmitting and receiving axes of the array transmitting and receiving light path in the horizontal and vertical directions. Figure 3.14 shows the technical principles.

3.4.1.1 Optical Path Calibration

One of the laser beams emitted by the array is transmitted to the first semi-transparent mirror which then splits the incident laser beam into two paths: the transmitted light in one path and the reflected light in the other path. The transmitted light, served as the reference light, successively enters into the neutral density filter and the collimator and then is focused on the CCD camera; while the reflected light enters into the simulated echo generator which is composed of an optical fiber coupler, a multimode fiber, an adjustable fiber attenuator and a fiber collimating mirror. The fiber collimating mirror outputs a beam of simulated laser echoes which are successively transmitted by the angle adjuster, reflected by the second semi-transparent mirror and the first semi-transparent mirror successively. Then, the laser echoes enter into the neutral density filter and the collimator and then are converged on the CCD camera. By adjusting the angle adjuster and observing the CCD camera, when the reference light and the simulated laser echo are overlapped at the same point on the focal plane of the collimator, the simulated laser echo is parallel to the reference light. At this time, the initial value of the beam angle adjuster is recorded. The transmitted light of the simulated laser echoes after being reflected by the second semi-transparent mirror directly enters into the receiver of the array optical path.

3.4.1.2 Registration Test of the Transmitting and Receiving Axes

By adjusting the fiber attenuator, the simulated echo energy is attenuated to that equaling to the receiving sensitivity of the array transmitting and receiving light path; the laser echoes are scanned horizontally and vertically by adjusting the beam angle adjuster; according to the scale of the beam angle adjuster, we can calibrate the direction and size of beam deflection. Meanwhile, based on the output of the array transmitting and receiving light path, the reading of the beam angle adjuster is recorded when there are no beams output in the light path. In the mean time, the reading data in the horizontal and vertical directions are written down at the moments when the output appears and disappears. Then, the average values of the two reading numbers in the vertical and horizontal directions are calculated respectively. By comparing these average values with the initial value of the beam angle adjuster, we can acquire the deviations of the transmitting and receiving axes of the light path in the horizontal and vertical directions. On this basis, the transmitting and receiving light path is calibrated through optical adjustment. We can calibrate the coaxiality of all channels of the transmitting and receiving light path successively.

3.4.2 Calibration for Ranging Precision

It aims to calibrate the ranging for each path of multi-path laser ranging and make the obtained ranging data not be influenced by the measurement system and reflect the real range information [17]. The emergent light in one path of the laser imaging system is transmitted to the simulated echo generator where laser signals with certain time delay can be generated. Afterwards, these signals are coupled to the receiving telescope by using a fiber coupler. By using the angle adjuster, the output

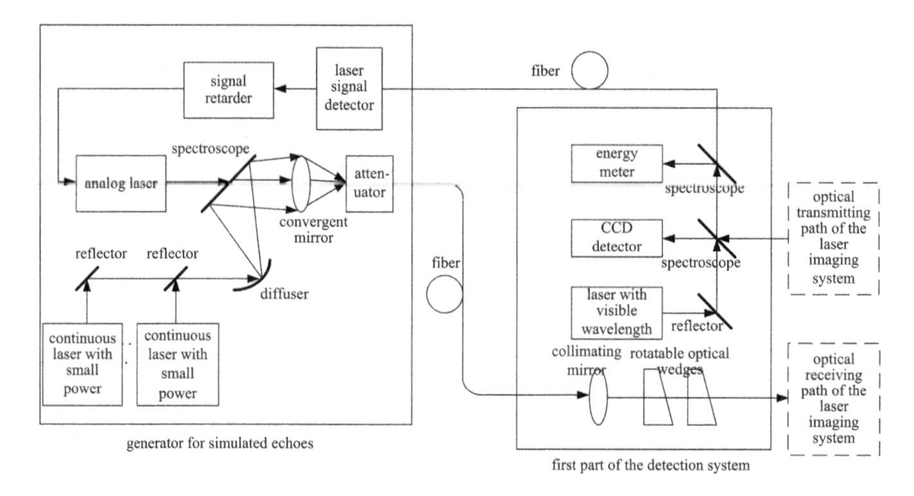

Fig. 3.15 Technical principle of the calibration for the accuracy of array ranging

signals in the corresponding channel are adjusted to be maximal while those in the other channels are the minimum. At this time, the signal intensity is measured. Afterwards, the ranging precision is calibrated by using the ranging error correction algorithm based on the echo intensity. The technical principle of the calibration for ranging precision is presented in Fig. 3.15.

After being semi-reflected by the first spectroscope, the emergent light in one path of the laser imaging system becomes a reflected laser beam, which is then semi-transmitted again to form a transmitted beam to be served as laser signals. These signals are transferred by fibers to the detector for transmitting laser signals which then changes the signals to electric signals. The retarder postpones the electric signals output by the detector and inputs them into the simulated laser. The simulated laser receives and triggers the electric signals output by the retarder and then transmits laser beams, that is, simulated echo signals. After being converged by the convergent mirror and then attenuated by the attenuator, the laser signals are output to the fiber for transmitting the simulated echo signals. Afterwards, they are coupled to the receiving telescope. By adjusting the angle adjuster, the output signals in the corresponding channel is adjusted to be the maximum, while the laser signals coupled to other channel is controlled to be the minimum. At this time, the intensity of the echo signals is measured. Finally, the ranging precision is calibrated by using the ranging error correction algorithm based on the echo intensity.

3.4.2.1 Echo Intensity Detection for Laser Echo Pulses

There exists a corresponding relation between the peak value of laser echoes and the error of laser detection when using the frontier identification method. Therefore, while ranging, ranging errors are possibly compensated as long as the peak voltage

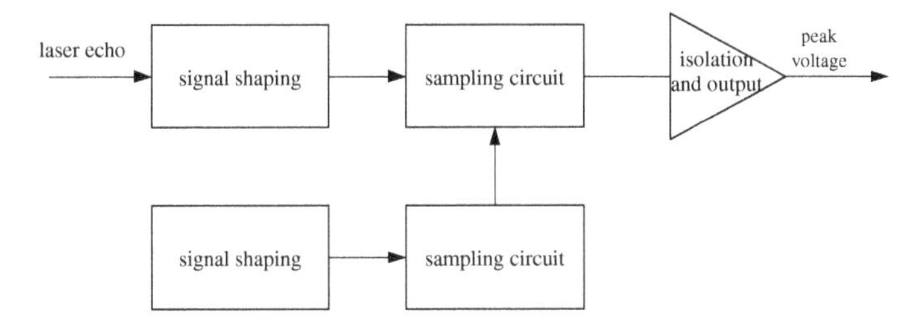

Fig. 3.16 Detection circuit for echo intensity

of laser echoes can be acquired. The peak voltage of laser echoes corresponds to the received laser energy. By combining the ranging information, we can deduce the gray information of targets.

Figure 3.16 shows the detection circuit for the echo intensity. After the signals of laser echoes are shaped, those with high frequency are filtered so that signals are smoothed and broadened. By doing so, the overshoot of the output signals caused by the delayed feedback are prevented. After the identification of the rising edge, the sampling circuit is used to sample the rising edges of echo signals with the falling edge remained the same so as to detect the peak value. After being isolated, the peak voltage is output. Meanwhile, the sequential circuit is utilized to control the bleeder circuit and the sampling circuit is discharged after each test to prevent the influence of backscattering at laser transmission.

The target detection systems with the function of echo intensity detection split the laser echoes into two paths based on the original ranging circuit using frontier identification method: one is the circuit for comparing the given thresholds, which generates stop count pulse for time interval measurement circuit and records the measured time interval values; while the signals of laser echoes in the other path are sent to the measuring circuit for the echo intensity. Meanwhile, the peak value of laser echoes is recorded after analog-digital conversion. The recorded ranging data and detection value of echo intensity need to be processed so as to compensate the ranging error.

3.4.2.2 Ranging Error Correction Method Based on Echo Intensity

The target detection system based on laser imaging developed by the author is capable of detecting echo intensity and therefore can be used to obtain target distance and laser echo intensity. By processing the tested echo intensity and the laser detection value using specific algorithms, this system can be utilized to correct ranging errors.

(i) System calibration

Calibration data can be obtained by using a target detection system based on laser imaging on the ground to measure known targets. By continuously adjusting the emergent laser energy, we can select (M + 1) energy levels of laser radiation including φ_0, φ_1, …, φ_M in a dynamic range. By doing so, the corresponding sampling value of peak values $V(\varphi_i)$ ($i = 0, 1, 2 \ldots M$) and ranging error $\Delta L(\varphi_i)$ ($i = 0, 1, 2, \ldots, M$) are obtained. In this way, the tables describing the relation between the time interval error and the laser echo energy, and that between the laser echo energy and the sampling value of peak values can be acquired.

(ii) Two-point multi-section linear correction through two times of look-up table

To reduce the nonlinearity between the detector and the sampling circuit of echo intensity, it is necessary to perform nonuniformity correction, the core of which is to select a suitable algorithm.

Different from the correction of general infrared systems, the target detection system based on laser imaging is corrected to obtain not only target gray, but also target distance information. Therefore, aiming at the detection system, a new method, that is, two-point multi-section linear correction, is proposed by combining the existing nonuniformity correction methods to correct target gray. The method can ensure obtaining the laser energy reflected by targets; while the two-point multi-section linear correction through two times of look-up table is applied to correct the distance. The method is firstly performed to correct the corresponding relation between the energy received and the sampling values of echo intensity so as to obtain the received energy; during the second time, the corresponding relation between the received energy and the ranging compensation value are corrected. By substituting the energy received obtained at the first time, we can obtain the corresponding ranging compensation values. Then, by compensating the ranging data, the precise target distance can be acquired. The specific method is presented as follows.

Firstly, it needs to judge which interval the peak detection value $V(\varphi)$ belongs to in $V(\varphi_i)$ (i = 0, 1, 2, …, M). Suppose that

$$V(\varphi_i) > V(\varphi) \geq V(\varphi_{i+1}) \tag{3.17}$$

Then, the received laser energy is:

$$\varphi = \frac{V(\varphi) - V(\varphi_{M+1})}{V(\varphi_M) - V(\varphi_{M+1})} \cdot (\varphi_M - \varphi_{M+1}) + \varphi_{M+1} \tag{3.18}$$

Formula (3.18) displays the two-point multi-section linear correction formula for calculating the received laser energy.

After performing the two-point multi-section linear correction for the received laser energy, we can obtain the ranging error $\Delta L(\varphi)$ by conducting the method for a second time. The interval that φ belongs to in φ_i (i = 0, 1, 2, …, M) is determined first, assume that

$$\varphi_m > \varphi \geq \varphi_{m+1} \tag{3.19}$$

Then, the ranging error caused by using frontier comparison method is:

$$\Delta L(\varphi) = \frac{\varphi - \varphi_{m+1}}{\varphi_m - \varphi_{m+1}} \cdot (\Delta L(\varphi_m) - \Delta L(\varphi_{m+1})) + \Delta L(\varphi_{m+1}) \tag{3.20}$$

The finally obtained ranging value is as follows.

$$L'(\varphi) = L(\varphi) + \Delta L(\varphi) \tag{3.21}$$

where, $L'(\varphi)$ shows the finally obtained ranging value; $L(\varphi)$ and $\Delta L(\varphi)$ present the ranging value acquired by using the frontier comparison method and the ranging error caused by using the frontier comparison method, separately.

Meanwhile, the calibrated gray information can be calculated based on the measured laser energy, energy for emitting laser and distance by using relevant formulas.

The laser imaging technology to detect echo intensity has been applied in target detection system based on laser imaging. This system adopts an APD array with 25-element to parallel receive laser echoes from multi-path. The original data obtained in each path are required to be calibrated and corrected to acquire precise data.

By using the proposed method, the laser echo signals in a path is calibrated and corrected. For the same observation target, the emergent laser energy of the laser imaging system is continuously adjusted to record the ranging values and peak voltages of the laser echoes. Figures 3.17 and 3.18 demonstrate the change of original data obtained in the laser detection caused by the variation of amplitude value of the received laser echoes. As can be seen, the peak-to-peak value of the time interval error reaches 6 ns, with the standard deviation of 1.42 ns. Moreover, the variation of the received laser energy also results in the periodic change of the sampling values of the peak voltages.

After the laboratorial calibration, the collected data are processed using the two-point multi-section linear correction through two times of look-up table method. The time interval values after calibration are shown in Fig. 3.19. As can be seen, the peak-to-peak value and standard deviation of the time interval error are 2 and 0.47 ns, respectively.

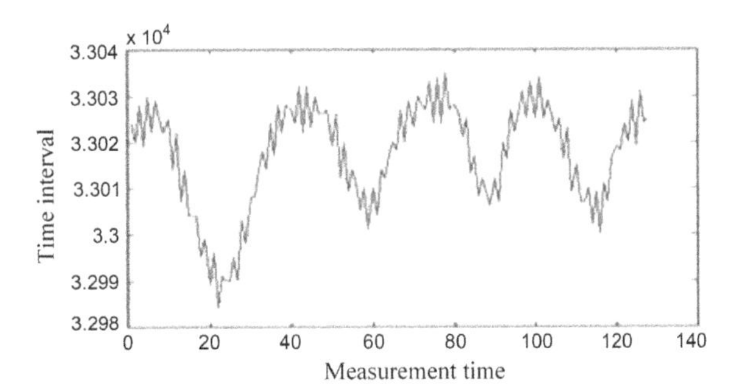

Fig. 3.17 Variation of ranging values caused by the continuous adjustment of the attenuation coefficient of the attenuator

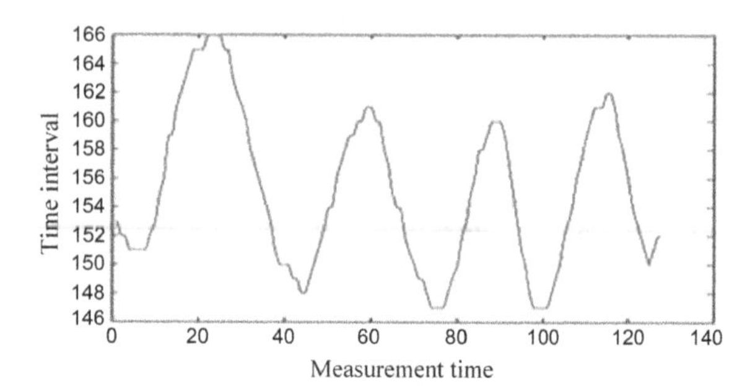

Fig. 3.18 Variation of the sampling values of the echo intensity caused by the continuous adjustment of the attenuation coefficient of the attenuator

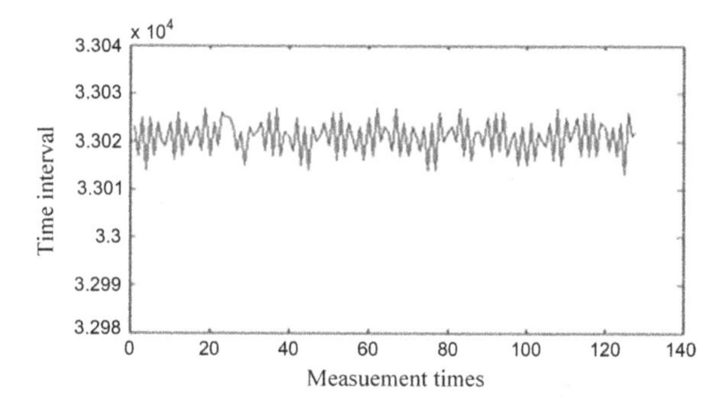

Fig. 3.19 Final ranging results obtained after two times of two-point multi-section nonuniformity correction

Thus it can be seen that owing to the difference of energy received, there is a quite obvious ranging error caused by the utilization of the frontier comparison method. After carrying out the two-point multi-section linear correction for two times, the ranging values are well corrected.

3.4.3 Calibration of Channel Consistency

The target detection system based on laser imaging is performed by using linear array APD devices. Owing to the unbalanced response of APD devices in each channel, and the imbalance of the circuits and in different channels of the time interval measurement circuit, the frontier identification method for laser pulses is used. Due to the usage of a fixed comparable threshold, the variation of the waveform of laser echoes can result in so called "range walk", thus leading to the ranging error.

The time interval value of ranging in each circuit can be calculated using the following formula.

$$T_i = t_i + \Delta t_i + \Delta t_{ij} + \Delta t_n + \Delta t_d \tag{3.22}$$

where, T_i is the measured result; t_i shows the expected measured value of time; Δt_i presents the system errors for ranging, including the system errors of time interval and channel time delay measurement circuits; Δt_{ij} denotes the error caused by the amplitude change of laser echoes when the received energy equals to E_j; Δt_n and Δt_d show the error caused by noises and resolution error for ranging, respectively.

System calibration can eliminate the system errors, but cannot remove random errors. For the convenience of analysis, random errors are not taken into consideration here.

$$T_i = t_i + \Delta t_i + \Delta t_{ij} \tag{3.23}$$

By conducting experiments for sampling and calibrating the echo intensity, we can reduce the system error for ranging and the error caused by "range walk". Figure 3.20 shows the devices used in the calibration experiment. At first, the target detection system based on laser imaging is placed on the horizontal turntable with high precision. Then, laser beams are split horizontally to irradiate remote targets. The specific experimental steps are shown as follows [18].

(i) Under certain energy for emitting the laser, the sampling value of the echo intensity and the distance value in the first channel are recorded.
(ii) The turntable is rotated to an angle (e.g. 2.5 mrad) which enables the laser beams to sample the target, so as to locate the second laser spot on the position of the first spot before rotation. At the time, the sampling value of the echo intensity and the distance value in the second channel are recorded.
(iii) The sampling values of the echo intensity and the distance values in all channels are written down by rotating the turntable continuously. If a 24-element detector is used, totally 24 groups of data are obtained.
(iv) By regulating the energy for emitting the laser, the steps (i) to (iii) are repeated, followed by the record of the sampling values of the echo intensity and the distance values in each channel.

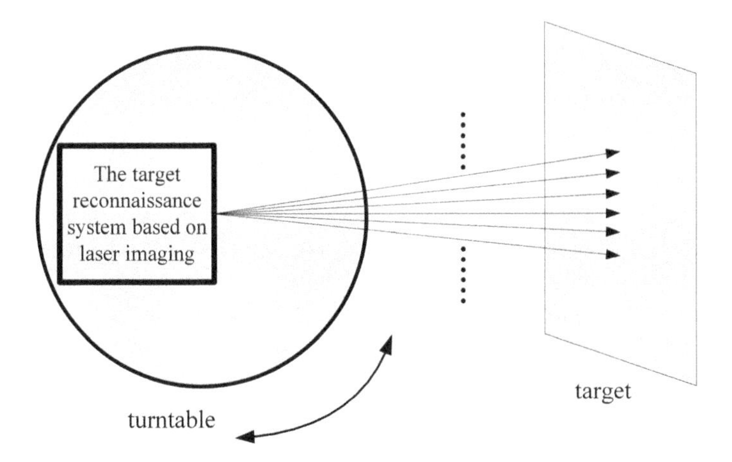

Fig. 3.20 Principles of calibration

The above steps are realized by selecting (M + 1) energy levels of laser radiation including φ_0, φ_1, …, φ_M in a dynamic range, thus obtaining the peak sampling value $V_i(\varphi_j)$ (j = 0, 1, 2, …, M; i = 1, 2, …, 24) and the time interval value $T_i(\varphi_j)$ (j = 0, 1, 2, …, M; i = 1, 2, …, 24) of the *i*th channel. In this way, we can acquire two tables which demonstrate the relative relation between time interval errors and the laser echo energy, and that between the laser echo energy and the peak sampling values.

For the *K*th energy level, as the channels are not saturated, it can be served as a benchmark energy level for comparison.

$$T_{ik} = t_i + \varDelta t_i + \varDelta t_{ik} \tag{3.24}$$

$$T_{1k} = t_1 + \varDelta t_1 + \varDelta t_{1k} \tag{3.25}$$

At the energy level K, the difference between the obtained values in the other channel and the first channel is calculated, as

$$t_i = t_1 \tag{3.26}$$

Then

$$\varDelta T_{ik1k} = T_{ik} - T_{1k} = (\varDelta t_i - \varDelta t_1) + (\varDelta t_{ik} - \varDelta t_{1k}) = (\varDelta t_i - \varDelta t_1) + \varDelta t_{ik1k} \tag{3.27}$$

When different energy levels are applied in the same channel, the difference between the acquired values at the energy level K and other energy level is obtained as follows.

$$\varDelta T_{imik} = T_{im} - T_{ik} = \varDelta t_{im} - \varDelta t_{ik} = \varDelta t_{imik} \tag{3.28}$$

For real targets, suppose that the received energy is φ_x, then

$$T_{ix} = t_i + \Delta t_i + \Delta t_{ix} \tag{3.29}$$

To eliminate errors Δt_i and Δt_{ix}, T_{ix} is corrected as follows.

$$T_{ix} = t_i + \varDelta t_1 + \varDelta t_{1k} + (\varDelta t_{ix} - \varDelta t_{ik}) = t_i + \varDelta t_1 + \varDelta t_{1k} + \varDelta t_{ixik} \tag{3.30}$$

In Formula (3.30), $\Delta t_1 + \Delta t_{1k}$ is a fixed value and can be removed by measuring the targets. The value of $\Delta t_{ix} - \Delta t_{ik}$ varies with the change of the received laser energy and is the key to the elimination of errors.

Firstly, the two-point multi-section method is utilized to obtain the irradiance φ corresponding to the sampling value A of the echo intensity. The sampling value is a monotone function for the laser radiation level.

$$A = V_i(\varphi) \tag{3.31}$$

That is,

$$\varphi = V_i^{-1}(A) \tag{3.22}$$

Based on the calibration results, it can be known that:

$$\varphi_m = V_i^{-1}(A_m)\ (m = 0, 1, 2, \ldots, M) \tag{3.33}$$

To begin with, the interval that the peak detection value $V(\varphi_x)$ belongs to in $V(\varphi_i)$ (i = 0, 1, 2, …, M) is judged. Suppose that

$$V(\varphi_m) > V(\varphi_x) \geq V(\varphi_{m+1}) \tag{3.34}$$

Then, the received laser energy is:

$$\varphi_x^{'} = \frac{V(\varphi_x) - V(\varphi_{m+1})}{V(\varphi_m) - V(\varphi_{m+1})} \cdot (\varphi_m - \varphi_{m+1}) + \varphi_{m+1} \tag{3.35}$$

In this way, the two-point multi-section linear correction is realized for the received laser energy.

By conducting the two-point multi-section linear correction again, the error of time interval measurement can be obtained.

First is to determine the interval that φ_x belongs to in φ_i (i = 0, 1, 2…, M). Suppose that

$$\varphi_m > \varphi_x^{'} \geq \varphi_{m+1} \tag{3.36}$$

Then, the ranging error caused by using frontier comparison method is:

$$\Delta T_{ixik}(\varphi_x) = \Delta t_{ixik}^{\prime} = \frac{\varphi_x - \varphi_{m+1}}{\varphi_m - \varphi_{m+1}} \cdot (\Delta T_{imik} - \Delta T_{i(m+1)ik}) + \Delta T_{i(m+1)ik} \tag{3.37}$$

The final ranging value is shown as follows.

$$T_{ixend} = T_{ix} - \Delta t_{ixik}^{\prime} = t_i + \Delta t_1 + \Delta t_{1k} + (\Delta t_{ixik} - \Delta t_{ixik}^{\prime}) \tag{3.38}$$

where, Δt_1 and Δt_{1k} are fixed values, namely,

$$\Delta t_{ixik} - \Delta t_{ixik}^{\prime} \approx 0 \tag{3.39}$$

$$T_{ixend} \approx t_i + \Delta t_1 + \Delta t_{1k} \tag{3.40}$$

Then, $\Delta t_1 + \Delta t_{1k}$ can be eliminated by comparing with the precise ranging value.

3.5 POS System

The POS system generally includes two parts: a dynamic differential GPS (DGPS) positioning system and an INS-based orientation measurement system. This system is mainly used to obtain the time of the target detection system based on laser imaging, calibrate the position of the system platform and collect the position and orientation information of the platform. In this way, it can determine the directions of laser beams and receiving fields. Moreover, it can determine the geographical coordinates of laser spots by combining the laser detection results.

3.5.1 Dynamic Differential GPS Positioning System

This system plays an important role in the target detection system based on laser imaging with high precision. It is used to measure the state parameters of a moving carrier including position, time, orientation, velocity and acceleration by jointly using the GPS receiver installed on the moving carrier and the receiver placed at one or more datum points on the ground [19]. Therefore, dynamic differential positioning is also named as relative dynamic positioning. According to the different data and methods used, the dynamic differential positioning can be separated into position differential, pseudo-range differential, comprehensive differential of phase and pseudo-range, and carrier phase differential.

The carrier phase differential technique is also called real time kinematic (RTK) technology, which is based on the real time processing of the observed data concerning the carrier phase at two observation stations. It can real-timely provide the 3D coordinates of the observation points with a high precision of centimeter-scale. By using the RTK method, we can eliminate the influence of clock errors of satellites and receivers, the satellite orbit errors, and atmospheric refractive errors, thus improving the positioning precision to centimeter-scale. The key of the RTK technology is to solve the integer ambiguity and detect and restore cycle slips under dynamic conditions.

The values observed by the dynamic differential GPS positioning system are influenced by various errors, including clock offset, tropospheric and ionosphere refractions, ephemeris errors, cycle slips, multipath, disturbance and other noises. In the target detection based on a laser imaging system, the factors influencing the positioning precision of the dynamic differential GPS system contain the cycle slips in the dynamic GPS data, the influence of atmospheric delay errors, multipath errors caused by the machine and electromagnetic interference in the operation areas. In addition, the lock-loss of GPS signals triggers the insufficient number of visible satellites and the long distance between the equipment platform and base station reduces the spatial correlation of errors. Both the factors can influence the positioning precision of the dynamic differential GPS system as well.

In the whole target detection system based on laser imaging, the DGPS can ① provide the accurate position data of the laser imaging sensor in the space, which generally are the 3D geographic coordinates of the rear node of the optical system at the time of pixel imaging in the center of the laser imaging sensor; ② combine with the INS-based orientation measurement system to form a GPS/INS compound orientation determination device so as to jointly filter the GPS and INS data, thus improving the measurement precision of three-axis orientation; and ③ offer the 3D position data of the system platform to the navigation display system, through which the detection platform can be moved accurately and high precise navigation can therefore be realized.

To precisely provide the 3D geographic coordinates of the rear node of the optical system at the time of pixel imaging in the center of the laser imaging sensor, GPS receivers that can receive external pulse and form GPS time mark record in corresponding data strings are required. As the GPS time system and the time system of the laser imaging sensor are not uniform, the time data provided by the GPS cannot be synchronized with the imaging time of the laser imaging sensor. Therefore, electric pulses are simultaneously emitted at the imaging time of the laser imaging sensor. Under the effect of these electric pulses, the GPS receivers form time stamps according to the GPS time serials. During the post-processing of the GPS data, the positioning data (3D coordinate data X, Y, Z) on the time markers are calculated using interpolation method. Afterwards, these data are served as the geographic coordinates of rear node of the optical system at the moment of laser imaging, thus realizing the time matching of laser imaging data and providing key reference systems for laser imaging based target positioning, as shown in Fig. 3.21.

By combining the dynamic differential GPS positioning system with the INS-based orientation determination system, a GPS/INS compound orientation determination device is formed. The GPS receiver provides the position and velocity information for the INS-based orientation determination system. In this way, the course is corrected by the INS system based on the provided information, thus improving the orientation determination precision.

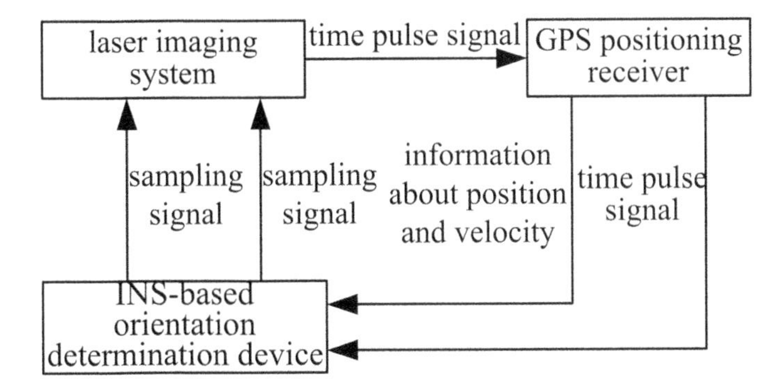

Fig. 3.21 Relationship between the dynamic differential GPS positioning system and other devices

3.5.2 INS-Based Orientation Determination System

The INS-based orientation determination system is mainly used to real-timely provide the attitude angles of the main optical axis of the target detection system based on laser imaging, including pitch angle, roll angle and course angle. In this way, the data of the attitude angles are combined with the differential GPS positioning data to locate the targets. The precision of the orientation determination plays a leading role in the obtaining of high precise positioning for laser spots. Based on the Newton's mechanics laws, the INS establishes the base of space coordinates, that is, the navigation coordinate by using a gyroscope. Then, by using an accelerometer to measure the motion acceleration of carriers, the INS converts the motion acceleration to the navigation coordinate system. After two times of integral operations, the motion parameters including position and velocity of carriers can be finally determined [20]. The INS is characterized by strong anti-interference ability, good concealment, integrity of navigation information and high data update rate. However, it also shows the disadvantages that the error of inertial devices can result in the accumulation of navigation error with time. Therefore, pure inertial navigation systems can not satisfy the requirement of high precise navigation for moving carriers in a long distance for a long time. The core component of the INS is an inertial measurement unit (IMU) composed of various inertial elements including a gyroscope and an accelerometer. Figure 3.22 shows the appearance of a Litton LN-200 inertial measurement unit.

Conversion and integration algorithms for the initial position and orientation as well as acceleration are needed for the INS. Thereinto, the initial position of carriers in the detection system can be obtained with the help of GPS, while the orientation information is acquired by using an initial alignment procedure with the alignment time in a range of 10–15 min. The initial level state is output by the accelerometer

Fig. 3.22 A Litton LN-200 IMU

in absolutely still mode, while the initial azimuths are calculated by monitoring the rotation velocity of the earth using a gyroscope.

Though the INS system is simple in principle, it is difficult to make each component works efficiently and precisely. There are two factors restricting the precision of the INS system: ① the measurement precisions for the angle and the acceleration and ② time measurement precision. Though currently the manufacturers for INS systems can provide complex inertial platform systems with high precision, these systems are also influenced by other working conditions after being installed on the moving carriers. For instance, when the moving carriers are excessively inclined or turned, the systems are possibly be suspended owing to the effect of gravitation and therefore cannot work normally. In addition, other factors such as gyroscope drift and error accumulation are likely to deviate the observed values. The GPS/INS compound orientation determination systems can significantly improve the determination precision of orientation parameters and overcome the error accumulation of the INS systems. Currently, the precision of the carrier orientation determined by high precise INS systems has exceeded 0.01°.

3.6 Typical Target Detection Systems Based on Laser Imaging

Based on the research results of the target detection technology based on laser imaging, the author put forward the idea of staring laser imaging. In addition, a novel target detection approach based on a laser imaging system was designed by combining the detection methods applying the linear array staring and the push-broom imaging with variable frequency. Moreover, by developing the first target detection system based on laser imaging in China, the author realizes precise 3D detection for targets through staring and perspective modes.

3.6.1 Compositions and Functions of the System

Figure 3.23 presents the compositions of the target detection system based on laser imaging when using direct detection with laser echoes developed by the author. This system contains a pulse laser source, a laser controller and scanning mechanism, an optical transmitting system, an optical receiving system, APD arrays, a circuit for amplifying and processing echoes, and a processing circuit for echo signals. In addition, a measuring circuit for echo delay, units for controlling, monitoring and recording, a differential GPS positioning system, an INS system, a target detection and processing system, and a calibration system for the imaging path of the laser array are also included.

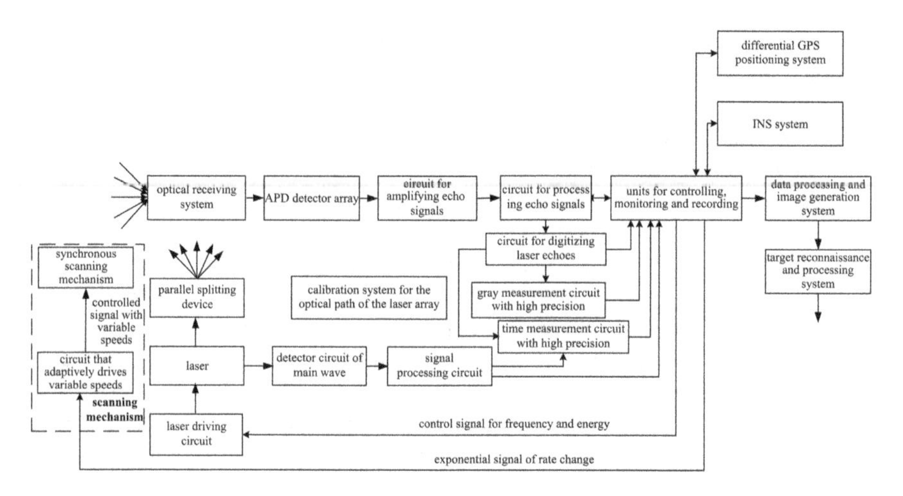

Fig. 3.23 Compositions of the target detection system based on laser imaging

Based on the direct laser imaging detection which jointly combines linear array staring and push-broom with variable frequency, the system detects the information carried by the echo pulses, including distance, reflection intensity and waveform characteristics of targets. In this way, the pixel distance, reflectivity data and waveform images of targets can be acquired. By combining differential GPS positioning data and the orientation data obtained by the INS system, we can directly obtain the range images and gray images of targets. Moreover, the generated laser images completely match with the 3D coordinates so that no additional reference control points are needed. By processing the generated images of targets, the 3D targets can be identified and classified, thus realizing active, direct and stereoscopic laser imaging based target detection with high resolution, efficiency, and precision.

This system realizes several aims including adaptive target detection with variable resolution which combines effective searching and high precision, and imaging detection and 3D positioning for targets with high precision. In addition, characteristics identification for the fine structure of targets and anti-shadowing detection, and on-line calibration for the optical paths and circuits of staring laser imaging systems are also achieved. These functions are realized by using various key technologies including intelligent and adaptive scanning for laser imaging, laser splitting emission with variable angles, array laser detection, and multi-dimensional image generation. Meanwhile, the following technologies are also used: generation of the laser with variable repetition pulses, calibration for array laser imaging systems, target 3D detection and processing, and multi-level information acquisition for targets using laser imaging radars.

3.6.2 Main Performance Parameters

The target detection system based on laser imaging developed by the author employs a novel optical detection method. The system firstly performs array staring detection in one dimension with high sensitivity and precision and then samples targets with variable resolution by scanning at variable speeds. In addition, this system combines the 3D positioning of laser imaging, information acquisition with high resolution of staring detection, dynamic sampling of targets of adaptive scanning with variable frequencies, and 3D detection and processing of multi-source information fusion. Therefore, it presents various functions including push-broom staring, laser imaging and 3D detection together. Based on these functions, it can conduct multi-beam staring, adaptively scan targets at variable speeds, and acquire and process the 3D information with high precision and sensitivity for targets of multiple layers. In this way, the rapid discovery and location of the 3D coordinate and identification of the 3D structures of targets with high precision are realized. The system is characterized by quick sensing, high precise 3D positioning and accurate identification for targets.

The main performance parameters of the system are presented as follows: by using a 24-element linear array detector, the system performs push-broom scanning every 2–20 times per second in a range of 45°. With a scanning field angle of 24 × (0.125–1.25 mrad), the targets are scanned at a frequency of 24 × (2–20) line/s to present the gray quantization of 8 bits and the sampling rate of targets can be equivalent to 65–650 kHz. The scanning speed and frequency, field angle and sampling rate can be automatically adjusted.

References

1. Rongrui W (1999) The state of arts on solid state LADAR technology. Laser Infrared 29(6):323–326
2. Cole TD, Cheng AF et al (2000) Flight characterization of the NEAR laser rangefinder. SPIE 4035:131–142
3. Xiping C, Lijie Q, Libao L et al (1999) State of the art of CO_2 imaging laser radar. Laser Infrared 6:327–329
4. Qiming W (2010) Breakthroughs and developments of semiconductor laser in China. Chin J Lasers 37(9):2190–2197
5. Sujie G, Jie W (2003) The semiconductor and its applications in military. Laser Infrared 33(4):311–312
6. Zhihui S, Jiahao D, Xiaowei G (2008) Progress and current state of the development of laser imaging detection system and its key techniques. Sci Technol Rev 26(3):74–79
7. Richard MM, Brian FA, Richard MM et al (2001) Three-dimensional laser radar with APD array. SPIE 4377:217–228
8. Richard MM, Timothy S, Robert EH et al (2003) A compact 3D imaging laser radar system using Geiger-mode APD arrays: system and measurements. SPIE 5086:1–12
9. Xinju L, Guobiao H, Yunan Z et al (1983) Laser technology. Hunan Science & Technology Press, Changsha, pp 56–82

10. Jianjun Q (2011) Design of the chaotic optical transreceiver units for chaotic lidar. Taiyuan University of Technology, Taiyuan
11. Yongchao Z, Mingjun Z, Wenping Z (2006) Trend of laser radar technology development. Infrared Laser Eng 35:240–246
12. Haihong Z, Jianyu W, Yihua Hu (2006) Effect of solar radiation on the receiver of near infrared laser altimeter. J Infrared Millimeter Waves 25(4):426–428
13. Genghua H, Junhua O, Rong S et al (2009) Influence of background radiation power on the signal-to-noise ratio of space-borne laser altimeter. J Infrared Millimeter Waves 28(1):58–61
14. Yuwei C, Zhiping H, Yihua H et al (2006) The study of the technology of computer aided calibration on active optical system. Laser Infrared 36(1):54–57
15. Huige D, Test on the registration of the receiving and transmitting axes of the laser detection system: China, 200710040397.7.2007-10-24
16. Yuwei C, Calibration devices and method of laser ranging system: China, 200410025632. X.2007-08-29
17. Haihong Z, Laser altimeter with multiple functions: China, 200510029959. 9.2008-03-12
18. Yihua H (2003) Key technologies of airborne push-broom laser 3D imaging. Shanghai Institute of Technical Physics, the Chinese Academy of Science, Shanghai
19. Renqian W (2004) Theory of kinematic positioning of GPS. Central South University, Changsha
20. Wei Q, Baiqi L, Xiaolin G et al (2011) INS/CNS/GNSS integrated navigation technology. National Defense Industry Press, Beijing

Chapter 4
Acquisition of Detection Information

The acquisition of detection information is considered as the basic function of the target detection based on a laser imaging system. It is designed to obtain relevant information for generating laser images and therefore realizing target detection and provide all kinds of data needed. The acquisition of detection information includes target sampling, beam scanning, echo detection, echo processing, delay and intensity measurement, waveform acquisition, collection of orientation and positioning data, synchronous labeling, storage and record of data, and so on. Among them, target sampling and beam scanning play the important roles in effectively acquiring information. In this chapter, the author states the basic theories of the acquisition of detection information for the target detection based on laser imaging system designed by the author. Moreover, the author explains the concrete realization methods for target sampling, beam scanning, and the processing and acquisition of the detection data.

4.1 The Target Sampling and Beam Scanning

Any imaging mode cannot be realized before sampling targets. The combination of a large number of sampling points forms an image. When taking pictures using films, the minimum sampling parameter for targets is based on the particle size of silver halide. As for digital photography, the sampling parameter of targets depends on the focus of the camera and the size of detector elements. In passive imaging, the sampling of targets is called receiving sampling. The imaging of a whole target area is generally realized by the scanning of a single-detector for the receiving field of view or the separate sampling of each element of an array detector for the targets. Laser imaging, as an initiative detection way, apart from receiving the sampled information, its major task is to perform laser radiation, namely, emission sampling, for the sampling points while receiving the information. In the process, a laser spot

Y. Hu, *Theory and Technology of Laser Imaging Based Target Detection*,
DOI 10.1007/978-981-10-3497-8_4

is formed. So, the synchronization of laser emission and receiving of sampled information has to be paid more attention, particularly the sampling effects in dynamic scanning or when the platform is moved. While detecting a specific target, it is necessary to sample the target with variable resolution by scanning it at an adaptively varying speed. In view of the emission sampling and scanning of single beams, as well as the receiving of the sampled information and the scanning of single-detectors, this section analyzes their sampling characteristics and effects of scanning and detecting. The analysis results are suitable for the target sampling and scanning based on the emission and receiving of array lasers.

4.1.1 Characteristics of Laser Spot in Emission Sampling

Emission sampling is to radiate a laser beam to directionally illuminate a target and therefore form a laser spot on a point of the target. The point is the sampling point for emitting the laser. The characteristics of the laser spot such as the shape and size are related to the divergence angle and emergent angle of the laser, the structure and dip angle of the target, and so on.

For a fundamental Gaussian laser beam, suppose that the divergence angle is α. If the surface normal of a planar target overlaps the emergent axis of the laser, then the formed laser spot is a standard ellipse. While in fact, as the conditions cannot be so standard, the laser spot is basically an ellipse with the long axis lying in the deflection angle of the laser beam. For example, when a laser beam scans along cross-track direction of the airborne platform, the long axis is in the cross-track direction, while the short axis is in the flight direction. As shown in Fig. 4.1, the dip angle of the laser beam is ϕ (the angle between the laser beam and the nadir) and the receiving field angle is β. Suppose that the long and short axes of the laser spot are a_L and b_L respectively, which are calculated in the following situation. The characteristics of the formed laser spot are analyzed by taking the air-to-ground laser imaging based detection as an example.

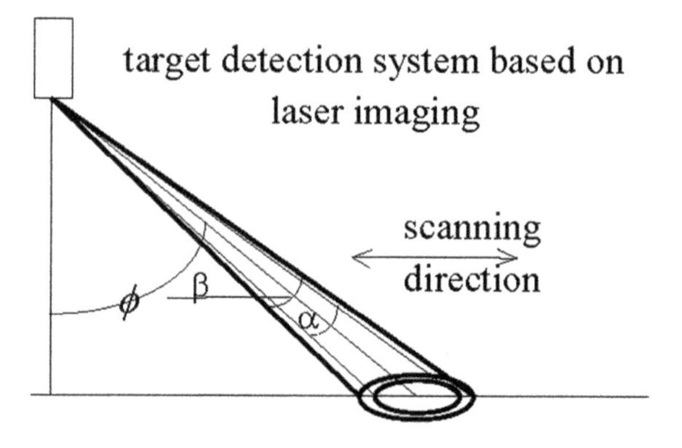

Fig. 4.1 The formation of the laser spot

(i) When the detection platform is horizontal with an elevation of H and the laser beam irradiates vertically the horizontal target, the long axis a_L and the short axis b_L are expressed as:

$$a_L = D + H\ \tan(\alpha/2) \quad b_L = H\ \tan(\alpha/2) \tag{4.1}$$

In general cases, when the detector aperture D and the divergence angle of the laser beam are relatively small, that is, $\tan(\alpha/2) \approx \alpha/2$, formula (4.1) can be simplified as

$$a_L = H \cdot \alpha/2 \quad b_L = H \cdot \alpha/2 \tag{4.2}$$

(ii) In case that the detection platform is horizontal with a vertical distance H to the ground and the laser beam irradiates the horizontal target with a dip angle ϕ, the long axis a_L and the short axis b_L are separately presented as:

$$a_L = \frac{H}{2}\left[\tan(\phi + \frac{\alpha}{2}) - \tan(\phi - \frac{\alpha}{2})\right] \quad b_L = \alpha H\ \sec\phi/2 \tag{4.3}$$

(iii) When detection platform is horizontal with an elevation H and the laser beams illuminates the inclined target (presenting a slope of δ) with a dip angle φ, the long axis a_L and the short axis b_L are presented as:

$$a_L = \frac{H\ \sec\phi}{2}\left(\frac{\sin\frac{\alpha}{2}}{\cos(\phi - \delta)} + \frac{\sin\frac{\alpha}{2}}{\cos(\phi + \alpha - \delta)}\right) \quad b_L = \alpha H\ \sec\phi/2 \tag{4.4}$$

4.1.2 The Detection Results Based on Dynamic Beam Scanning

The coaxiality of emitting and receiving is the basis for realizing laser imaging based detection. It generally requires the laser spot to be located in the receiving footprint of targets and therefore the transmitting field of view can be ensured to be within the receiving field of view in static conditions, namely, not scanning, through systematic design. However, when transmitting lasers to scan and receiving echoes in the dynamic situations (that is, the platform is moved or scanning mirrors are rotated continuously), the dynamic coaxiality of emitting and receiving is difficultly to be realized. Especially, this situation is more serious while scanning using a satellite for ground targets at a high speed or with a long operating range.

When the scanning detection based on dynamic beams is carried out, the laser beam is emitted at a certain moment with a directional angle of ϕ (the included angle between the laser beam and the direction of the nadir). After a certain time delay of the echoes reflected from the target, the echoes enter in the optical

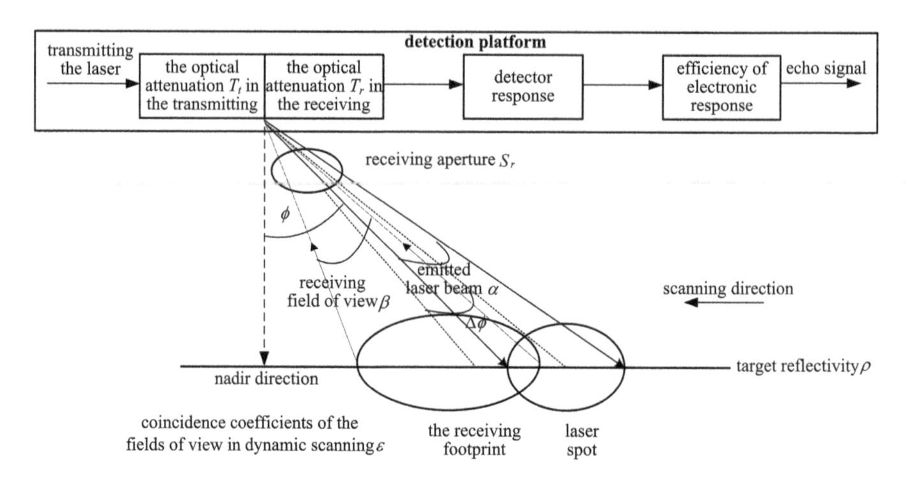

Fig. 4.2 The receiving footprint and the spot of the laser footprint

receiving system by penetrating through the scanning mirror because of the continuous rotation of the scanning mirror and the moved detection platform. Obviously, the deflection of the scanning mechanism at this moment is inconsistent with that while transmitting the laser. That is, the instantaneous receiving field of view presents an angle difference $\Delta\phi$ with the transmitting direction of the laser beam, which causes the dynamic deviation of the field of view, as shown in Fig. 4.2. This phenomenon shows that the coincidence degree between the laser spot and the receiving footprint is small. This dynamic deviation of the transmitting and receiving fields of view cannot be eliminated even if the fields of view are completely overlapped in static condition through precise adjustment. As a result, there are errors in the scanning detection results based on dynamic beams.

The scanning detection results based on dynamic laser beams can be described as the relatively effective receiving powers of a standard reflector and presented by the dynamic coincidence coefficients. The dynamic coincidence coefficient is defined as the ratio of the overlapped area between the laser spot and the receiving footprint to the area of the laser spot. Meanwhile, the effective receiving power of the laser detector is directly proportional to the dynamic coincidence coefficient [1]. Obviously, if the dynamic coincidence coefficient is very small, it is difficult to obtain preferable detection results through dynamic scanning even if favorable detection results can be obtained in static conditions. Figure 4.3 presents the relative relations concerning the definition of the dynamic coincidence coefficient and the overlapped area is presented by the shadow. Only the backscattering laser signals reflected from this area can be received by the detector.

According to Figs. 4.2 and 4.3, if the deviation angle $\Delta\phi \geq (\alpha + \beta)/2$, the spot is entirely outside the footprint, as showed in Fig. 4.3a. Under such circumstance, the coincidence coefficient is 0. In contrast, if the deviation angle $\Delta\phi \leq (\beta - \alpha)/2$, the spot is entirely within the footprint, as showed in Fig. 4.3b. In this case, the coincidence coefficient is 1.

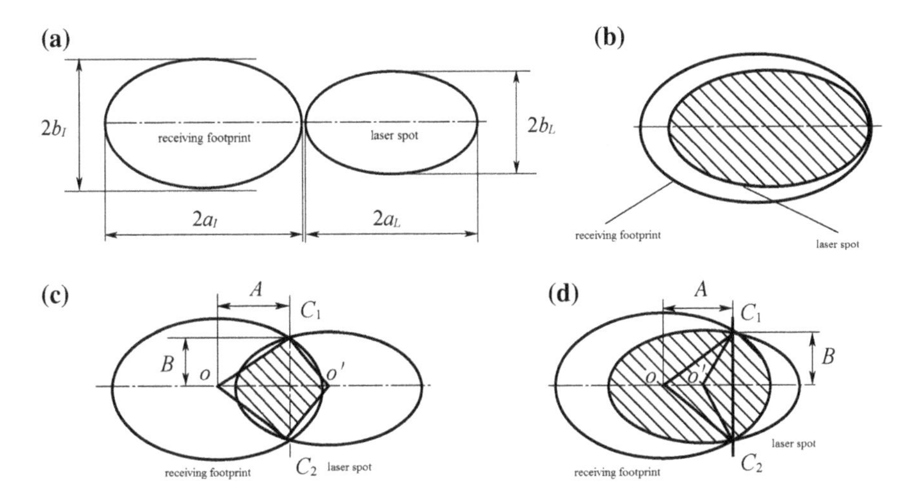

Fig. 4.3 The relative relationships between the laser spot and the receiving footprint. **a** the laser spot is outside the footprint, **b** the laser spot is inside the footprint, **c** and **d** part of the laser spot is inside the footprint

Suppose that the long and short axes of the laser spot are a_L and b_L respectively and those of the laser footprint are a_I and b_I, they are expressed as

$$a_L = \alpha H \sec^2 \phi / 2 \tag{4.5}$$

$$b_L = \alpha H \sec \phi / 2 \tag{4.6}$$

$$a_I = \beta H \sec^2 \phi / 2 \tag{4.7}$$

$$b_I = \beta H \sec \phi / 2 \tag{4.8}$$

In the case of linear scanning, the long axes of the two ellipses are overlapped. In this case, the distance between the centers of the two ellipses is

$$\Delta l = oo' \approx H \Delta \phi / \cos^2 \phi \tag{4.9}$$

Suppose that the distance from one of the intersection points ($C1$ or $C2$) of the two ellipses to the long axis is B, the distance from the line $C1C2$ connecting the two intersection points to the center o of the receiver is A, then:

$$\begin{cases} \dfrac{A^2}{a_I^2} + \dfrac{B^2}{b_I^2} = 1 \\ \dfrac{(\Delta l - A)^2}{a_L^2} + \dfrac{B^2}{b_L^2} = 1 \end{cases} \tag{4.10}$$

Based on formula (4.9), Eq. (4.10) can be acquired as follows:

$$A = \frac{\Delta l}{2} + \frac{a_L^2}{2\Delta l}\left((\beta/\alpha)^2 - 1\right) \tag{4.11}$$

$$B = b_L\sqrt{(\beta/\alpha)^2 - A^2/a_L^2} \tag{4.12}$$

If $\Delta l - A \geq 0$, then $\Delta\phi \geq \sqrt{\beta^2 - \alpha^2}/2$. The area of the shadow region in Fig. 4.3c is

$$S_{LI} = a_L b_L \arccos\frac{\Delta l - A}{a_L} + a_I b_I \arccos\frac{A}{a_I} - \Delta l \cdot B \tag{4.13}$$

If $\Delta l - A \leq 0$, then $\Delta\phi \leq \sqrt{\beta^2 - \alpha^2}/2$. The area of the shadow region in Fig. 4.3d is

$$S_{LI} = a_L b_L\left(\pi - \arccos\frac{A - \Delta l}{a_L}\right) + a_I b_I \arccos\frac{A}{a_I} - \Delta l \cdot B \tag{4.14}$$

The area of the laser spot is

$$S_L = \pi \cdot a_L \cdot b_L \tag{4.15}$$

According to the definition, the dynamic coincidence coefficient can be summarized as [2]

$$\varepsilon = S_{LI}/S_L = \begin{cases} 1 & (\Delta\varphi \leq (\beta - \alpha)/2) \\ 1 - \dfrac{\arccos\left(\dfrac{A - \Delta l}{a_L}\right)}{\pi} - \left(\dfrac{\beta}{\alpha}\right)\dfrac{2\arccos\left(\dfrac{\alpha A}{\beta a_L}\right)}{\pi} - \dfrac{\Delta l B}{\pi a_L b_L} & \left((\beta - \alpha)/2 < \Delta\varphi \leq \sqrt{\beta^2 - \alpha^2}/2\right) \\ \dfrac{\arccos\left(\dfrac{\Delta l - A}{a_L}\right)}{\pi} - \left(\dfrac{\beta}{\alpha}\right)\dfrac{2\arccos\left(\dfrac{\alpha A}{\beta a_L}\right)}{\pi} - \dfrac{\Delta l B}{\pi a_L b_L} & \left(\sqrt{\beta^2 - \alpha^2}/2 < \Delta\varphi \leq (\beta + \alpha)/2\right) \\ 0 & (\Delta\varphi > (\beta + \alpha)/2) \end{cases} \tag{4.16}$$

According to the above formula, the dynamic coincidence coefficient ε is related to multiple factors. These factors include the divergence angle α of the laser spot, the instantaneous angle β of the receiving field of view, the height H of the platform where the data acquisition system is located, the directional angle ϕ of the laser beam, the scanning rate of scanning mirror, and so on. It directly affects the size of the laser detection power in the form of a multiplication factor, and also determines the detection results of the laser scanning.

Above is the analysis of scanning conditions in the cross-track direction. Similar analysis can be conducted for the dynamic deviation of the fields of view caused by the platform motion along the flight direction. If the imaging detection system is carried by an aircraft, the influence of the movement can be neglected due to the smaller height and the variable velocity of targets, while the influence cannot be ignored in satellite-to-ground detections. For example, suppose that H = 450 km and the velocity is 8000 m/s, then the time difference from laser emission to the receiving of echoes is 3 ms, during which the satellite has moved 24 m.

4.1.3 Adaptively Variable-Speed Scanning for Target Sampling with Variable Resolutions

In laser imaging, to ensure that the entire target area is sampled, the rates for laser transmitting and receiving of sampled information need to be high enough. Under such circumstance, if the repetition frequency of the laser is high enough, the target sampling can be achieved at a high density. Afterwards, 3D images of targets can be generated by processing the signals of laser echoes. The denser the sampling of the targets is, the smaller the laser spot and the receiving footprint and therefore the higher the resolution of the generated images.

When laser scanning for sampling and generating images is performed on a moving platform, the size of the sampling interval for targets (reflecting the resolution of the target) is inversely proportional to the repetition frequency of the pulse laser. In contrast, it is proportional to scanning rate and flight speed platforms. Therefore, to reduce the sampling interval and make the sampling points denser, it is necessary to improve the laser pulse repetition frequency and reduce flight and scanning rates. However, the laser pulse repetition frequency is difficult to be improved to a high level due to the limitations of various technical conditions. Furthermore, slowing down the flight speed of the sensor platform is likely to bring disadvantages such as increasing the detection time and low efficiency, which make the real-time performance difficult to be realized. So, only reducing the scanning rate can solve this problem. However, the decline of scanning rate can bring the decrease of the number of scanning lines within a certain operating time of the platform. This is supposed to increase the sampling interval in the flight direction. That is, the decline of scanning rate can only improve the sampling resolution in the direction perpendicular to the moving direction of the platform carrying the detection system. While the sampling resolution in the direction parallel to the moving direction of the platform is likely to be decreased. Obviously, this result is not desirable. Then, what means can be utilized to achieve satisfactory sampling interval? This problem can be solved by a compromise way, that is, scanning at an adaptively varying speed [3]. According to the method, the scanning rate can be reduced to decrease the sampling intervals for interested targets; while in the non-target area, the scanning rate can be increased. By using the method, the target

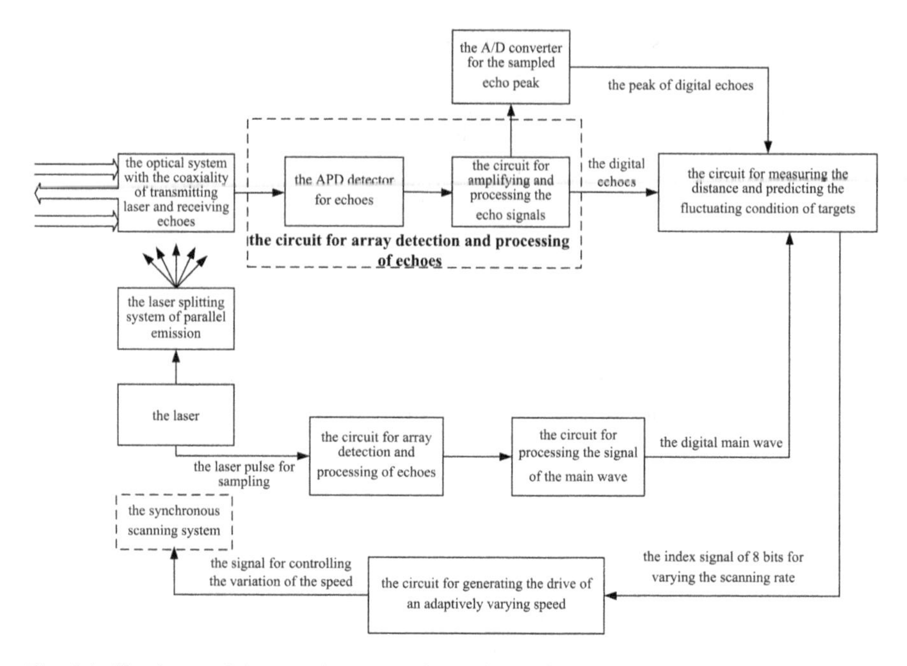

Fig. 4.4 The theory of the scanning at an adaptively varying speed

sampling with variable resolution can be realized, with probably slight change in the total equivalent scanning rate.

According to Nyquist Sampling Theorem, the sampling rate which can reflect target features is twice faster than the variable velocity of targets. Therefore, to obtain better results of the target detection using a laser imaging system, an appropriate sampling rate needs to be employed. No doubt that employing a high sampling rate can acquire favorable results, while it can result in waste and the high rate is hard to be realized at some times. While sampling at a very low rate is likely to lose the 3D features of targets. In the case of limited resources, real-time adjustment of sampling rate based on the predictions of the fluctuating condition of targets is expected to provide favorable laser imaging effects [3]. On the promise of fixed repetition frequency of laser and flight speed of the sensor platform, the scanning at an adaptively varying speed can ensure the generated images of targets presenting high spatial resolution and gray scale fidelity. In this way, the staring detection for interested targets can be realized, so as to achieve the detection with high resolution and the efficient scanning.

Figure 4.4 shows the specific theory of the scanning at an adaptively varying speed. The emission laser beam is divided into the sampling laser beam and the detection laser beam by the laser beam splitter. Then the sampling laser beam is converted to digital main wave through the circuit for the photoelectric detection and the signal processing of main wave and transmitted to the input terminal of the circuit for measuring the distance and predicting the fluctuating condition of targets.

The backscattering echo signals reflected from targets form the digital echo signals through the circuit for detecting and processing echoes. The echo signals are transmitted to two circuits: the another input terminal of the circuit for measuring the distance and predicting the fluctuating condition of targets, and the A/D circuit for sampling echo peeks. According to the predicted fluctuating conditions of targets, the former circuit sends an index signal of 8 bits for varying the scanning rate to the circuit for generating the drive of an adaptively varying speed. Then the circuit offers the driving signals to the stepping motor with variable scanning rates to drive the scan mirror to scan at an adaptively varying speed.

The scanning at a varying speed is realized by a loop which controls the step-scan motor to drive the scanning mirror. The loop is composed of optics devices and a laser detection system, a variable-speed stepping motor, the circuit for measuring the distance and predicting the fluctuating condition of targets (including prediction algorithm), and the circuit for generating the drive of an adaptively varying speed. The key technologies of adaptively variable-speed scanning are the circuit for measuring the distance and predicting the fluctuating condition of targets (prediction algorithm) and the circuit for generating the drive of an adaptively varying speed. The former is the core of variable-speed scanning and has to calculate the fluctuating characteristics for the structure and gray scale of targets in the next area based on the currently measured results regarding the distance. These fluctuating characteristics including fluctuation frequency of spatial structure, reflectivity, and distance fluctuation, are a basis for controlling the scanning rates. The circuit for measuring distance and predicting the fluctuating condition of targets is shown in Fig. 4.5. This circuit comprises the circuits for converting the pulse of the main wave and the echo pulse, generating the distance gate, and measuring the

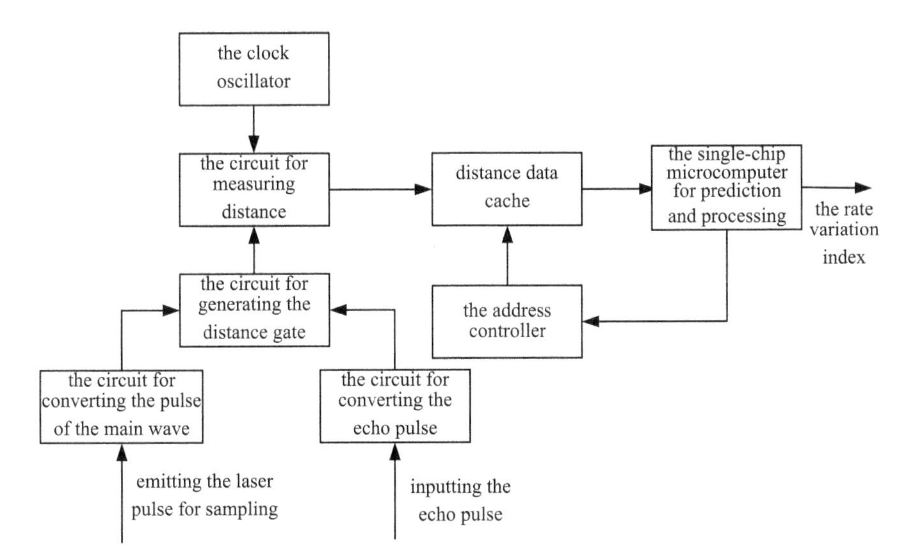

Fig. 4.5 The composition of the circuit for measuring distance and predicting the fluctuating condition of targets

distance, a clock oscillator, a distance data cache, an address controller, the single-chip microcomputer for prediction and processing, etc.

The working process of circuit for measuring the distance and predicting the fluctuating condition of targets is as follows: the data provided by the circuit of time delay of the echoes are converted into the target distance. Then these distance data are transmitted into the cache for distance data, which stores all the data in the scan lines prior to the current sampling point. The amount of the data is expected to meet the requirement of the circuit for prediction and processing. Moreover, the single-chip microcomputer for prediction controlled by the program calls and processes the previous distance data and thereby estimates the fluctuating condition of targets. Then the single-chip microcomputer predicts the fluctuating condition of the subsequent targets, so as to provide the variation index of the scanning rate, which is sent to the circuit for generating the drive of an adaptively varying speed. The rate variation index is a binary number of 8 bits. Then the rate variation can be divided into 256 grades.

The prediction procedure is shown in Fig. 4.6, the data source of which come from the data cache. The basic method for the prediction is as follows:

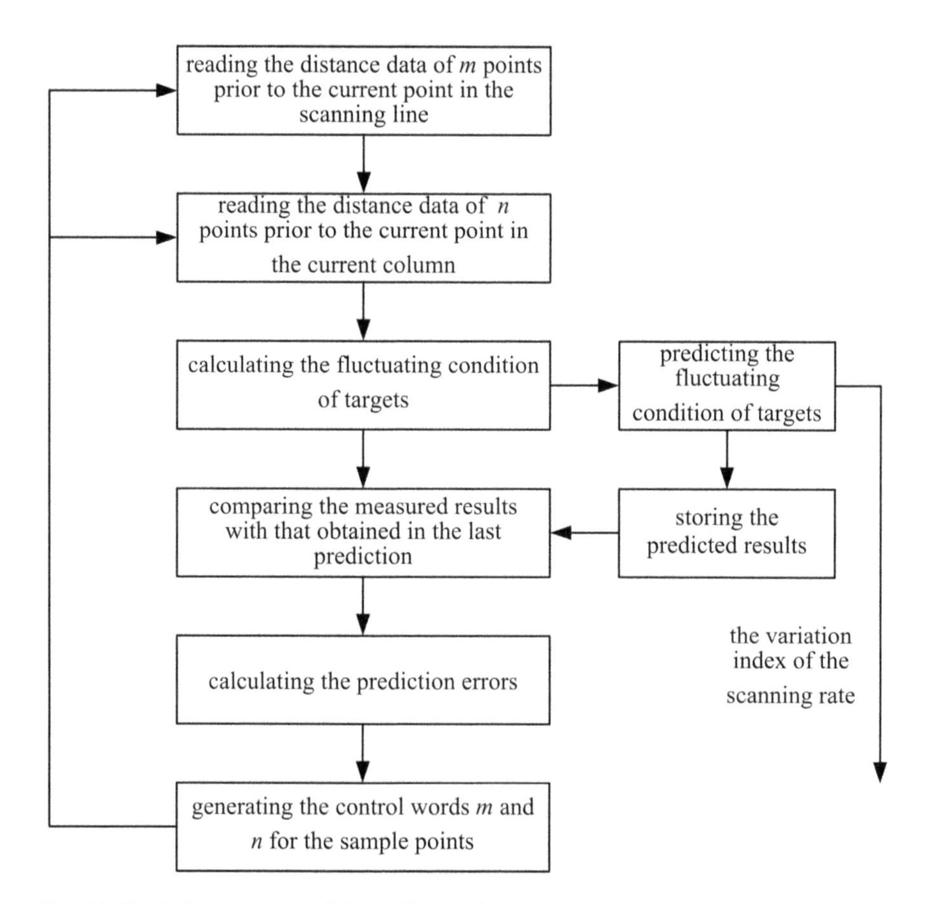

Fig. 4.6 Prediction procedure of the prediction circuit

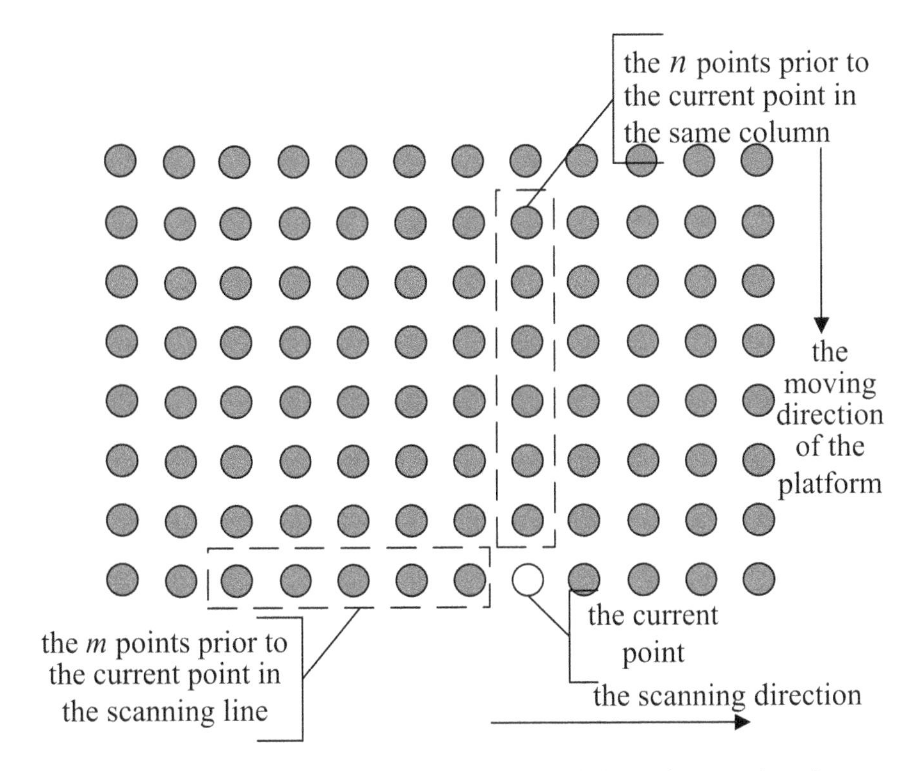

Fig. 4.7 The relation between the current sampling point and the previous m and n points

(i) To begin with, the distance data of m points located in the current scan line and those of n points located in the current scan column (along the track direction) obtained prior to the current moment are selected, as shown in Fig. 4.7. The target points corresponding to these two types of data are intersected at the current point and basically reflect the conditions surrounding the targets.

(ii) Then, the fluctuating condition which refers to the spatial frequency and amplitude of the fluctuation in target height of previous targets is calculated. Through the linear prediction operations on them, fluctuating condition of the target surrounding the current sampling point can be inferred.

(iii) Next is to provide a variation index for the fluctuating condition of targets. As the basis for adjusting the rate of the signal for controlling the variable speed, the index is transmitted into the circuit for generating the drive of an adaptively varying speed. Then, the scanning rate for the sequential points in this line and the points in the columns intersected with the subsequent scanning lines is supposed to be controlled according to the predicted results.

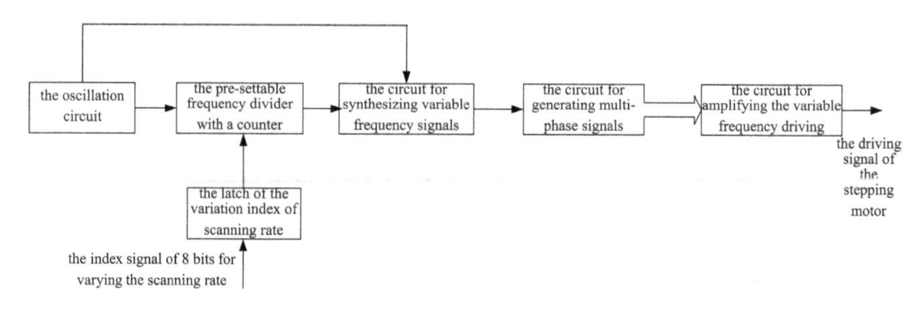

Fig. 4.8 The composition of the circuit for generating the drive of an adaptively varying speed

In the prediction process, to ensure the high prediction precise, it is important to compute the errors between the predicted and the measured results of the last time. Then control words m and n for the number of sample points are generated as the basis for selecting the number of the current sample points predicted. The number of the predicted sample points is the number m of the distances data of target points in front of the current point in the scanning line and the number n of the distances data of target points prior to the current point in the current column. The two numbers influence the prediction error.

The circuit for generating the drive of an adaptively varying speed is illustrated in Fig. 4.8. The circuit receives the variation index of scanning rate transmitted from the circuit for measuring distance and predicting fluctuating condition of targets, and then sends the variation index to the corresponding latch. By applying the data to the preset terminal of pre-settable frequency divider with a counter, the frequency division ratio of the signal frequency output by the oscillation circuit can be changed.

Because the frequency division ratio is varied regularly among the integers from 1 to 256, the interval of frequency conversion output by the frequency divider changes in the double manner. So it is necessary to use a circuit for synthesizing variable frequency signals to synthesize the signals undergoing and not experiencing frequency division according to certain a rule so as to vary the frequency in accordance with the arithmetic law. Therefore, the output pulse frequency of the circuit can be changed at same intervals. After being output and passing through the circuit for generating multi-phase signals, the signal outputs several signals with variable frequency and different phases to the circuit for amplifying the variable frequency driving. The number of the signals is determined according to the requirements of the stepping motor. The circuit for amplifying the variable frequency driving improves the driving ability of the signals by amplifying their power and then sends the signals to the stepping motor. Because the driving signal of the stepping motor is pulse trains, the pulse interval of which determines the rotation rate. So the major function of the circuit for controlling the frequency of the driving signal used to drive the stepping motor is to real-timely alter the pulse interval. This circuit is realized by the frequency dividing circuit in which the frequency dividing ratio is pre-settable. The frequency dividing ratio is the rate variation index sent from the last circuit. As the pulse train frequency output by the oscillation circuit is high, about dozens of times of the actual frequency required by the stepping motor, the time interval is many times smaller than the smallest interval of the driving signal of the stepping motor. At this time, after the rate variation index is received, the time interval between the current and last pulses output by the circuit for controlling the frequency of the driving signal can be changed immediately, thus outputting the signal for controlling the variation of the speed.

The stepping motor for scanning at a variable speed is the executor for realizing the scanning at an adaptively varying speed. Although the function can be realized by an ordinary stepping motor, the selection of the stepping motor has to consider the parameters of the scanning mirror and the speed of the laser scanning. This is because the selection of load capacity of the stepping motor is determined by the rotational inertia and the starting torque of the scanning mirror. Moreover, the selection of the rotation rate of the stepping motor depends on the speed of the laser scanning for targets. Meanwhile, the stepping interval is selected according to the requirement of the detection system for the minimum interval of sampling based scanning. To change the scanning rate, it needs to alter the frequency of the driving pulse. Therefore, the scanning at a variable speed can be realized by applying driving signals with variable speeds to the stepping motor.

4.2 The Information Acquisition for the Target Detection Based on a Laser Imaging System

The information acquisition in the target detection based on a laser imaging system involves the formation of the laser beam to be emitted for sampling, the reception and detection of laser echoes, the processing of echo signals, and the digitization of echo signals. Information acquisition is performed in the same principle in both element and array sampling and detection, while it shows specific characteristics in the array sampling and detection. This section mainly focuses on the information acquisition in array sampling and detection. It discusses the parallel splitting of laser, the laser beam emission with variable angles, the selection of parameters of the sampling laser beam, the multi-element array detection of echoes, the digitization of echo waveforms, and so on.

4.2.1 The Formation of Transmitting Laser Beam in Array Detections

In the scanning-style target detection based on a laser imaging system, the target sampling is to generate images showing the spatial distribution of targets. It is realized by carrying out the two-dimensional and ergodic scanning for detected areas or targets point by point, so as to obtain information of targets at high resolution in the space and time. Therefore, the resolution of the sampling in laser scanning significantly influences the spatial resolution of multi-dimensional images of targets, and further affects the detection precision based on these images. For array detections or the laser imaging in the mode of transmitting and reception of array laser, the sampling for the targets is achieved by the separate sampling of each element in an array detector for the targets. In the condition, scanning is not necessary or it merely needs to scan in certain a dimension. In this way, the sampling rate for the targets can be improved, therefore greatly improving the spatial resolution of images generated through laser imaging. To realize this, it is necessary to explore the methods for splitting and parallel transmitting laser beams for sampling targets.

1. **Parallel splitting of the laser beam**

In the information acquisition, the sampling rate of the transmitting laser for targets depends on the repetition frequency and scanning rate of the radiated laser pulse. When the repetition frequency and scanning rate of the laser pulse are low, to improve the sampling rate for targets, breakthroughs and innovations are expected to be made in technology. Therefore, there are many equivalent methods of improving the sampling rate of targets including the liner array irradiation of laser beams and point-to-point array detection and receiving. In addition, the angle between the transmitting lasers and the receiving field of view needs to be adjusted according to practical conditions of targets, so as to conduct the staring detection in a linear array for targets.

The radiation of laser beams in a linear array can be realized by forming multiple laser beams in the transmitting end of the laser according to the linear array, area array or a certain pattern. In this process, an array transmitting program or a parallel splitting device are needed. The array transmitting program with multiple laser sources can arrange light sources in a certain pattern through the optical fiber array. However, it presents a disadvantage that it is hard to guarantee the consistency of multiple channels, which is likely to cause errors of the distance measurement in each channel. As a result, the channels are imbalanced, thereby influencing the precision of multi-dimensional images of targets. At present, the relatively effective method is to split a laser beam produced by a laser into required patterns by using a parallel beam splitting device. The commonly used parallel splitting devices include diffraction and transmission of raster splitters, optical fiber splitters, reflecting mirror assemblies, laser diode splitters, prismatic decomposition splitters, and so on [4].

In the parallel splitting method studied by the author, a diffraction and transmission of raster splitter is adopted to conduct linear array splitting for high-power laser beams. The magnitude and the angle among the split laser beams are determined by the parameters of the optical raster. The diffraction raster divides an input laser beam into multiple homenergic laser beams. The diffraction angle of the laser beams output by the diffraction raster is calculated using Formula (4.17).

$$\sin\,\theta_m = \pm m\lambda/d_1 + \sin\,\theta_0 \tag{4.17}$$

where m is the diffractive order and θ_m is the diffraction angle, namely, the angle between the center light beam and the diffracted beam in the m order. In addition, θ_0, d_1, and λ are the angle of incidence, the raster constant, and the wavelength, respectively.

According to formula (4.17), when the wavelength is fixed, different diffraction angles can be obtained by varying the raster constant d_1. As the incident angle θ_0 changes, the diffraction angle θ_m is distributed in different diffractive orders. The working principle of the diffraction raster is illustrated in Fig. 4.9.

What is worth noting is that a beam expander is needed to compress the divergence angle before the laser enters into the diffraction raster. The beam expander is an inverted Galilean telescope, as shown in Fig. 4.10. The negative lens with the focal length being 16.2 mm is faced with the incident laser beam to make it divergent. Then the divergent beams are collimated by passing through the positive lens with the focal length of 32.4 mm. In the process, high energy of the laser beam is not concentrated in the optical path. The laser beam diameter is enlarged by 2 times, while the divergence angle is reduced by 2 times. Meanwhile, the laser wavelength is 1.064 μm, with the diameter and the divergence angle of the input beam being ϕ6 mm and less than 3 mrad, respectively. It can be obtained that the realized beam-expanding multiple is 2.

As shown in Fig. 4.11, a He–Ne laser is used to irradiate the light-splitting device and the results of laser beam splitting are observed. It is worth noting that the

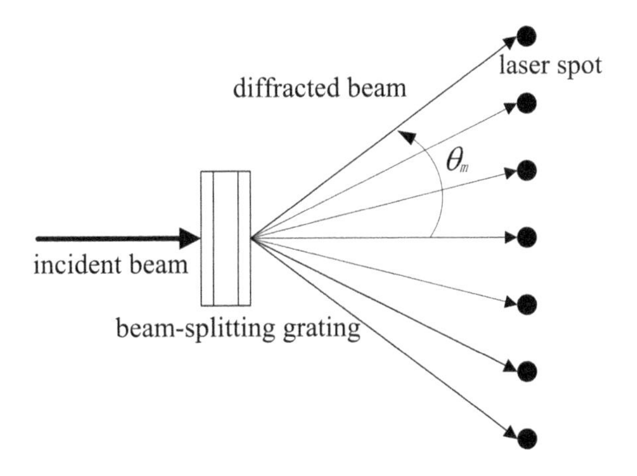

Fig. 4.9 The splitting of the laser beam using a diffraction raster

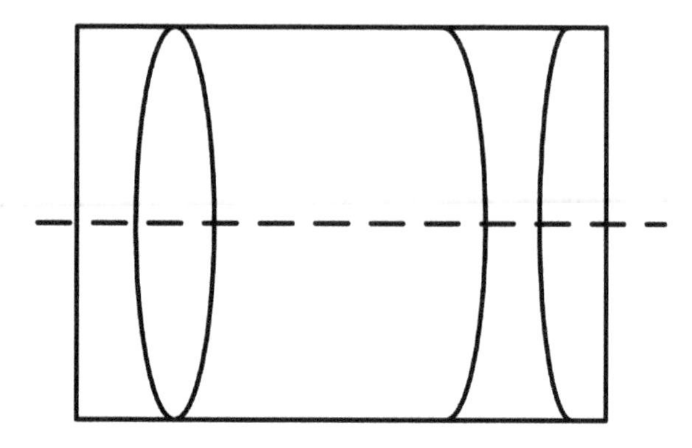

Fig. 4.10 The beam expander

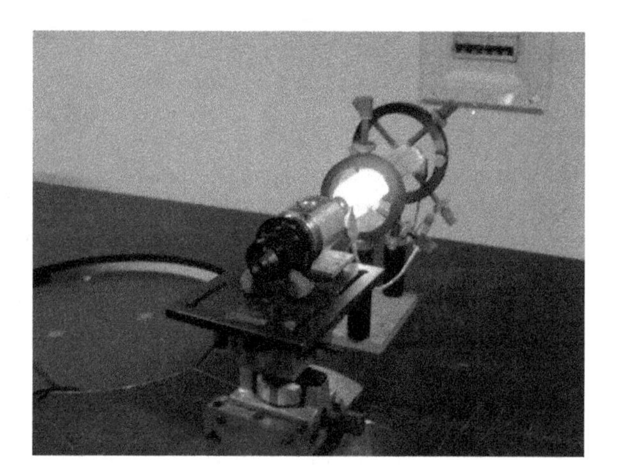

Fig. 4.11 The experimental components for parallel beam-splitting

beam-splitting angle needs to undergo the wavelength conversion. The experimental result of the beam splitting is shown in Fig. 4.12. According to the figure, there are other laser spots apart from the diffracted beams of 33 points. The existence of these laser spots is inevitable in splitting laser beams using a diffraction raster, which can lead to loss of the laser energy. Preliminary estimation suggests that the energy loss, namely, loss in splitting laser beams, accounts for about 15% of the total laser energy. Therefore, the corresponding laser splitting efficiency is 85%.

2. **Transmitting the laser beam with variable angles**

If the incident angle $\theta_i = 0$, the angle among adjacent beams spilt by the transmission-mode diffraction raster is

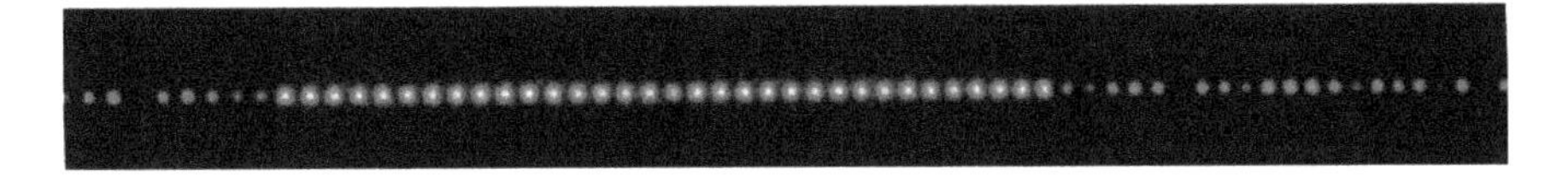

Fig. 4.12 The experimental result of the device for parallel splitting laser beams

Table 4.1 The angles of the diffracted beam in each diffractive order with the adjacent and the central beams

Diffractive order	Angle between each beam and the central beam θ_m (mrad)	Angle between each beam and the adjacent beam $\Delta\theta_m$ (mrad)
0	0	
1	2.497	2.497
2	4.994	2.497
3	7.491	2.497
4	9.988	2.497
5	12.486	2.498
6	14.983	2.497
7	17.480	2.497
8	19.978	2.498
9	22.476	2.498
10	24.973	2.497
11	27.471	2.498
12	29.969	2.498
13	32.467	2.498
14	34.966	2.499
15	37.465	2.499
16	39.964	2.499

$$\Delta\theta_m = \arcsin[(m+1)\lambda/d_1] - \arcsin(m\lambda/d_1) \tag{4.18}$$

Suppose that the above diffraction raster outputs the beams in 33 grades, that is, the diffractive order m is 33, together with 1064 nm of wavelength and 426.1 μm of the raster constant d_1. Under such condition, the angles of the diffracted beam in each diffractive order with the adjacent and the central beams are presented in Table 4.1 (as the diffracted beams in both sides of the central beam are symmetric, merely the angles of the beams in one side are given and those of the other side are the same).

It can be seen from Table 4.1 that in the 33 laser beams, which are bilaterally symmetric to the center beam, the angles between each beam with their adjacent beams basically range from 2.497 to 2.499 mrad, with a deviation less than

0.002 mrad. The deviation occurs because of the existence of calculation errors in the above results. Therefore, the angles between each beam and their adjacent beams are regarded as identical after performing parallel splitting for the laser beam using the diffraction raster.

Without considering other factors, when laser beams in a linear array with the uniform angle irradiate targets, the interval of the sampling points is constant, which leads to the unchanged spatial resolution for targets. In fact, this is not conducive to the high-resolution imaging for targets. Therefore, the angles of different output beams are changed to make the transmitted multiple laser beams illuminate targets with different densities on the basis of the fixed angle of the parallel split laser beams. In this way, it is also capable of sampling targets with varied resolutions in linear array staring detection, which is particularly suitable for conducting laser imaging for an interested target with high resolutions.

By connecting a zoom telescopic system behind the transmission-mode diffraction raster, the transmitted laser beam can be split in a linear array with variable angles. Therefore, linear array staring irradiation with variable interval of points can be achieved so as to realize the high-resolution sampling with variable intervals. Figure 4.13 shows the optical structure of the optical transmitting device for split laser beams with variable angle [5]. The zoom telescopic system with controllable adjustment is composed of the front lens group and the rear lens group which are confocal. Thereinto, the latter is zoomable. According to the features of the telescope system, $\gamma = f1'/(-f2')$, that is, the angular magnification γ of the telescope system only depends on the ratio of the focal lengths of the two lens groups. When one of the lens groups is a zoom system, the angular magnification γ is expected to be changed with the focal length of the lens group.

As shown in Fig. 4.13, a transmission-mode diffractive raster 1 and a zoom telescopic system 2 are located in the propagation direction of the beam. Thereinto,

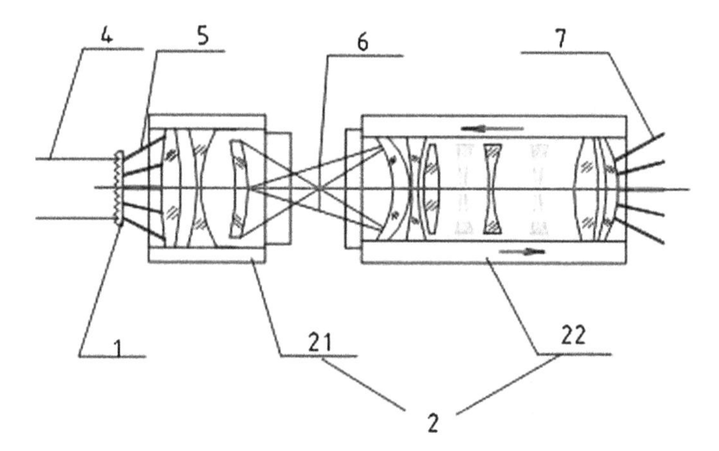

Fig. 4.13 The optical structure of the optical transmitting device for split laser beams with variable angles

the zoom telescopic system 2 is composed of the front lens group 21 and the rear lens group 22 which are confocal. Thereinto, the rear lens group 22 presents a variable focus. By adjusting the focal length of rear lens group, the angular magnification of the system can be changed. By doing so, the emergent angle of the multiple laser beams being split can be adjusted.

The whole working process is as follows: ① After the laser beam 4 is split into multiple beams by the diffraction raster 1, the multiple beams 5 are emitted to different directions according to the designed angle interval; ② different laser beams passing through the front lens group 21 are converged on the focal plane 6 to form a point array of equal intervals. Then the laser beams are transmitted to different directions with equal angular intervals after passing through the rear lens group 22 which is confocal with 21. At this moment, if the focal length of the zoomed lens group 22 is equal to that of the front lens group 21, the emission direction of the emergent laser beams 7 is the same as that of the split beams 5. If the focal length of the zoomed rear lens group 22 is greater than that of the front lens group 21, the angle interval of the emergent laser beams 7 is less than that of the spilt laser beams 5. In case that the focal length of the zoomed rear lens group 22 is less than that of the front lens group 21, the angle interval of the emergent laser beams 7 is larger than that of the split laser beams 5. In this way, the emergent angle of multiple laser beams being split can be adjusted.

Because the focal length the zoomable rear lens group 22 can be changed continuously in a wide range, the angle interval of the split laser beams 7 can be changed in a considerable range as well. Therefore, dynamic adjustments of the target sampling resolution can be achieved by imaging the target with varied focal length in the case of not increasing the repetition frequency of the laser. This, to some extent, meets the requirements for improving the resolution of imaging detection.

3. **The selection of parameter in linear array emission**

(1) The scanning width for sampling

If the angle of the laser beams in a linear array is $\Delta\theta_m$ and the divergence angle of the laser beam is α, then the laser spots irradiating targets have the characteristics illustrated in Sect. 4.1.1. Suppose that, in the air-to-ground laser imaging based detection, the height of the platform is H and the angle between a beam and the nadir is ϕ_i (i is the number allocated to a laser beam from the nadir point). When $\Delta\theta_m$ is smaller, the interval of the laser spots can be expressed as

$$d_{Ti} = H \cdot \Delta\theta_m \cdot \sec^2 \phi_i \tag{4.19}$$

If n (n is generally an odd number) laser beams are transmitted parallel along the track direction, the scanning width d which enables the laser to sample the target is expressed as

$$d_T = 2\sum_{i=1}^{(n-1)/2} d_{Ti} \tag{4.20}$$

where, d is the maximum width of the parallel emitted laser beams, and is an important basis for determining the maximum receiving field of view when designing the parallel receiving system.

(2) The repetition frequency of transmitted laser pulses

While emitting laser beams in a linear array, each laser pulse is emitted after being divided into n laser beams. The laser beams distributed in a line array are transmitted for scanning in another dimension (also known as scanning along the track dimension). It can be push-broom scanning with the movement of the platform, or realized by driving the scanning mirror using the scanning mechanism. Assume that the scanning rate is V, in the case that the laser spots in each scanning line need to be connected so that no line is missed, the requirement for the laser repetition frequency f_p is

$$f_p = \frac{V}{H\alpha \cdot \sec\phi} \tag{4.21}$$

According to the above equation, the elevation of the detection platform and scanning rate show a specific requirement for the repetition frequency of laser pulses, that is, the requirement for velocity to height ratio in remote sensing detections. It is also called velocity to distance ratio in target detections.

(3) Laser emission power

Suppose that the beam splitting efficiency of a diffraction raster is η, the number of detector elements equals to that of split beams n, and the instantaneous receiving field of view is larger than the laser divergence angle. Under such condition, for the maximum range z_{max}, the requirement for the peak power P_T of laser emission is

$$P_T = \frac{\pi \cdot nP_{r\min}Z_{\max}^2}{\eta S_r T_T T_R A(z)\rho T_1} \tag{4.22}$$

where S_r is the area of the receiver telescope, T_T and T_R are optical transmissivities of the transmitting system and the receiving system, respectively, and $A(z)$ is the transmissivity of the two-way atmospheric transmission. Besides, ρ represents the target reflectivity, T_1 is the center transmissivity of the optical filter, and $P_{r\min}$ is the minimum measurable input power.

According to Eq. (4.22), when other conditions remain constant, there are square-root relations between the power and the maximum range. Suppose that the diameter of a telescope is 80 mm, with T_T, T_R, and $A(z)$ being 0.7, 0.5, and 0.4, respectively, and the transmissivity T_1 of the central wavelength of the band pass filter is 70%. In addition, other conditions such as the minimum measurable power

$P_{r\min}$, the minimum reflectivity ρ of the targets, the number n of the split beams, and the beam splitting efficiency are set as 1×10^{-7} W, 0.1, 25 and 85%, separately. Then, the maximum power P_T needed is 1.9×10^{7} W to measure a target 10 kW away from the detection system. When the pulse width of the laser is $\tau = 10$ ns, the pulse energy is 190 mJ.

4.2.2 Array Reception and Detection of Laser Echo

There are two major methods for simultaneously receiving multi-channel laser echo signals. One is to arrange element devices as an array (linear array or area array) and the other is to use a linear array detector or an area array detector. The former is realized through beam splitting using beam splitters or the import of fiber arrays to lead an optical signal into different element detectors. In other words, the echo detection circuits for single-point lasers are integrated into a linear array circuit or an area array circuit matching with the sampling pattern of laser emission. The array detector method is to employ a linear or an area array APD detector. In this way, each element of the receiving detector corresponds to a laser spot for sampling targets and receives reflected echo signals. This array detection for laser echoes has been paid close attentions by researchers abroad in recent years. The MPLAB developed by National Aeronautics and Space Administration (NASA) of America designed a linear array APD receiving system of 128 elements by adopting this receiving technology using the modular method. In addition, the small receiving field of view also can be enlarged to show a much larger scanning width through modular method [6]. In addition, the application of the area-array APD chips [7] and focal plane arrays are two rapidly developed technologies of array detection for laser echoes. The former integrates APD area arrays and processing circuits as a chip to simplify the peripheral circuits and decrease the volume and power of the system. Based on the fiber-optic focal plane technology, the latter transmits the multi-channel echoes received by a reception telescope to the photosensitive surfaces of APD elements through optical fibers. Apart from transmitting light, the optical fiber technique can be used to transform light beams as well. This chapter mainly discusses array detection.

1. **The selection of receiving parameters in linear array detections**

The following aspects need to be considered while selecting the receiving parameters of linear array detections. First, a linear array detector is selected with the consideration of the number and size of elements as well as the interval between every two elements. Besides, the operating wavelength, responsivity, bandwidth and quantum efficiency of the detector have to meet the requirements. Second, the diameter and focal distance of the telescope are determined. Generally, the transmission-type telescope is used. Its diameter is related to the laser power, the maximum range, etc. The selection of the focal distance is mainly affected by the instantaneous receiving field of view and the total field of view. On the one hand, the instantaneous receiving field of view needs to be larger than the laser divergence angle, that is, the size of the receiving footprint needs to be larger than that of the laser spot. On the other hand, the interval between the receiving fields of view of every two elements has to equal to the interval $\Delta\theta_m$ of laser beam emission so that each laser beam can be covered by the corresponding receiving field of view.

Suppose that the size of the elements of the linear array APD is γ_0, the interval between every two elements is γ_1 ($\gamma_1 > \gamma_0$), and the focal distance of the telescope is F, then the angles of the instantaneous receiving field of view of each element and its adjacent field of view are respectively

$$\beta = \gamma_0/F \tag{4.23}$$

$$\beta_1 = \gamma_1/F \tag{4.24}$$

The interval between every two receiving footprints of the target is expressed by the following formula. Where, ϕ_i is the angle between the instantaneous receiving field of view and the nadir.

$$d_{Ri} = H \cdot \beta_1 \cdot \sec^2 \phi_i = H \cdot \gamma_1 \cdot \sec^2 \phi_i/F \tag{4.25}$$

The corresponding scanning width d of the n receiving elements is

$$d_R = 2 \sum_{i=1}^{(n-1)/2} d_{Ri} \tag{4.26}$$

In order to match the transmitting and receiving fields of view, the following requirement has to be satisfied:

$$\begin{cases} \beta_1 = \Delta\theta_m \\ d_R = d_T \\ \beta > \alpha \end{cases} \tag{4.27}$$

By solving formula (4.27), the focal distance of the optical receiving system, the size of detection element, the interval between every two elements, the splitting angle and the divergence angle can be obtained.

2. **APD linear array detection**

In the acquisition of detection information in laser imaging based detection, parallel reception is realized based on a linear array circuit, namely, multiple laser echoes are detected precisely based on an APD array detection circuit. Additionally, an automatic balancing circuit for the bias of the APD linear array is designed to automatically adjust the bias high-voltage with the changes of ambient temperature, background noise and circuit noise to guarantee detection sensitivity [8]. Meanwhile, the receiving circuits of echoes in each channel are adjusted to be consistent with each other to guarantee the consistency of gains and time delays of multiple echo signals.

Figure 4.14 shows the components of the processing circuit for array detection. It can be seen that echo signals are converged on the APD array of the focal plane through the receiver telescope system. After the signal of each detection element undergoes photovoltaic conversion, the echo signals are split into three parts by the splitter and the amplifying module. The first part enters into the sample-and-hold circuit for peaks through the isolation amplifier, and then transmitted to and converted by the next analog to digital converter (ADC) through the deserializer. In this way, original data of the gray image of the target are obtained. After being changed to digital echoes by passing through the forming module for digital echoes, the second part of the echo signals are imported into the module for measuring the time interval as the termination signals for distance measurement and counting. The ranging results are input into the computer to store. After being smoothed by going through the noise extraction circuit, the third part is employed as the feedback signal for bias control. Due to the strong back scattering in laser emission, the sequential circuit is employed to control the amplifier gain and APD bias to avoid the spurious triggering caused by back scattering. Besides, the sequential circuit can output synchronizing sequential information to be used in real-time processing and post-processing. Because APD gain decreases with temperature rise, temperature compensation needs to be conducted to APD bias so as to keep the consistency of echo signals. Thus in the design of the circuit, a temperature sensor is adopted to

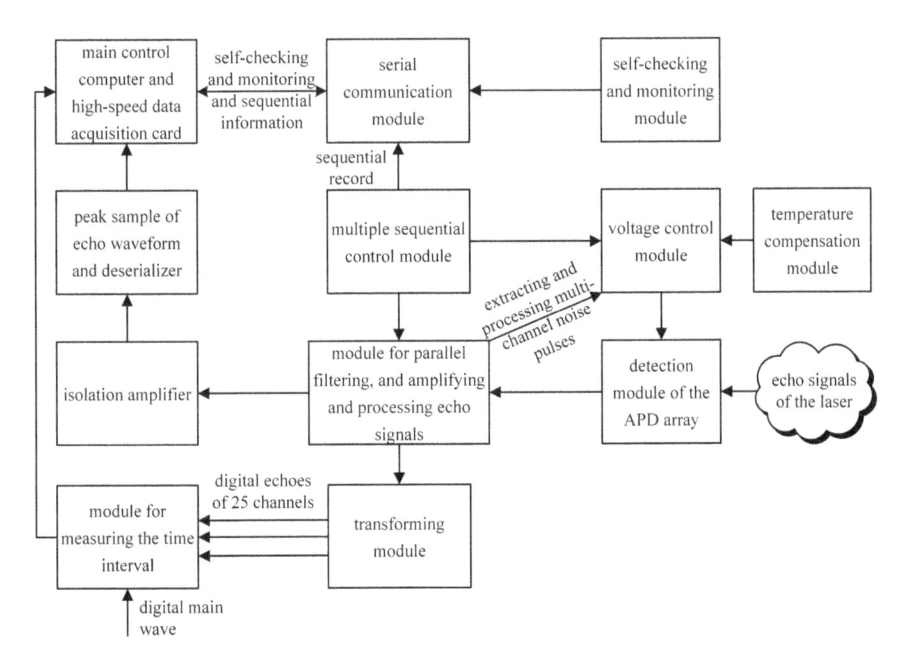

Fig. 4.14 The components of the processing circuit for array detection

perform feedback control on APD bias so that the echo gains of each element are consistent in a wide range of working temperature.

The multiplication coefficient of an APD is closely related to the bias voltage applied, and the output signal-to-noise ratio (SNR) is greatly affected by the multiplication coefficient and temperature. Therefore, a regulator circuit is designed to adjust the bias voltage of the APD linear array to make it change with the variations of the output noise and temperature so as to obtain the optimum SNR and the lowest false alarm rate.

The following matters need to be considered when controlling the bias. First, there is strong back scattering in laser emission. Therefore, the gain of the amplifier module needs to be controlled in addition to decreasing the bias of the APD array to prevent the strong echo signals from generating echo signals and therefore falsely triggering the subsequent circuits. Second, because APD gain decreases with temperature rise, the temperature needs to be compensated for the APD bias so as to maintain the consistency of the echo signals. Thus in the circuit design, a temperature sensor is adopted to perform feedback control on APD bias to ensure that the echo gains of each element are consistent in a wide range of working temperature. Third, the bias is affected by the background noise as well. As oversized background noise is likely to lead to a false alarm, the APD bias is decreased to control the noise. In the system, the feedback control of the bias is realized by using the method of extracting and integrating multi-channel noise. That is, the bias is controlled according to the average level of noises. Finally, because of different manufacturing processes of the device, the gains of APD elements are likely to be inconsistent even in the same bias voltage. Hence, before operating the system, the system needs to be calibrated and the calibration data are applied as the basis for calibrating the system when the gray images are generated.

3. **Consistency adjustment of multi-channel echo signals**

The method of laser array detection can simultaneously obtain laser echoes of all the sampling points. However, when the transmitted beams not accord with the reception channels, detection results of laser echoes cannot reflect the real features of targets. As a result, the laser images obtained are anamorphic so that the real properties of the targets cannot be obtained in the detection. In an array receiving system, the consistency of gains and time delays in all channels determines that of the multi-channel echo signals. Therefore, the consistency is the crucial issue that needs to be solved including the consistency of gains and time delays of channels and lags in transmitting laser beams and receiving echo signals. By ensuring the consistency of the above aspects, the accurate gray images, range images, and spatial relation of targets can be obtained, respectively. These need to be considered in system design and data processing.

(1) The consistency adjustment of gains in multi-channel

The parallel reception of echo signals in multi-channel is to obtain accurate gray images of the targets. Considering the purpose, from the emission of the splitting laser beams to the output of echoes signals, the relative gains (responding gains normalized by the emission energy) of each channel are supposed to be basically consistent or can be measured accurately. To obtain the consistency of gains in multi-channel, devices selected need to be basically same while designing the detection and processing circuit for multi-channel signals. What is more important is to adjust the gains of each channel after designing the detection system. It is generally realized by using two methods, namely, actual measurement and adjustment based on diffuse reflection in laboratories.

(i) The method of actual measurement and adjustment. The system is set beside the window of a laboratory to transmit multi-beam lasers outside and irradiate targets with uniformity such as a lawn and a building. In the process, it needs to ensure that all the receiving footprints are located in the selected targets. Next, lasers are emitted to detect the targets and then measure the echoes in each channel using a high-speed data acquisition circuit. Afterwards, the laser energies obtained are normalized using the ranging data of each channel. The results obtained basically reflect the differences of gains in these channels, namely, the relative gains. Based on this relative gains, the overall gains of the detector and the amplifier in each channel are adjusted respectively to realize the consistency of gains in multi-channel.

(ii) The adjustment based on diffuse reflection in laboratories. A laser is applied to irradiate a diffuse reflection plate in a laboratory, and then laser signals scattered by the diffuse reflection plate are received by an array detection circuit. Because the brightness of the scattering lasers of diffuse reflection plates is basically consistent in various directions, the power that each detection element received is theoretically the same. At this moment, the values of the output signals of each detector are collected and can be used as references for relative gains in the following data processing to adjust the gains of the channels. Nevertheless, it is worth noting when the laser power cannot saturate the detector, the laser beam needs to be split into more beams with the application of a proper incidence angle.

(2) Consistency adjustment for time delays of multi-channel

Apart from generating gray images of targets, another purpose of receiving multi-channel echoes is to obtain accurate range images of targets, which depends on the consistency in time delays of echo signals received by each channel. Therefore, the response delays of each channel to laser echoes need to be basically consistent or can be measured accurately. However, it is difficult to achieve the consistency of response delays. In the beginning of designing a processing circuit of array detection, wiring design, selection of channel bandwidth and impedance matching need to be appropriate. Besides, the differences between time delays of channels are supposed to be accurately measured after developing the processing circuit. In addition, consistency adjustment for time delays of channels needs to be performed according to the measured data.

At a position several meters away from the detection system in a laboratory, a miniature pulsed laser is employed to irradiate a diffuse reflection plate. Under such condition, the receiving system is turned towards the diffuse reflection plate to receive the scattered signals of the pulsed laser, and then a laser pulse is expected to be output from each channel. At this moment, one of the laser pulses (generally, the central channel) is employed as the reference to measure the differences of leading-edge delays among laser pulses from different channels using a two-channel high speed circuit for data acquisition. After obtaining delay differences of laser pulses, the channels are calibrated based on the measured data to realize high consistency of delays in multi-channel. It is worth noting that temporal resolution needs to meet the requirement of range resolution for laser range imaging. For example, to guarantee the range images reaching the required resolution of 18 mm, the sampling rate needs to be 10 GHz. A comparing threshold value needs to be set to measure the differences of leading-edge delays, and it generally equals to the threshold value for detecting laser echo pulses.

(3) Consistency adjustment for the lags in transmitting laser beams and receiving echoes

According to Sect. 4.1.2, in the dynamic scanning of beams and receiving field of view, the uncoaxiality of emitting lasers and receiving echoes is likely to occur. This phenomenon becomes more obvious while detecting at a higher scanning rate or for targets located further. Therefore, in the dynamic scanning of array beams and multi-element receiving field of view along track direction, the uncoaxiality also occurs in the corresponding element. The further the distance from an array element to the central point is, the more distinct the phenomenon becomes, which leads to the lag angle effect, as shown in Fig. 4.15. Where $\theta_{sc,n}$ is the tilt angle of the field of view corresponding to the nth element, and $\theta_{lag,n}$ is the lag angle

Fig. 4.15 The lag angle

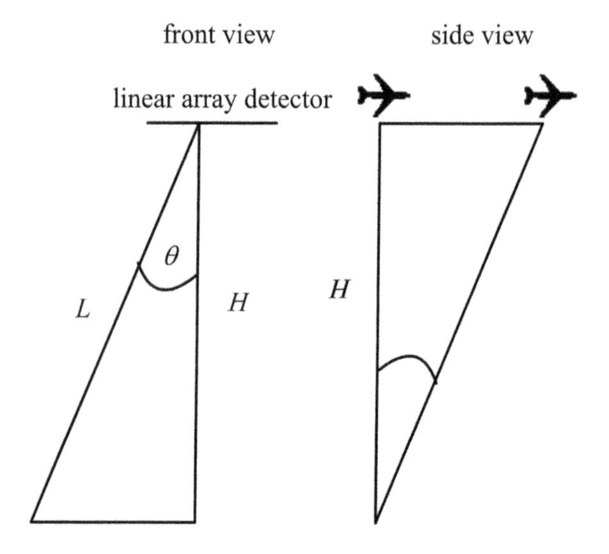

corresponding to the nth beam. When the flight speed is V at a height of H, the lag angle corresponding to each laser signal is

$$\theta_{lag,n} = \mathrm{Arctan}(2VH/(Hc\cos(\theta_{sc,n}))) \tag{4.28}$$

When $\theta_{lag,n}$ is small, it can be regarded that

$$\theta_{lag,n} = 2V/(c\ \cos(\theta_{sc,n})) \tag{4.29}$$

When $\theta_{lag,n}$ is large, it needs to be considered while calculating the ranging data of each channel and analyzing echo waveforms. In this case, the receiving footprints of target sampling for array detection constitute an arc, instead of a straight line.

After the array receiving echo signals, the pulse signals of the echoes obtained are rectified by utilizing the above lag angel relation, so that the collected gray, range and waveform data of laser echoes can correspond to the right positions. By doing so, laser images with accurate position relations can be generated.

4.2.3 Amplifying and Processing Laser Echo Signals

There are many methods for amplifying and processing laser echo signals. This section mainly introduces the measures applied in analog receiving channel of laser echoes including parallel high-speed amplifying and automatic control of the bandwidth according to the requirements of the target detections based on a laser imaging system [9].

1. **Signal processing of an ordinary ranging receiver**

The function of a general ranging receiver for laser pulses is to detect laser echo pulses from the noise. The general requirement is to obtain optimal output SNR regardless of pulse fidelity and measure time delay of the laser echoes. The major components of the general ranging receiver include a photoelectric detector, its corresponding load and matched filtering amplifier, and a pulse-detecting circuit. Generally, laser signals collected by a telescope are transformed into current pulses by a detector. According to the characteristics of emitted laser signals, the pulse currents and frequency spectrum of Gaussian waveform can be expressed as:

$$i_s(t) = I_s exp(-kt^2/\tau_R^2) \tag{4.30}$$

$$I_s(\omega) = \frac{\tau_R I_S}{2\sqrt{\pi}\ \ln 2}\exp[-\ln\sqrt{2}(\frac{\omega}{\omega_{ci}})^2] \tag{4.31}$$

where, $t = t' - t_R$. t_R refers to the time delay of echo pulse at peak points relative to the transmitted dominant wave and t' denotes the time delay of any point in the echo pulses. Moreover, I_S is the peak value of the pulse signal, and τ_R is the half-peak width of the current of echo pulses at the input port of the detector. In addition, $k = 4\ \ln\ 2$, and $\omega_{ci} = 2\sqrt{2}\ \ln\ 2/\tau_{R0}$, which is the 3 dB cut-off frequency of the signal.

The noise in signal is regarded as additive white noise, when the input noise and equivalent noise in the detector are considered, the root mean square of the current density of the noise is denoted as i_n^2. At the input port of the detector, there is distributed with equivalent load that also can transform the signal and noise output by the detector. Afterwards, the signals of preceding stage are filtered and amplified by the matched filtering and amplifying circuit, which is designed according to the principle of obtaining the optimal output SNR based on the known input signal. For laser echo pulses of Gaussian waveform, suppose that the matched filter is designed with the peak value and width of the current pulse being I_{S0} and τ_{R0}, respectively. Then while processing the detection signals given in Eq. (4.30), the optimal SNR is obtained from the output port of the detector under the conditions of $I_S = I_{S0}$ and $\tau_R = \tau_{R0}$. However, in general, echo pulses are not likely to meet the above standard conditions and input signals not always match with the filter, thus the signal voltage and the root mean square of noise voltage at the output port of the matched filtering amplifier are

$$v_s(t) = k_1 I_s(\tau_R/\tau_{RV}) exp[-4\ \ln\ 2(t/\tau_{RV})^2] \tag{4.32}$$

$$V_s(\omega) = \frac{k_1 \tau_R I_s}{2\sqrt{\pi}\ \ln\ 2} exp[-\ln\sqrt{2}(\frac{\omega}{\omega_{cv}})^2] \tag{4.33}$$

$$V_n^2(\omega) = k_1^2 i_n^2 exp[-2\ \ln\ \sqrt{2}(\omega/\omega_c)^2] \tag{4.34}$$

Equation (4.34) shows the root mean square of noise voltage. Where ω_c is the 3 dB cut-off frequency of the standard input signal considered in the design of the matched filter, $\omega_c = 2\sqrt{2}\ \ln\ 2/\tau_{R0}$, and k_1 is the overall gains of the circuit. All these parameters are related to the standard signal and circuit system set earlier.

According to Eqs. (4.32) and (4.33), the half-peak width and 3 dB cut-off frequency of the signal at the output port of the amplifier are

$$\tau_{RV} = \sqrt{\tau_R^2 + \tau_{R0}^2} \tag{4.35}$$

$$\omega_{cv} = \frac{\omega_c \omega_{ci}}{\sqrt{\omega_c^2 + \omega_{ci}^2}} = \frac{2\sqrt{2}\ \ln\ 2}{\tau_{RV}} \tag{4.36}$$

Accordingly, the SNR at the output port of the amplifier can be obtained by calculating the average power and the noise, as shown in

$$SNR_V = 1.133\tau_R \frac{I_s^2}{i_n^2} \cdot \frac{\sqrt{2}\tau_{R0}}{\sqrt{\tau_{R0}^2 + \tau_R^2}} \quad (4.37)$$

Finally, the existence of echoes is detected by the constant false alarm rate (CFAR) decision circuit set behind the matched filtering amplifier with the application of a certain threshold level. When there are echoes, the pulse signal of a representative echo is output to be subsequently processed.

While ranging using an airborne laser by radiating laser beams towards the ground, suppose that amplitude and pulse width of the laser echoes in the nadir direction are I_{S0} and τ_{R0} respectively. With the same energy of radiation and a tilt angle of ϕ, the pulse signal enters into the detector is broadened and the signal amplitude decreases to

$$\tau_R = \tau_{R0} + 2\alpha Htg(\phi)/(c \cdot \cos\phi) \quad (4.38)$$

$$I_S = \frac{\tau_{R0}}{\tau_R} I_{S0} \cdot \cos\varphi \quad (4.39)$$

where, α is the divergence angle of a laser beam, and H represents the elevation of the laser device and detector in the air, ϕ is the angle between the beam axis and the nadir, and c represents the velocity of light.

When the amplifier of the receiver is designed based on the echo in the nadir direction, the width of the pulse voltage at the output port can be obtained according to Eqs. (4.35) and (4.38).

$$\tau_{RV} = \sqrt{2\tau_{R0}^2 + 2\tau_{R0} \cdot (2\alpha Htg\varphi/c \cdot \cos\varphi) + (2\alpha Htg\varphi/c \cdot \cos\varphi)^2} \quad (4.40)$$

The output SNR can be obtained according to Eqs. (4.37), (4.38) and (4.39).

$$SNR_V = (1.133\tau_{R0}\frac{I_{s0}^2}{i_n^2}) \cdot (\frac{\tau_{R0}}{\tau_R}\cos^2\varphi) \cdot (\frac{\sqrt{2}\tau_{R0}}{\sqrt{\tau_R^2 + \tau_{R0}^2}}) = SNR_0 \cdot k_\varphi \cdot k_\tau \quad (4.41)$$

where, $SNR_0 = 1.133\tau_{R0}\frac{I_{s0}^2}{i_n^2}$ is the optimal SNR in the nadir direction. In addition, $k_\varphi = \frac{\tau_{R0}}{\tau_R} \cdot \cos^2\varphi$ is the SNR loss factor caused merely by the beam pointing angle, $k_\tau = \frac{\sqrt{2}\tau_{R0}}{\sqrt{\tau_{R0}^2 + \tau_R^2}}$ is the SNR loss factor caused by the mismatch between the width of the input signal pulse of the detector and the structure of the matched filter. Therefore, $k_\varphi \cdot k_\tau$ is the SNR loss factor.

When $\alpha = 3$ mrad, $\tau_{R0} = 7$ ns, and $H = 2000$ m, the output SNR loss factor $k_\varphi \cdot k$ of the amplifier and output pulse width vary with the change of the scanning angle ϕ, as shown in curves A and B in Fig. 4.16, respectively.

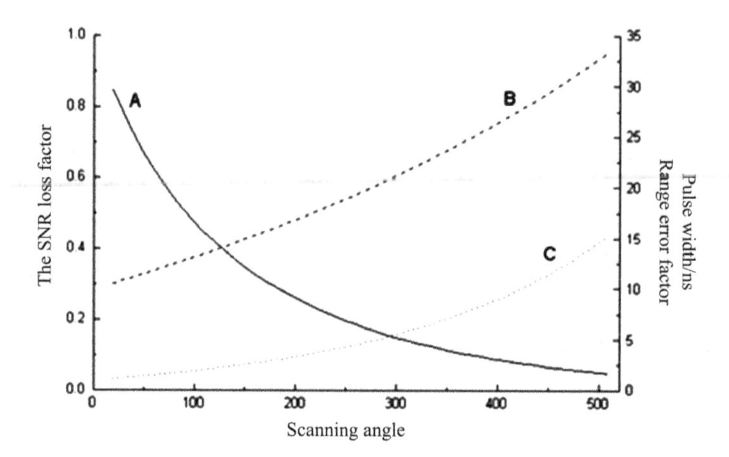

Fig. 4.16 SNR loss factor, pulse width and range error factor at the output port of the amplifier

As shown in Fig. 4.16, the output SNR obviously varies with the scanning angle, and deviates from the optimal value in the meantime. Such change is more apparent especially when the aircraft flies at high altitudes to scan with a large laser divergence angle, which makes it difficult for pulse determination using the CFAR. Besides, the detection ability for laser echo pulses is also directly affected with the decrease of detection sensitivity in the tilt direction and the possibility of losing range data, which is intolerant for airborne detection systems. In addition, the width of the output pulse of the amplifier changes in beam scanning, which causes pulse trigger errors in the pulse-detecting circuit with a fixed threshold. As a result, the ranging error varies in the detection with different SNRs and pulse widths and the range precision decreases. In this case, the ranging error shows the following relations to the SNR and the pulse width:

$$\Delta z = \frac{c \cdot \tau_{RV}}{2\sqrt{SNR}} = \frac{c \cdot \sqrt{2}\tau_{R0}}{2\sqrt{SNR_0}} \cdot k_z \tag{4.42}$$

where $k_z = \frac{\tau_{RV}}{\sqrt{2}\tau_{R0}} \cdot \frac{1}{\sqrt{k_\varphi k_\tau}}$ is the range error factor in the nadir direction. Curve C in Fig. 4.16 presents that it varies with the scanning angle. For example, when the scanning angle is 300 mrad, SNR decreases to 1/10 of that in the nadir direction and pulse width is 28 ns, then, the range error factor is 6, that is, the ranging error increases by 6 times. Thus, for an airborne target detection system based on a laser imaging system, echo characteristics vary with the changes of the beam pointing angle when it is assumed that the influences of targets and transmission media on the echo characteristics are the same in all the directions. Under such condition, it is hard for general ranging receivers to meet the requirements for high reliability and high precision of ranging.

2. **Parallel high-speed amplification**

Each output element of an array detector has a high-speed amplifying circuit. The major functions of the circuit include realizing impedance transformation and providing a large output SNR, an appropriate bandwidth to obtain distortionless laser echoes, and a gain setting interface to adjust the consistency of gains in different channels.

The following aspects require to be noted in the design of amplifying circuits: First, the bandwidth, time delay and gain of each channel need to be as consistent as possible. In addition, the bandwidth is able to be adjusted automatically while and the time delay and gains can be regulated artificially. Second, echo signals are not saturated, which requires a large dynamic range. Third, the functions including automatic gain control and logarithm amplifying are not needed. Fourth, there are effective anti-interference measures among the circuits.

In general, APD outputs signals at high frequencies. When the load is 50 Ω, the rise time of the output signal is 2 ns. In order to amplify the signal without distorting it, the signal bandwidth can be calculated according to the following formula:

$$\Delta B = 1/(\pi t_r) \tag{4.43}$$

where ΔB is the signal bandwidth and t_r is the rise time, the corresponding spectrum width is 160 MHz when the rise time is 2 ns.

For the weak output signals of general detectors, they need to be amplified using low-noise amplifying technology to increase the SNR of the system. In practical application, ideal distortionless amplification is difficult to be realized. Generally, the signal bandwidth of an amplifier can be determined according the formula below.

$$\tau_R \cdot B = k_B \tag{4.44}$$

where τ_R is the width of echo pulses, and k_B is the constant coefficient under optimal bandwidth. In the target detections using a laser imaging system, when echo waveforms need to be collected, the maximum distortionless amplification requires to be realized. In this case, $k_B = 4$ can be adopted. If merely laser ranging is performed, the amplification for the maximum output SNR is needed and $k_B = 0.5$ satisfies the requirement. If $\tau = 20$ ns, when the bandwidth of the amplifier is 200 MHz, the signal can be considered as undistorted.

3. **Automatic control of the bandwidth**

In the design of the parallel amplifying circuit mentioned above, bandwidth is an important parameter. If the bandwidth is not appropriate, the distortionless echo or optimal output SNR cannot be obtained. Therefore, the bandwidth in the amplifying circuit needs to be regulated automatically and varies with the parameters such as pulse width. In the laser scanning based detection, the scanning angle is known, and the scanning angle is related to the width and amplitude of echo pulses. Based on this, the parameters of the filtering amplifier, especially the bandwidth, are able to synchronously change with the scanning angle in an automatic control manner. In this way, the maximum distortionless echo or optimal SNR required can be output.

According to the design principles of a filter, the bandwidth B of the amplifier requires to be inversely proportional to the pulse width of the input signal, namely, $B = k_B/\tau_R$. If the bandwidth of the amplifier in the nadir direction is B_0, it is expected to be changed in the following rule in beam scanning:

$$B = \frac{k_B}{\tau_{R0} + 2\alpha tg\phi/(c \cdot \cos\phi)} = 1/\left(\frac{1}{B_0} + \frac{2\alpha \tan\phi}{k_B \cdot c \cdot \cos\phi}\right) \tag{4.45}$$

where, k_B is the constant coefficient with the optimal bandwidth and is supposed to be 4 and 0.5 so as to output the maximum distortionless and the optimal SNR. Besides, τ_{R0} is the pulse width of a laser echo in the nadir direction,α is the divergence angle of a laser beam, ϕ is the angle between the beam axis and the nadir, and c represents the velocity of light.

When the amplifier of the receiver is designed based on the echo in the nadir direction, the pulse width at the output port is

$$\tau_{RV} = \sqrt{2\tau_{R0}^2 + 2\tau_{R0} \cdot (2\alpha H tg\varphi/c \cdot \cos\varphi) + (2\alpha H tg\varphi/c \cdot \cos\varphi)^2} \tag{4.46}$$

In practical operation, given that the scanning of laser beams is discontinuous, a set of amplifiers with different bandwidths are supposed to be utilized to meet each scanning angle in the equation above, respectively. By using electronic switches, the corresponding amplifier can be easily connected with the signal processing circuits. In the working mode of automatic control of the bandwidth, the structure of the signal processing circuits matches with input signals, so that approximately optimal outputs are obtained at the output port of the amplifier. The width of the output pulse is approximately

$$\tau_{RV} = \sqrt{2}\tau_R \tag{4.47}$$

where τ_R is the half-peak width of the echo pulse at the input port of the detector.

In this case, the SNR is

$$SNR_V = 1.133\tau_R \frac{I_S^2}{i_n^2} = SNR_0 \cdot k_\varpi \tag{4.48}$$

where I_S is the peak value of the pulse signal, and i_n^2 is root mean square of the current density of the noise, SNR_0 is the SNR in the nadir direction, and k_ϖ is the SNR loss factor caused by beam pointing angles.

The analysis above merely considers the changes of beam pointing angles existing in beam scanning while neglecting the inclination of targets. A method for estimating the tilt angle of the target in advance is expected to be used to decrease the influence of the inclination on ranging. When the resolution is large enough, the change of the inclination of targets are related to that of the beam pointing angles. While measuring ranges in a ranging period, signals are sent to each filtering and amplifying channel. The amplitudes and widths of echo pulses can be estimated by comparing the signal intensities output from filtering amplifiers with different bandwidths. The estimated values are used as references for selecting filtering and amplifying channels in the next ranging period to determine the bandwidth of the amplifier together with the influences caused by scanning angles.

4.2.4 Waveform Digitization of Laser Echoes

The waveform, range and gray images of targets are all generated by processing data relating pulse waveforms, time delays and power of laser echoes. However, influenced by the changes of range, surface slope and reflectivity of targets, the structural fluctuation, etc., the waveforms of pulse echoes are likely to be widened even distorted. Under such condition, the pulse energy fluctuates remarkably, together with the apparent change in the time delays of echoes relative to the dominate wave. Therefore, the entire waveform data of laser echoes received need to be collected to obtain the echo pulses with high fidelity, so as to ensure the effectiveness of the high-resolution measurement for laser echo waveforms.

The acquisition of the entire waveform data of laser echoes is realized mainly by digitizing echo waveforms. The digitization of waveforms means to sample weak signals of laser echoes reflected from targets output by detectors at a high speed, so as to obtain the discrete digital waveforms of laser echoes. Through a series of processing such as pre-processing, filtering, peak value detection and intensity integral, these discrete digital waveforms are expected to generate data files with trigger periods of lasers as the variables. The data files include the filtered echo waveforms, the distributions of time delays, powers, pulse widths, and amplitudes of the echoes, and the distribution of backward scattering rates of targets [10].

Figure 4.17 shows the principle of the digitizing circuit for echo waveforms. It can be seen that the components of the digitizing circuit mainly include an emitter follower for echoes, a time delay unit, a circuit for generating sampling clocks, a high-speed analog to digital (A/D), a high-speed data random-access memory (RAM), a circuit for generating sampling gates, etc. After going through the detection and isolation circuit for dominate waves, one of the dominate wave signals is transmitted into the circuit for generating sampling gates and then to the A/D card as the control signal for the record option of waveform sampling. In actual measurements, the wave gate is expected to cover all the possible target echoes, and the data output from the A/D are input in the high-speed RAM in the validity of the wave gate. Afterwards, the data are restored in a computer. The width of the wave gate and sampling rate determine the number of sampling points in single sampling periods. While, the leading edge of the sampling gate is controlled by the pulses of

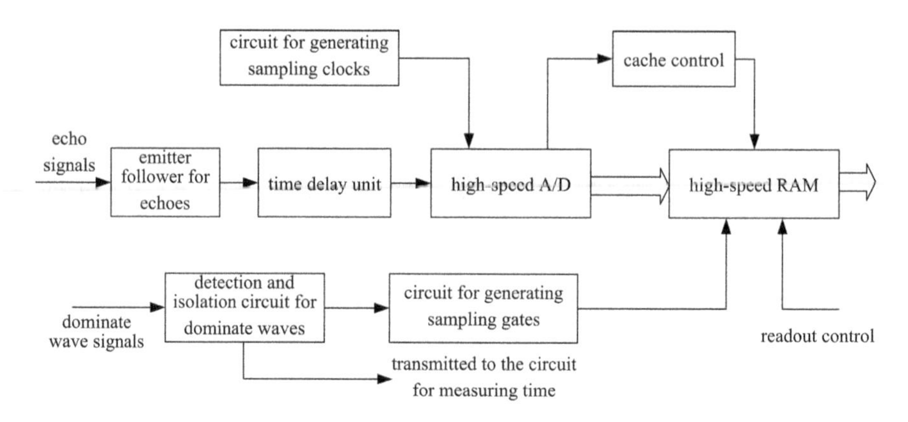

Fig. 4.17 The principle of the digitizing circuit for echo waveforms

the dominate wave and its time delays relative to the dominate wave can be set as fixed values relating the minimum distances of all the possible targets. In the meantime, all target echoes need to appear after the leading edge of the sampling gate. In order to reduce data rate and save memory space, floating sampling gates are applied. In this case, the sampling gate is generated by echo pulses and does not show a fixed time delay relative to the dominate wave. As the echo signals are delayed before being sent to the A/D, they are in the wave gate. Under such condition, the width of the wave gate equals to the sum of the broadened width of echo signals (about 20–50 ns) and redundancy time (about 50 ns). According to the width, forty points need to be sampled in a ranging period at a sampling rate of 240 MHz.

Laser echo signals are sampled with the laser triggering period as the working period, the magnitude of which is basically kilohertz at present. Owing to the data sampling results are expected to be stored in the hard disk of a computer, it is impossible to store the data synchronously with the analog-to-digital conversion in real time. So the data are stored by means of bulk storage in a laser detection period. In laser detections, it merely needs to obtain partial results around echo signals, which are real-timely stored in the high-speed RAM and then transferred and saved in the hard disk of the computer in the rest time of the period.

Apart from the digitized echo waveforms, to generate laser images of targets, position and attitude data of the platform need to be collected as well. Meanwhile, in the initial test, reference images of the target require to be obtained through other methods to serve as references in the construction of a relation model. That is, other data have to be collected while collecting the digitized information of echo waveforms. All these data collected need to match with the sampling points of targets, which is the basic requirement for the 3D image generation and identification of targets. Therefore, the data collection to be conducted in a synchronous control sequence which is the synchronizing pulse in scanning lines and signals

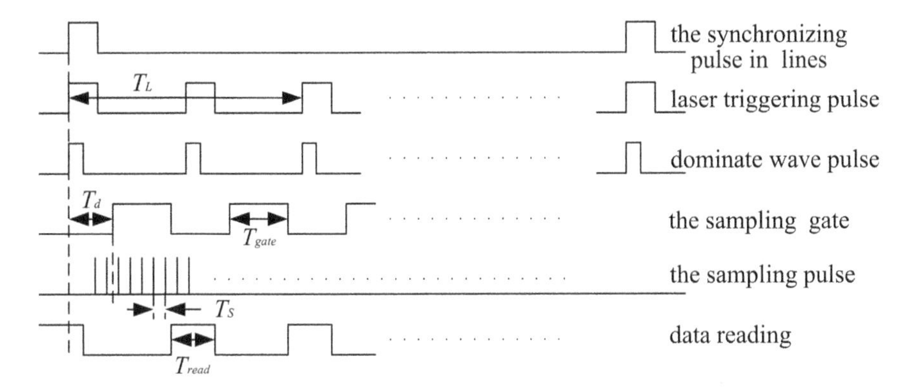

Fig. 4.18 The major sequential relationships of waveform digitization

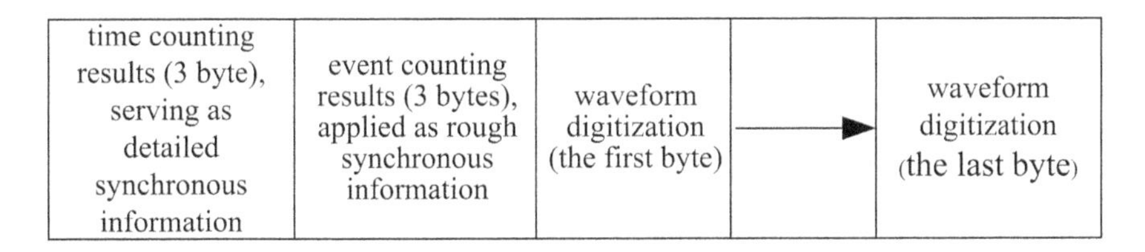

Fig. 4.19 The storage structure of waveform digitization in each laser period

produced by the pulse such as laser triggering pulse. Figure 4.18 demonstrates the major sequential relationships.

The synchronizing pulses in a scanning line are used to trigger an event counter and the counting results are employed as the rough synchronous information of waveform digitization. For laser triggering pulses in a scanning line, the time of data reading of counting results of a latch time counter is applied as the detailed synchronous information. The two parts of synchronous information obtained are recorded before the data of echo waveforms in the corresponding period, respectively, so as to match with other data in the subsequent data processing. The storage structures of waveform digitalization in each laser scanning period are shown in Fig. 4.19.

By digitizing echo waveforms, the digitized echo waveforms obtained are merely discrete digital data based on laser echo pulses, which contains noise, disturbance, etc. These factors are not beneficial for the accurate, sufficient and convenient application of the information in the subsequent processing.

Thus, on the basis of echo waveforms digitization, the digitization results need to be pre-processed to provide data files of different types and characteristics. The pre-processing includes:

(i) Filtering the quantized data. On account of the presence of the noise and disturbances in waveforms digitization, stray data not belonging to the laser echo pulses exist in the data. Therefore, filtering and relevant processing are required to purify the data.

(ii) Calculating the peak value V_m of the target echo and the serial number L_m of its sampling point.

(iii) Computing the equivalent pulse width. The numbers n_1 and n_2 of points whose amplitudes of target echo are 0.707 V_m to the peak point in two sides of L_m are calculated respectively. By multiplying the sum of n_1 and n_2 by the interval T_S of waveform digitization, the result obtained is approximately equal to the echo pulse width τ

$$\tau = (n_1 + n_2) \cdot T_S \tag{4.49}$$

(iv) Calculating echo delays. The voltage being certain proportions smaller than the peak voltage (0.5 Vm, for example) is selected as the floating threshold. Then, the number of sampling points k from the point of the floating voltage to the leading edge of the sampling gate is calculated and multiplied by the interval T_S of waveform digitization. By adding the time delay T_d of the sampling gate relative to the pulse of the dominate wave, the result obtained equals to time delay of laser echoes relative to the dominate wave, that is, $T_r = T_d + k \cdot T_s + T_e$. The time parameters are shown in Fig. 4.18. In the above equation, T_e is the time difference between the first sampling pulse and the leading edge of the wave gate. This method overcomes the error in estimating time delays caused by the range walk phenomenon when the fixed threshold is large [11].

(v) Calculating the echo power. Within the equivalent pulse width of laser echoes, integral operation is performed for the digital quantity of laser echoes in terms of the intensity. Through normalization, the integral result is used as the laser echo power [12].

4.3 Data Collection for the Detection Based on a Laser Imaging System

The multi-dimensional laser images needed for target detection are generated based on the data such as echo waveform, time delay, and gray of targets. Therefore, apart from the information mentioned in the previous section, other data also need to be collected including echo pulses with high fidelity, the time delay and gray with high resolution of laser echoes, etc. For the acquisition of echo pulses with high fidelity, based on the echo waveforms digitization, the entire laser echoes can be collected under the limited storage capacity. Thus, obtaining time delay and gray with high resolution of laser echoes is another important aspect to improve the precision of the multi-dimensional images of targets. In this section, data collection for the detection based on laser imaging system is discussed to measure the time delays and the gray data of laser echoes with high resolution.

After multi-channel laser echo signals received by the elements of an APD linear array detector are parallel amplified, the signal in each channel is divided into two parts. After going through a leading-edge timing discrimination circuit, one part of the signal generates a pulse of time discrimination which is transmitted into a circuit or measuring time intervals of high resolution. This measuring circuit can measure the time interval between the starting pulse and the stop pulse, that is, the time delay with high resolution is measured. In this way, the information of time delays of laser echoes is collected. The other part of the signal passes through a peak sampling circuit to collect the gray of laser echoes [8].

4.3.1 *Discrimination of the Arrival Time of Echo Pulses*

Range information is obtained based on the round-trip time of the laser pulse radiated to and reflected from targets, which is acquired mainly by measuring the time interval between transmitting pulses and receiving echo pulses. Hence, the arrival time of echo pulses has to be discriminated accurately. The commonly used methods for discriminating the arrival time include leading-edge timing discrimination, Leading and trailing edge timing discrimination, self-correcting constant ratio timing discrimination and high-pass capacitance-resistance timing discrimination [13].

1. **Leading-edge timing discrimination**

Leading-edge timing discrimination is to determine the start and end times through fixing a threshold, that is, the arrival time of the point whose pulse intensity is equal to the set threshold in the leading edge of the pulse is used as the start time. The specific theory is shown in Fig. 4.20.

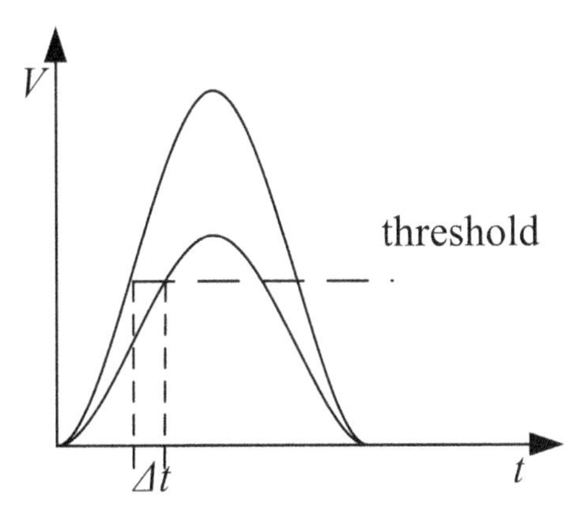

Fig. 4.20 The theory of the leading-edge timing discrimination

In general, suppose that laser echoes are in standard Gaussian waveforms and the comparing threshold is $th1$, then the leading-edge timing discrimination refers to detecting the moment of echo signals passing $th1$ to be used as the arrival time of the echo pulses. This method is simple to be operated. However, its biggest problem is that the changes of the echo amplitude are likely to lead to the error of frontier identification. Suppose that the echo amplitude of Gaussian pulses is A, then the time error Δt_A induced by variations of echo amplitude is

$$\Delta t_A = \sqrt{-2\sigma_p^2 \cdot \ln(th1/A)} \tag{4.50}$$

where, σ_p is the root mean square of the pulse width.

Simultaneously, owing to the existence of noise, the echo amplitude fluctuates and the error of timing discrimination is generated. The time error Δt_N induced by the noise can be expressed as:

$$\Delta t_N = \frac{t_r}{\sqrt{SNR}} \tag{4.51}$$

where t_r is the rise time of laser echo pulses and is defined as the time during which the waveform amplitude of echo pulses rises from 10 to 90%. SNR is the signal-to-noise ratio of the system.

The final time error induced by the leading–edge timing discrimination is

$$\Delta t = \Delta t_A + \Delta t_N = \frac{t_r}{\sqrt{SNR}} + \sqrt{-2\sigma_p^2 * \ln(th1/A)} \tag{4.52}$$

In the application of the leading-edge timing discrimination, the comparing threshold and the root mean square width of the echo pulse are fixed. Meanwhile, echo amplitude is expected to change constantly with the variation of target features. According to formula (4.50), the changes of echo amplitude can give rise to a large time error.

2. **Leading and trailing edge timing discrimination**

In practical applications, due to changes in the target distance and the performance of components, the amplitude of echo pulses is impossible to be kept constant while prone to fluctuates. On account of this, errors are likely to occur in the leading-edge timing discrimination based on the leading threshold. Therefore, it is necessary to improve the leading-edge timing discrimination so as to reduce the influence of the echo amplitude.

The leading and trailing edge timing discrimination is a new timing discrimination method which is put forward on the basis of the principle of the leading-edge timing discrimination. This method is used to identify the pulse time simultaneously using the information of leading and trailing edges to eliminate the drift error caused by the changes of the amplitude and shape of laser pulses and finally improve the measurement resolution of time values. By using the method, the measurement resolution of time values can exceed 120 ps.

The theoretical basis of the method is that if the laser echo is a symmetric Gaussian pulse, then the centroid of laser echo waveforms is considered as the ideal starting and ending time of laser ranging. The steps are as follows: (1) The dominate waves of lasers pass through the fixed threshold to generate the starting count pulse which is then input into the time measurement circuit to make the two channels measure the pulse time simultaneously. (2) The amplified laser echoes are input into two channels of the time measuring circuit to measure time simultaneously. Among them, the first time measurement channel uses a rising edge trigger, while the second one applies a falling edge trigger. (3) The values measured by the two channels are averaged to obtain the measured values of pulse time. By doing so, the drift error caused by changes of the amplitude and shape of the echo pulses is eliminated followed by obtaining the gray value of targets. Moreover, this circuit is simple and easy to be realized.

3. **Self-correcting constant-fraction discrimination**

Constant fraction timing employs the constant-fraction trigger to dynamically adjust the threshold level according to the amplitude of echo pulses, so as to avoid the occurrence of the drift error induced by the fluctuation of the echo amplitude [14]. The employing processes are as follows: the signal output by the filter is divided into two signals. Among them, one signal becomes $f_a(t)$ after the attenuation, and the scale factor is k_a. The other signal changes into $f_c(t)$ after delay of T_d and phase reversal. At last, the sum of these two signals is input into the zero-crossing timing discriminating circuit to discriminate the threshold. By doing so, the detected output pulses of the echo signal can be obtained. If the signal output by the filter is expressed as $v_{so}(t_n)$, and t' is the time difference between any point and the peak of the echo signal, obviously, the time delay for the occurrence of the peak relative to the dominate wave is stable when the signal amplitude changes. Therefore, if t' can be kept constant, then detection accuracy of echo pulses does not fluctuate with the variation of the amplitude. The triggering of the constant-fraction timing can be represented as:

$$f_a(t) - f_c(t) = 0 \tag{4.53}$$

$$k_a v_{so}(t) - v_s(t - T_d) = 0 \tag{4.54}$$

If the signal presents Gaussian waveform, then

$$v_{so}(t) = V_{so} \exp[-4\ln 2(t/\tau_{RV})^2] \tag{4.55}$$

Substituting formula (4.55) into formula (4.54), we can obtain

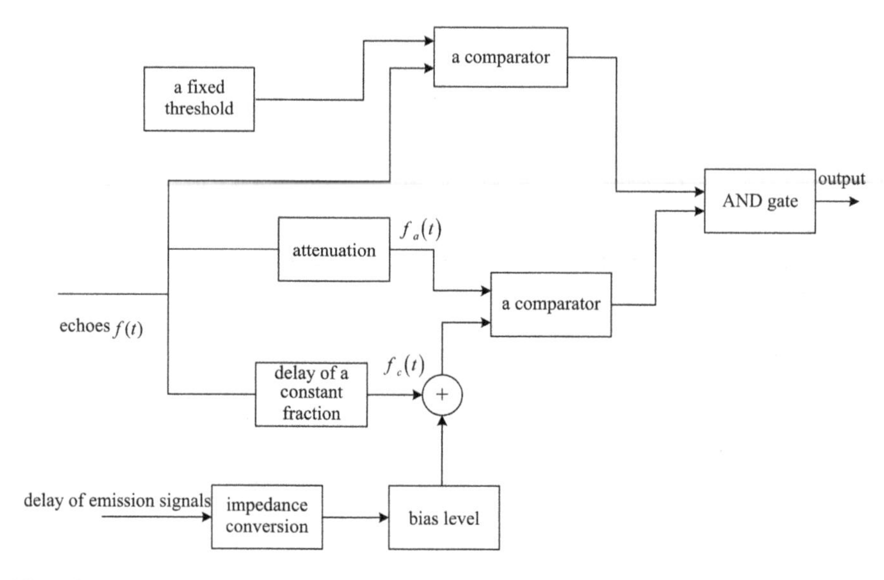

Fig. 4.21 The circuit of the constant-fraction timing discrimination

$$t = \frac{\ln k_a \cdot \tau_{RV}^2}{8 \ln 2 \cdot T_d} + \frac{T_d}{2} \tag{4.56}$$

It can be seen from formula (4.56) that the value of t' is not influenced by the fluctuation of the amplitude but by the pulse width τ_{RV}. When the pulse width is constant, the ranging error in timing discrimination based on the constant-fraction trigger is constant and thus can be eliminated as long as proper T_d and k_a are selected. This is an advantage of using the constant-fraction trigger to determine pulse time. The circuit of the constant-fraction timing discrimination designed based on above theories is shown in Fig. 4.21.

The premise of applying constant-fraction timing discrimination is that echo signals only show changes of the amplitude while the shape and width of pulses are constant. If the width of the pulses output by the matched filtering changes constantly, the estimation error of echo time delay changes as well, which cannot be solved by the constant-fraction timing discrimination. In the scanning laser detection, no matter applying the filter with fixed parameters or the variable-bandwidth filter introduced in the previous section, the width of output pulses changes with the width of input pulses. Thus it can be concluded that the detection errors of echo pulses still exist even constant-fraction trigger technology is utilized in scanning laser detections. Therefore, it is necessary to take certain measures to modify this technology, that is, self-correcting constant-fraction discrimination.

The so-called self-correcting constant-fraction discrimination means calculating the width of echoes entering the laser detector according to the scanning angle and

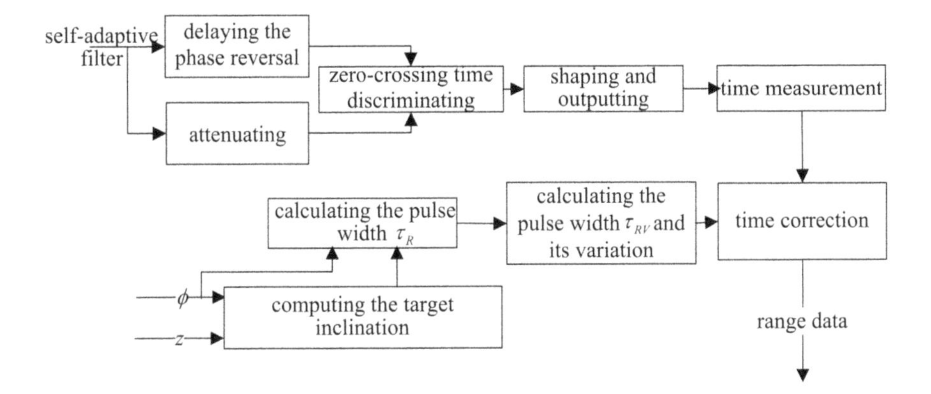

Fig. 4.22 The principle of the self-correcting constant-fraction timing discrimination

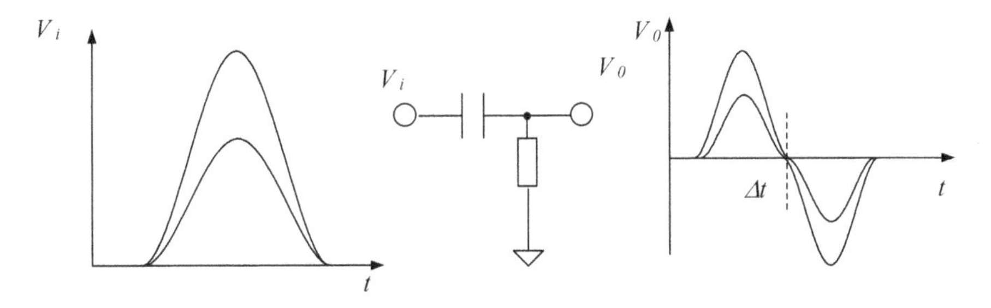

Fig. 4.23 The principle of the high-pass CR timing discrimination

the estimated inclination of targets based on the traditional constant-fraction timing discrimination. Then τ_{RV} and its variation are calculated, respectively. According to these results, the formula (4.40) is used to modify the subsequent ranging results. The principle of the self-correcting constant-fraction discrimination is shown in Fig. 4.22.

4. **High-pass CR timing discrimination**

The high-pass CR timing discrimination sends the starting and ending signal pulses output by the receiving circuit through the high-pass CR filtering circuit to transform the signals into the bipolar ones. The original extreme point is changed into zero point and serves as the starting time point. Here, the error induced by the variation of the amplitude is 0, and mainly influenced by the slope of signal pulses near to the extreme point. The principle of the high-pass CR timing discrimination is demonstrated in Fig. 4.23.

In the above methods for discriminating pulse times, the leading-edge timing discrimination is most easily to be realized while shows the largest errors. The leading and trailing edge timing discrimination is suitable for the cases that the shape and width of laser pulses are constant. Once the shape and width of laser pulses change, its time error increases as well. In comparison, the self-correcting constant-fraction discrimination has lower requirements for the shape and width of laser pulses and stronger adaptability for targets. The accuracy of high-pass CR timing discrimination is the highest among these methods, while it is hard to be realized.

4.3.2 The High-Resolution Measurement of Time Delay of Multi-channel Echoes

The high-resolution measurement of time delay of multi-channel echoes is the premise of the generation of the range images of targets with high resolution. In the data acquisition of the laser imaging, the most commonly used method is the time delay measurement based on the pulse counting. In addition, the analog interpolation method is generally combined with the time interval measurement to realize the high-resolution measurement of time delay of echo pulses.

1. **Delay measurement based on the pulse-counting**

The timing discrimination of laser pulses only determines the starting and ending times of receiving echo pulses and transmitting laser pulses. For the data acquisition of time delay of laser echoes, it needs to measure the time interval T_x between transmitting laser pulses and receiving echo pulses (that is the time delay T_r of echo pulses). The basic method for measuring the time interval of echo pulses is pulse-counting method. That is, the counter starts to count from the leading edge of the dominate wave of laser pulses and stops to count at the leading edge of echo pulses, and then the result of multiplying the counting result by the counting cycle is considered as the time interval T_x.

The relation of the dominate wave of transmitted laser pulses, received echoes and the counting clock is shown in Fig. 4.24. Suppose that the cycle of counting

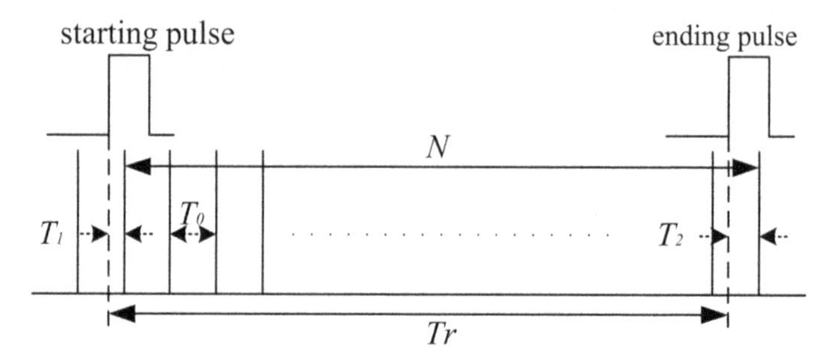

Fig. 4.24 The relation of the dominate wave of transmitted laser pulses, the received echoes and the counting clock

clock is T_0 (the counting frequency is f_0), and the actual time interval between the dominate wave and the echoes is T_x. Meanwhile, the time difference between the leading edge of the dominate waves and the first efficient counting pulse is T_1, the time difference between the leading edge of echoes and the last efficient counting pulse is T_2, and the counting result is N.

The time interval T_x can be computed using the following formula

$$T_x = N \cdot T_0 + T_1 - T_2 \tag{4.57}$$

where, the results of $T_1 - T_2$ changes randomly in a limitation of $\pm T_0$.

The maximum error of ranging data induced by above factors is

$$T_x = N \cdot T_0 + T_1 - T_2 \tag{4.58}$$

In fact, the pulse counting method only uses the leading edges of the dominate wave of laser pulses and echoes, while there are two error factors affecting the result of time measurement in terms of the amplitude and time of actual laser pulses [15, 16].

One factor resulting in the error of time discrimination is the different intensities of reflected echoes. Besides, the echo pulses are not standard square waves and have certain inclination at the leading edge. As a result, the instable trigger caused by the varying echo amplitude brings about errors in timing discrimination.

The other factor is the quantization error of counting. Owing to the fact that the appearing times of the dominate waves and echoes are not likely to synchronize with the counting clock, they present a random time difference with the counting clock. The time difference is the so-called quantization error of counting, that is, the above $T_1 - T_2$. In order to reduce the quantization error of counting, the most direct way is to improve the frequency of counting pulses. However, the improvement of pulse frequency is also limited by various factors, that is, to achieve a resolution of 0.5 ns in the measurement of the time delay, the frequency of the counter needs to reach 2 GHz.

2. **High-resolution time delay measurement based on an ADC conversion method**

In order to obtain the high-resolution of the time interval and acquire the time delay of echoes with high-resolution, it has to reduce the quantization error of counting. Two ways can be used to reduce the quantization error of counting. The first method is an ADC conversion method [15]. At the beginning of counting, an integrating capacitor is charged and discharged. The discharge is stopped at the end of counting, when the voltage of the capacitor is converted into the time interval value by employing A/D conversion. The second method is delay line interpolation method [16]. The high-resolution delay line is used to read the counter and the status value of delay line. Moreover, the sufficiently high resolution can be obtained through the interpolation and subdivision for the counter.

(1) The basic principle of A/D conversion

The arrival time of dominate waves is T_1 earlier than that of the first counting pulse. In order to real-timely measure the quantization error T_1 of counting, it has to convert the time interval T_1 into the electrical signal. In the A/D conversion, the leading edge of the dominate waves is used as the starting trigger to launch the constant current source I of first order, so as to charge the capacitor C. Here, the internal resistance of constant current source is R, and then the voltage V_C of the capacitor C is

$$V_C = RI(1 - \mathrm{e}^{-t/RC}) \quad (t \geq 0) \tag{4.59}$$

Then, by using the leading edge of the first efficient counting pulse as the control signal to stop charging the capacitor, the growth of the capacitor voltage also stops. Suppose that the voltage is V'_C at this moment, then the time delay at this moment corresponding to $V_C = 0$ (at the leading edge of the dominate wave) is T_1. Similarly, this method also can be utilized to obtain the charging voltage of the capacitor corresponding to T_2. In the following analysis, T' is used to represent T_1 or T_2, then V'_C is

$$V'_C = RI(1 - e^{-T'/RC}) \tag{4.60}$$

Connected with the charging capacitor is an isolation amplifier with preferable performance, which has high input impedance, generally tens of megohms. The function of the isolation amplifier is to isolate the influence of post-circuits on the charging capacitor, so as to keep the voltage of the capacitor for a long time. Meanwhile, it also has certain amplification functions but not affects the charging of the constant current source for the capacitor. When the charging of the capacitor was stopped by the first counting pulse, the voltage V'_C of the capacitor is input into the A/D circuit through the isolation amplifier to conduct the A/D conversion so as to acquire a digital code N'. It can be considered that the amplifier has a unit gain for ease of analysis. Suppose that the number of bits of A/D is m and the input voltage in the full-scale range is V'_{Cm}, then the digital code N' is

$$N' = 2^m (V'_C / V'_{Cm}) \tag{4.61}$$

According to formulas (4.60) and (4.61), N' shows one-to-one correspondence with V'_C and T'. By combining formulas (4.60) and (4.61), we obtain

$$T' = -RC \ln[1 - (N' V'_{Cm})/(2^m \cdot RI)] \tag{4.62}$$

where, R, C, m and V'_{Cm} are determined in the design of the circuit, and N' is given by the output value of A/D. Thus, the quantization error T_1 of counting can be determined, so does the measurement principle of T_2. The sequential relation of time measurement using A/D conversion is displayed in Fig. 4.25.

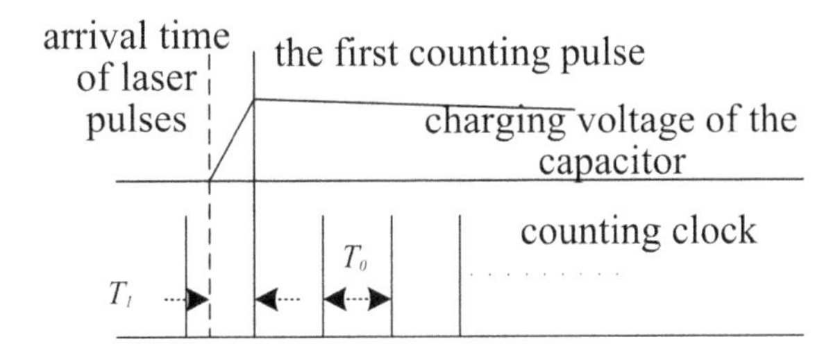

Fig. 4.25 The sequential relation of time measurement using A/D conversion

Formula (4.62) can be used to calculate exact T_1 and T_2, with the consideration of the nonlinear influence in the charging of the capacitor. Therefore, the result of time measurement obtained through A/D conversion is accurate. For example, suppose that R = 10 Kn, C = 1000 pF, V'_{Cm} = 10 V, m = 16, and I = 150 mA, when N' = 64, then T' = 6.5 ps. When $N' = 2^{10}$, then T' = 0.1 ns.

If $RC \geq T'$, by performing exponential expansion to formula (4.62) and neglecting higher order terms, the linear solution of V'_C can be obtained as

$$V'_C = (IT')/C \tag{4.63}$$

The linear approximation result of time measurement can be acquired as

$$T' = (C/I) \cdot (V'_{Cm} N'/2^m) \tag{4.64}$$

In the period within T_0, the nonlinear process of the capacitor charging can be shortened through reasonably choosing the charging capacitor and the constant current source. By doing so, the capacitor charging can be treated as a linear charging.

(2) Feasibility and precision of A/D conversion

The circuit for realizing the time measurement through A/D conversion is shown in Fig. 4.26. According to formula (4.64), corresponding changes occur to the results of time measurement with certain variations of I, C and V'_{Cm}. Therefore, while designing the circuit for time measurement based on A/D conversion, it needs

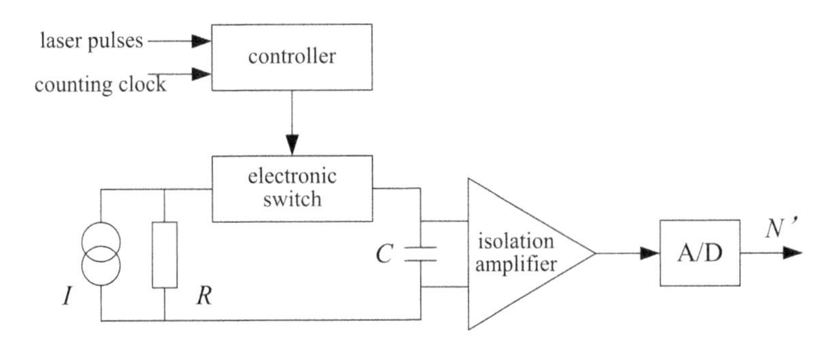

Fig. 4.26 The principle of employing the A/D conversion to realize time measurement

to select the highly stable circuit parameters such as the constant current source and capacitance. Meanwhile, a capacitor with low leakage can be selected as the charging capacitor. Moreover, it needs to consider the influence of the insulation absorption of the capacitor on keeping voltage stable. Capacitors with preferable performances contain polystyrene, polypropylene and polytetrafluoroethylene (PTFE). The A/D reference source needs to be kept as stable as possible because of the influence of the reference source on the full scale V'_{Cm}.

The floating input design of the isolation amplifier completely eliminates the coupling of the input port of signals with the output port of the amplifier and the end of the power supply. In the circuit, an instrumentation isolation amplifier is adopted, which requires to have high input impedance and common-mode rejection ability, as well as low offset voltage and temperature drift. The isolation amplifier is used to isolate and amplify the DC voltage of the capacitor which stops being charged. Therefore, the frequency characteristic is not specially required. Under such condition, it is not difficult to select and realize the isolation amplifier.

The performance indexes of A/D conversion mainly include conversion rate and resolution. Therefore, the selection of the A/D resolution (namely, the corresponding number m of bits) plays the key role in this method. In order to make full use of the A/D conversion, it generally makes $V'_C \approx V'_{Cm}$ when $T' \approx T_0$.

In linear approximation,

$$I/C = dV_C/dt \tag{4.65}$$

By substituting the above result into formula (4.64), it can be obtained that

$$\Delta T' = [(V'_{Cm}C)/(2^m \cdot I)] \cdot \Delta N' \tag{4.66}$$

The error of measuring time by applying A/D conversion is

$$\delta T = T_0/2^m \tag{4.67}$$

That is to say, the quantification error of counting in A/D conversion is reduced to $1/2^{\mathrm{m}}$ of the original error. Because the maximum value of $1/2^{\mathrm{m}}$ is T_0, the precision of time discrimination is improved by 2^m magnitudes using A/D conversion in the laser ranging.

If the input resistance of the isolation amplifier is R_i, the discharge time constant of the capacitor which stops being charged is R_iC, for m bits A/D, the quantification interval of the amplitude is $V'_{Cm}/2^m$. If the capacitor discharges in the linear approximation manner, the needed time for the capacitor voltage to decline by a quantification interval of amplitude is approximately

$$f_t \geq 1/T' = 2^m/(R_iC) \tag{4.68}$$

In order to guarantee that the minimum T' can be distinguished, A/D conversion requires to be accomplished prior to that the capacitor voltage declines by a quantification interval of amplitude while selecting A/D. That is, the A/D conversion rate has to meet the following requirement.

$$f_t \geq 1/T' = 2^m/(R_iC) \tag{4.69}$$

For example, when the frequency of the counting clock is 15 MHz, and the error of time discrimination using the pulse counting method to measure time delay is 66.7 ns. While, when the A/D of 16 bits is selected to perform the high-resolution measurement through the A/D conversion, according to formula (4.69), the theoretical error of time measurement can be reduced to 1 ps around. Because the maximal value of T' is T_0, the RC can be set as 10 μs to meet the linearization requirement of $RC \geq T'$. If R = 10 Kn, C = 1000 pF can be selected. If V'_{Cm} = 10 V and the voltage of the capacitor is supposed to be charged to approximate 10 V, the linear charging rate needs to reach $V'_{Cm}/T_0 = 1.5 \times 10^8$ V/s. According to formula (4.65), I = 150 mA can be obtained. If R_i = 20 M, it can be obtained from formula (4.69) that the conversion rate of A/D is:

$$f_t \geq 2^{16}/(2 \times 10^7 \times 10^{-9}) = 3{,}276{,}800\,\text{MHz} \approx 3.3\,\text{MHz} \tag{4.70}$$

It can be seen from the above analysis that A/D conversion presents low requirements for the A/D conversion rate. However, A/D conversion is favorable for improving the precision of time measurement.

3. **High-resolution measurement of time delay based on a delay line interpolation method**

The resolution of time measurement by measuring the time delay based on A/D conversion can reach dozens of picoseconds while it is hard to be realized. In this case of the resolution of time measurement is in the range of hundreds of picoseconds to 1 ns and the ranging precision varies from 10 to 20 cm, employing A/D conversion to measure time not only increases the difficulty in the design but also calls for excessive cost. Therefore, the delay line interpolation method can be used to measure time delay followed by realizing the precision requirement of the above time measurement employing high-precision tapped delay line.

The principle of the high-resolution measurement of time delay based on the delay line interpolation method is shown in Fig. 4.27. The pulses of the dominate

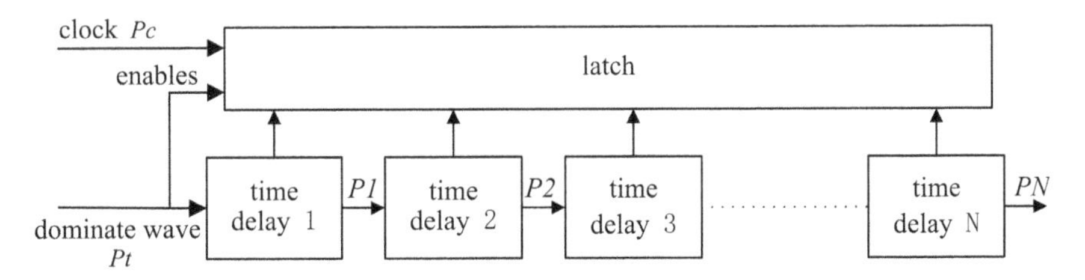

Fig. 4.27 The principle of using the tapped delay line to compensate time measurement

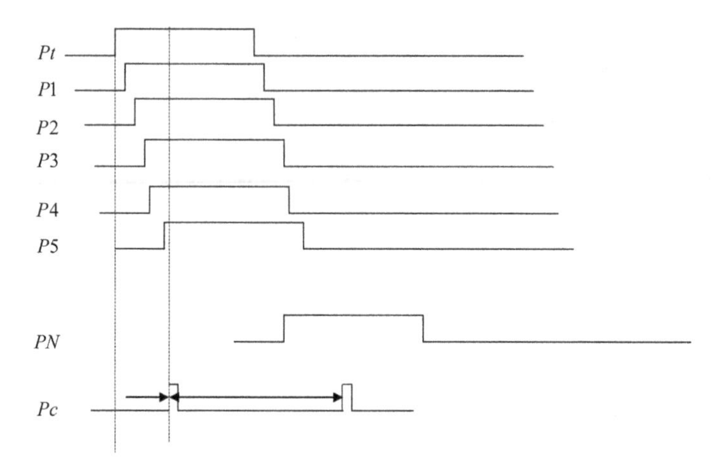

Fig. 4.28 The counting clock and the relation of the signals output by the tapped delay line

wave (or echo) are sent to the first stage of the tapped delay line. It is assumed that each tapped delay unit provides a time delay Δt. To measure *T1* or *T2* which is smaller than the counting cycle *T0*, Δt needs to be far smaller than *T0*. At this point, the time delays of the output *P1*, *P2*, *P3 ... and PN* by each delay unit are Δt, $2\Delta t$, $3\Delta t$... and $N\Delta t$, respectively. The sequential relationship output by the time delay units is shown in Fig. 4.28.

The logical state of the output port of all the time delay units is cut in the latch through the counting clock. If m units arranged from left to right in the results of latch are logic 1 and the following N-m units are logic 0, which shows 11…10…0. At this time, the high-precision time measurement can be realized with the minimum interval being Δt, that is, the measurement result of the time interval T_1 (or T_2) from the rising edge of pulses of the dominate wave (or echo) to the first counting clock is $m\Delta t$. The result of time measurement given by Fig. 4.28 is $5\Delta t$. The maximum range $N\Delta t$ of the effective time measurement in the precise time measurement depends on the delay interval and the stages of the tapped delay line. Because the maximum value of T_1 or T_2 is T_0, as long as $N\Delta t > T_0$, precise time measurement can be achieved with a resolution of Δt. For example, if N = 10 and $\Delta t = 1$ ns, the resolution of precise time measurement is 1 ns within the effective time measurement range of 10 ns.

The basic principle of delay line interpolation method is illustrated in Fig. 4.29. The method is realized by *n* delay units and an encoding unit with two input terminals and an output terminal. The upper input terminal is for the clock while the lower one is for starting or ending signals and the output terminal outputs binary time quantity t_a or t_b. The delay units are compromised of a buffer B and a trigger L. Every the starting and ending pulses passing through a delay unit, the time interval between the two signals is shortened by a LSB (least significant bit). After passing through the *i*th delay unit, if the starting and ending signals stay ahead of the clock, the Q of the *i*th trigger is high potential, while that of the front trigger is low

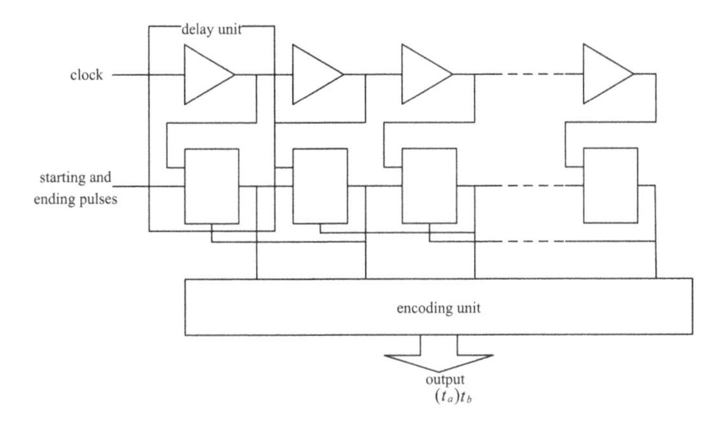

Fig. 4.29 The principle of delay line interpolation method

potential. Otherwise, the Q of the *i*th trigger is low level, while that of the front trigger remains unchanged. After the starting signals and clock signals all pass through the delay line, if the output of the *m*th delay unit is high level while others are low level, then the following relation exists, that is, $t_a \approx \mathrm{m} \times \mathrm{LSB}$ (more precisely, $m \times \mathrm{LSB} \leq t_a < (m + 1) \times \mathrm{LSB}$).

The prominent advantage of delay line interpolation method is that the structure is simple and can be realized through monolithic integration on monolithic field programmable gate arrays (FPGA) [17]. However, the method shows a disadvantage that the precision is supposed to be up to hundreds of picoseconds limited by the LSB. Therefore, the precision of time measurement is in the range of centimeter and decimeter levels. The error is mainly induced by the following four aspects: (1) Discrete process, that is, the continuous time variables t_a and t_b ($0 \rightarrow \mathrm{T}$) are discrete time output of $\mathrm{m} \times \mathrm{LSB}$ (m = 0, 1, 2…n), the highest resolution of time interval is only up to a LSB, which is the maximum error. (2) Nonlinear integration of delay lines. Because it is impossible to realize the complete consistency of each delay unit in the integration process, the time delay of each delay unit is different, which appears as the nonlinear effect. The correction methods include average method (the mean of the maximum and minimum values is adopted as the LSB) and vector method. Thereinto, the vector correction method shows more desirable performance. For example, while correcting the 6 bits linear output as 8 bits nonlinear one using the two methods separately, the nonlinear integral error drops from 4.2 LSB to 0.24 LSB. (3) Random change, which is caused by the change in the temperature and the power supply voltage of the delay unit, which vary in the ranges of +0.4%/°C and −30%/V, respectively. The variation of the temperature and the power supply voltage can be decreased to ±80 ppm/°C and ±0.5%/V by using the phase-locked loop. (4) Time jitter, which includes clock jitter and time jitter of using delay unit signals to trigger switches. The latter can lead to ranging errors of 7.6 mm.

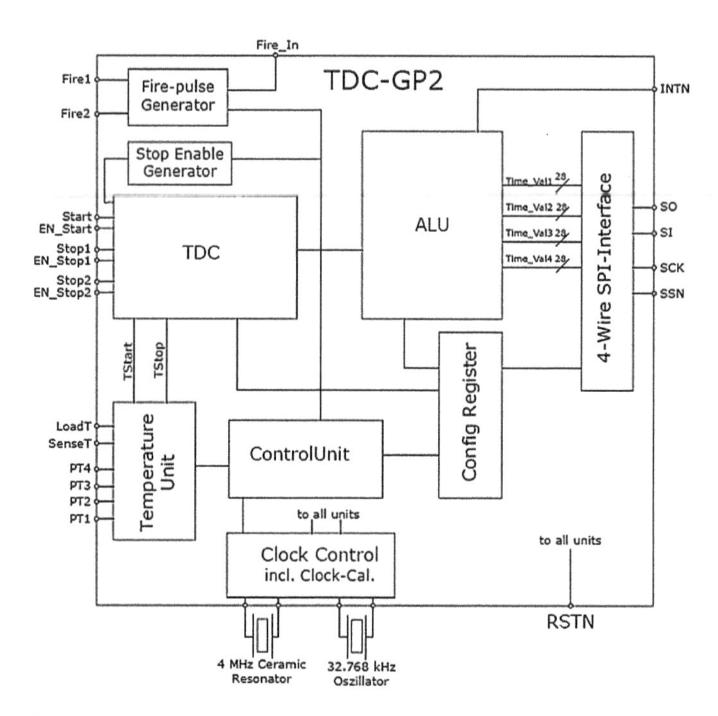

Fig. 4.30 The internal structure of the TDC-GP2

The high-integration time measurement chips made by using delay line interpolation technique mainly include the high-resolution time-to-digital conversion chips in the time-to-digital convert (TDC) series produced by German ACAM Company. Taking TDC-GP2 as an example, its time resolution, mean square error (MSE), and the range are 65 ps, 50 ps, and 0–1.8 μs, respectively. Figure 4.30 presents the interior structure of the TDC-GP2.

The author once achieved ranging with a resolution of 15 cm by using the discrete delay lines and registers. Meanwhile, a high-resolution device for measuring time intervals is developed using the time measurement chips and applied to achieve time delay measurement with a resolution of 120 ps (ranging resolution is 18 mm). The device uses a dedicated ASIC chip to improve the counting frequency based on the phase locked loop in the chip and achieves high-resolution measurement using delay line technology [2].

The device for measuring time intervals is connected with the PCI bus interface of a computer through the chip of the PCI bus interface. The hardware structure is shown in Fig. 4.31. The system transforms the PCI bus interface of the computer to the subscriber line which can be used to easily control the reading and writing through the chip PLX9052 of the PCI interface. Then, the configuration information is written into FIFO3 through the PCI interface module, and then MCU is supposed

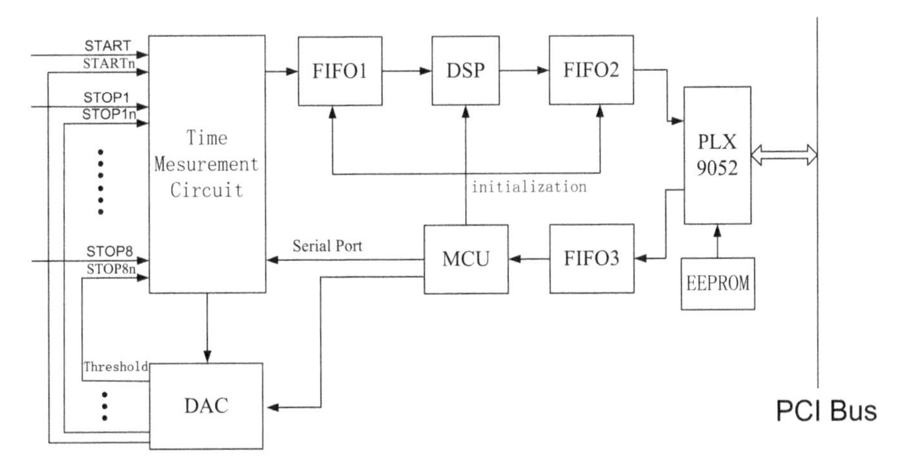

Fig. 4.31 The overall structure of the high-resolution device for measuring time intervals

to restore and initialize the system according to the information in the FIFO3. Meanwhile, the MCU sets the voltage of the comparing threshold for each measurement channel through the D/A converter. Furthermore, the module for measuring time intervals inputs the measured values into the FIFO1 to process the data by the DSP chip. After being processed, the data are sent to the FIFO2 and then transmitted into the computer through the chip PLX9052.

Figure 4.32 shows the work flow of the device for measuring time intervals. After the system is charged with electricity and reset, the host writes configuration command in the FIFO3. Then, the configuration command is read by the MCU to complete the initialization for the time measurement circuit and other circuits and set the parameters of the phase-locked loop in accordance with the command. Meanwhile, the comparing threshold of each measurement channel is determined by setting the DAC. After initialization, the system starts to measure the time interval. After the time measurement circuit writes the measured results in the FIFO1, the DSP processes the data in the FIFO1 and writes the processed results in the FIFO2. Then PLX9052 reads the data in the FIFO2 and sends them to the system to be displayed, recorded and analyzed.

The main technical indexes of the high-resolution device for measuring time intervals are as follows: (1) having 8 measuring channels; (2) measurement resolution of each channel is 120 ps; (3) each channel can record up to eight count values in each measurement to make the device has the ability of time measurement for multi-echo pulses; (4) the trigger threshold of each channel is programmable; (5) the rising or falling edge trigger of each channel can be programmed and selected; and (6) the measurement range is 5 ns–7.8 μs.

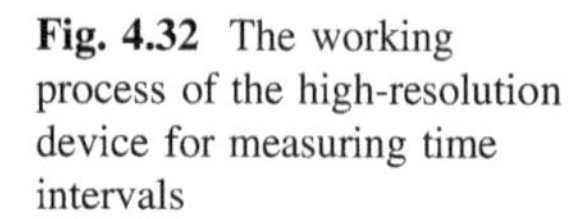
Fig. 4.32 The working process of the high-resolution device for measuring time intervals

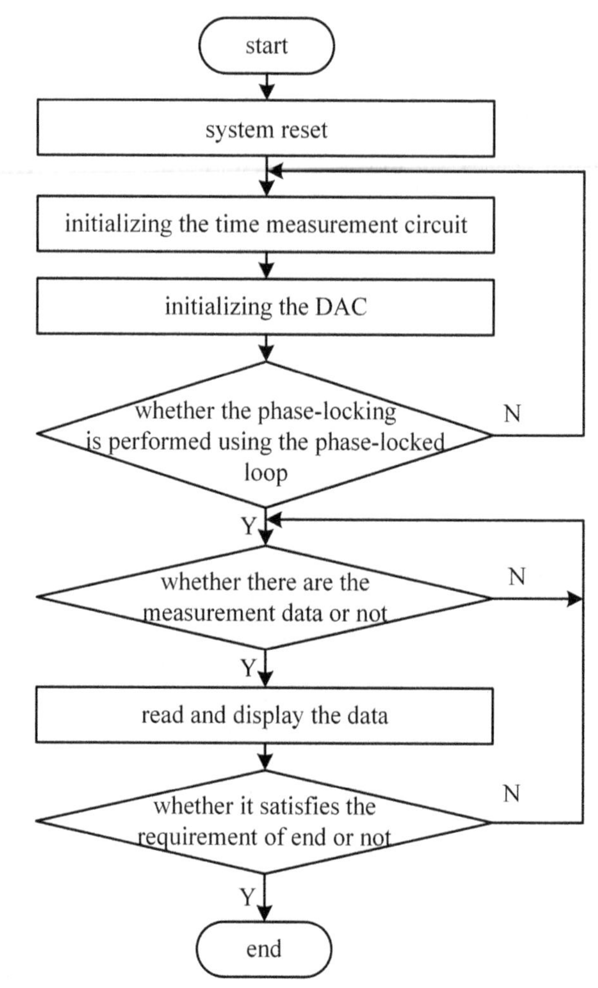

4.3.3 *The High-Resolution Measurement of the Gray of Multi-Channel Echoes*

The intensity of multi-channel echo signals reflects the reflectivity of corresponding sampling targets in array detections. After the consistency of the gains in each channel is adjusted, the reflectivity level of target points for the irradiating laser wavelength can be determined by detecting the peaks of pulse echoes. Therefore, the peak detection circuit can be adopted to obtain the gray information of targets so as to provide the data for generating the gray images of targets. Meanwhile, the peak of laser echoes also can be acquired by obtaining the echo energy through sampling pulse waveforms and integral operation.

1. **The gray measurement based on the peak detection**

Peak detection can be realized by means of charging and discharging capacitors, and the keys to the circuit design and selection of component parameters are: ① To ensure the echo signal of narrow pulse widths can be quickly charged to its peak; ② the results of peak detection can be rapidly transformed to digital quantity after the end of a pulse echo; ③ After the end of the echo pulse, the capacitor requires to release the charged signals at a high rate; and ④ The quantization bits of peak voltage sampling needs to meet the resolution requirements for the generation of gray images.

The coupling signals output by the parallel multi-channel receiving amplifier charge the capacitor through the isolation amplifier and the isolation diode. Assume that the capacitor of 220 pF is selected as the charging capacitor to detect the peak in the rising edge of echo signals. Here, the upper limiting frequency of the peak detection circuit is $f_{3\mathrm{dB}} = 1/2\pi RC = 36$ MHz, which can meet the requirement for tracing and detecting laser echo signals of 10 ns in the rising edge. In order to maintain the signals for peak detection in the falling edge of laser echo pulses, the signals have to be isolated and output by the operational amplifier and then input into the sampler using the operational amplifier with high-input impedance. The performance of the sampler, to a large extent, depends on the quality of the holding capacitor and the sampling speed. Thus, it is crucial to select a high-quality capacitor and a high-speed sampler. While selecting the holding capacitor, we need to pay much attention to the performances of the insulation resistance and dielectric absorption of the capacitor. After sampling, the stored charge of the capacitor needs to be released through a switch as much as possible so as to charge the capacitor in the next time. The release rate depends on the repetition frequency of pulses and has to be greater than the repetition frequency with a certain redundancy.

After being charged to a certain voltage, the capacitor undergoes short-circuit discharge and then the circuit is opened. Under such circumstance, the voltage of the capacitor is expected to grow slowly from zero to the original voltage. The restoration of the voltage of the capacitor is called the dielectric absorption. This feature is supposed to produce errors in the holding voltage. Table 4.2 presents some capacitors which meet the requirements of high-resolution peak detection.

Table 4.2 The high-resolution capacitors

Metalized polycarbonate	−55 to +125	3×10^5	4×10^3	0.05%
Polypropylene	−55 to +105	7×10^5	5×10^3 (105 °C)	0.03%
Metalized polypropylene	−55 to +105	7×10^5	5×10^3 (105 °C)	0.03%
Polystyrene	−55 to +85	1×10^6	7×10^4 (85 °C)	0.02%
PTFE	−55 to +200	1×10^6	1×10^5	0.01%
Metalized PTFE	−55 to +200	5×10^5	2.5×10^4	0.02%

2. **The gray measurement based on the waveform sampling**

The optical power, namely, average power of a continuous light source is the energy radiated by the light source per unit time. The average power $\overline{P}$ of pulsed light is the ratio of the output energy E of each monopulse to the repetition cycle T of the pulse. The peak power of the pulsed light P_p is the power of the monopulse in the persistence time, that is, the ratio of the output energy E of the monopulse to the pulse width τ. That is to say, the peak power of echo pulses is

$$P_p = E/\tau_{RV} \tag{4.71}$$

The sampling data of all waveforms of laser echo pulses are obtained by using the above-mentioned method. Then the sampling data of each pulse experiences integral operation to acquire the laser echo energy E of the pulse and calculate the pulse width. Finally, the value of the peak power of echoes can be calculated according to formula (4.71).

References

1. Hu Y, Fang K, Shu R et al (2001) Sounding effect of laser scanning imaging in earth observation. J Infrared Millim Waves 20(5):335–339
2. Hu Y (2003) The quantification technical research of airborne three-dimensional imaging. Shang Institute of Technical Physics, Chinese Academy of Sciences, Shanghai
3. Hu Y (2002) The laser imaging device based on scanning at an adaptive speed: China, 02157696.3. 2002-12-24
4. Chen Y, Zhang L, Hu Y et al (2004) Array detection technology of echo on earth observation laser imager. J Infrared Millim Waves 23(3):169–171
5. Fang K (2006) The laser beam splitting device with variable splitting angle: China, 03151057.3.2006-02-15
6. Christoph H (1999) Introducing the next generation imaging laser almimeters: concept of a modular airborne pushbroom laser altimeter (MAPLA). In: The fourth international airborne remote sensing conference and exhibition 21st Canadian symposium on remote sensing, Ottawa, Ontario, Canada, 1999: pp 21–24
7. Richmond R, Stettner R, Bailey H (1999) Laser radar focal plane array for three dimensional imaging. SPIE 2748:61–67
8. Hu Y (2009) The three-dimensional detection technology of targets in the target staring of laser imaging -engineering report. College of Electronical Engineering in Hefei, Hefei
9. Hu Y, Wei Q, Zhang L (1997) A study on the signal processing in an airborne laser scanning range find receiver. Chin J Quantum Electron 14(3):284–288
10. Hu Y, Wang J, Xue Y et al (2001) The waveform digitization of laser return in airborne laser remote sensing imaging. J Remote Sens 5(2):110–113
11. Hu Y, Shao H, Xue Y (1998) A study on the precision of the geo-referenced image acquired by remote sensing. Proc SPIE 3505:144–150
12. Hofton MA, Blair JB (2002) Laser altimeter return pulse correlation: a method for detecting topographic change. J Geodyn 34:477–489
13. Chen Q, Yang C, Pan Z et al (2002) The development of laser time-of-flight distance measurement technology. Infrared Laser Eng 32(1):7–10

14. Araki T (1995) Optical distance meter using a short pulse width laser diode and a fast avalanche photodiode. Rev Sci Instrum 66(1):43–47
15. Hu Y, Wei Q, Liu J et al (1997) Using A/D converter to improve precision of time interval measurement in pulse laser range finder. Laser Technol 21(3):189–192
16. Zhang L, Chen Y, Hu Y (2004) A high performance time interval measurement instrument and its application in laser imaging. Infrared Technol 26(3):71–74
17. Kalisz J (1997) Single-chhip interpolating timecounter with 200-ps resolution and 43-s range. IEEE Trans Instrum Measure 46(4):851–856

Chapter 5
Detection Data Processing and Image Generation

The multi-dimensional images of the target most directly and accurately reflect the characteristics of the target. Based on the multi-dimensional images, many characteristics of the target can be obtained such as the outline, shape, spatial distribution, structure, size, and fluctuation. In addition, the higher the precision of the images, the more accurate the characteristics are. Thus, after obtaining the multi-element detection data and the platform data, comprehensively processing these data so as to generate the multi-dimensional images of the target is of great significance in the target detection based on a laser imaging system. These data include the laser echoes reflected from the target, the sampled echo waveform, time delay, gray, multi-level relations in vertical direction, GPS and IMU data, etc. Aiming at the generation of multi-dimensional images, this chapter mainly discusses the methods for generating special images for target detection such as the range, gray, waveform and hierarchy images of the target through processing detection data.

5.1 Processing of Detection Data

A laser imaging based target detection system can be used to obtain the following detection data: time delay, gray, echo waveform, GPS, IMU, time, synchronization mark, distribution of scanning and sampling, etc. These data are a basis for obtaining the 3D coordinate of the laser footprint. Hence, these data require to be processed before being used as a basis for the generation of laser images of targets. Processing of detection data mainly includes data pre-processing and data file processing.

Y. Hu, *Theory and Technology of Laser Imaging Based Target Detection*,
DOI 10.1007/978-981-10-3497-8_5

5.1.1 File Processing

Data files processing mainly includes the integrity analysis of the data for each file, the generation of the uniform files containing the final distance, grey, and timeslice data by matching the data in each file, and the retrieval for the real distance and grey values.

1. **Integrity analysis of the file data**

In this step, the main task is to detect whether there are lost rows. Considering the sampling period of progressive scanning and each row of data are marked with its timeslices, the following method is applied for detecting:

(i) Obtaining the timeslices of each row of data from all the files recording the time delay, gray and waveform data, respectively.
(ii) Calculating the difference between two adjacent timeslices. In the ideal situation, namely, without consideration of frame losses and the influence of the software, the values of timeslices can be considered as the period of scan sampling.
(iii) Subtracting the period of scan sampling from the timeslice sequence. In the ideal situation mentioned above, the values of all timeslices are equal to 0.
(iv) All the timeslices are added together to obtain a sequence $T(n)$, that is, $T(n) = t(1) + t(2) + \cdots + t(n)$.
(v) Judging the $T(n)$. When there is no frame lost, the value of $T(n)$ is supposed to fluctuate around 0. Otherwise, the value of $T(n)$ is likely to grow by one period of scan sampling with each lost frame.
(vi) Detecting whether there are lost rows in data files.

The values for timeslice of the data file are showed in Fig. 5.1.

Figure 5.1 shows that the values of processed timeslices fluctuate in the range of −2 to 0, which is far less than one period of scan sampling (50 ms in the computation). This means that there is no lost line in the data files.

(vii) Analysis of the data files in which rows are lost.

It can be seen from Fig. 5.2 that, jumps occur in four periods of the scan sampling (50 ms in the calculation) in the data flow. It indicates that some rows are lost in the corresponding jump point, so the corresponding rows need to be marked.

1. Matching of data files

All the data files are supposed to be combined to generate a uniform file. While collecting data, the file of gray and waveform files is not expected to record data without being triggered by the main wave, while the range file can still do so. Thus, the matching for data files should be conducted on the basis of applying the timeslice of gray files as the benchmark.

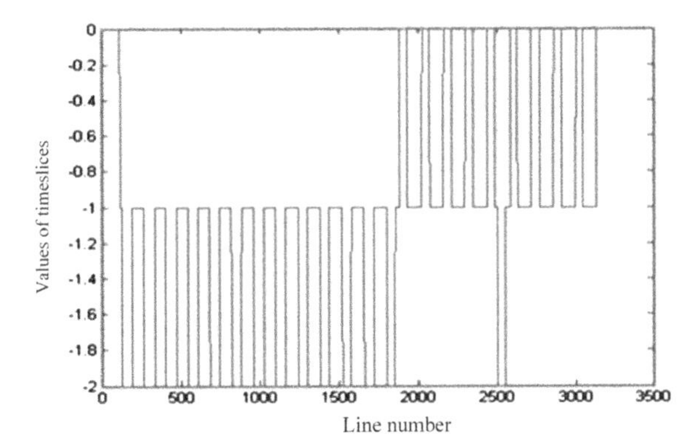

Fig. 5.1 The values of timeslices for data files

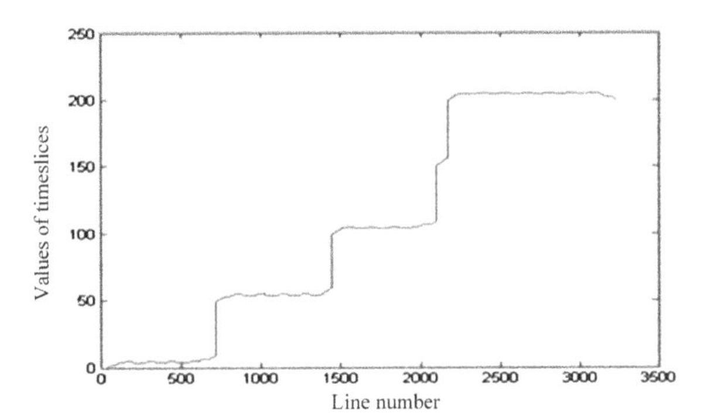

Fig. 5.2 The values of timeslices for gray files

At the beginning and the end of a file, some data are likely to be lost, causing the incomplete row data. Under such circumstance, these rows require to be deleted. In the matching process, the positions of the lost rows are expected to be marked in the premise of not influencing the overall data format, for later interpolation processing.

Two files are obtained after matching of data. The two files are corresponding to the revised file and the file with lost rows, respectively. The later is generated by writing the serial numbers successively into the file in the format of int32.

2. The calibration of data files

The system needs to obtain the gray values of laser echoes by analyzing the peak sampling value, so as to compensate the range error caused by the application of frontier identification method using the peak sampling value of laser echoes. Meanwhile, due to the imbalance of detecting channels, the measured data of time delay, gray, and waveform are inconsistent. Therefore, the original data need to be revised based on the acquired data by employing calibration methods.

5.1.2 Data Processing

The data processing contains many contents, while the author merely analyzes typical aspects of data processing in this book. These typical aspects include the efficiency analysis of ranging data and echo intensity, calculation of typical parameters of echo signals, determination of distributions of laser sampling points and interpolation processing of GPS and INS data.

1. **The efficiency of ranging data and echo intensity**

The laser echoes reflected from the targets need to be received in the process of laser imaging. However, the difference in the distances, the dip angles of reflector, the surface reflectivity and the structures of, targets, tends to result in the diverse intensities of the received echoes. In some conditions, echo signals are likely to be covered by the noise. For example, the echo signals from water surface are hard to be detected because of the reflection characteristics of water. Influenced by interferences, the recorded values are likely to become abnormal. Therefore, it is necessary to analyze the efficiency of laser ranging data and echo intensity prior to using this abnormal values. The basic method for analyzing the efficiency is to set a tolerance for each type of data, and then read all data to compare with the tolerance. If the data are within the tolerance, they can be used as valid data. If individual data are outside the tolerance, the positions of laser sampling points corresponding to these data are recorded and other data are used as usual. If a group of continuous data is beyond the tolerance, these data and data relating to them cannot be used.

2. **Calculating typical parameters of the echo signals**

The typical parameters of the echo signals mainly include results of quantization, peak, pulse width and power. The processing procedures of these parameters are shown as follows:

(i) Filtering processing for the quantitative data. Due to noises and interferences in the sampling based on the waveform digitization, stray data which do not belong to the pulse of the laser echo inevitably exist in the data. Therefore, the data need to be purified at first through filtering (for instance, median filtering) and related processing methods.

(ii) Calculating the peak V_m of target echoes and the serial number L_m of the position of the target sampling points.

(iii) Computing the equivalent pulse width. The numbers n_1 and n_2 of points from the point whose amplitude of target echo is equal to 0.707 V_m, to the peak point in two sides of L_m are separately calculated. Then the sum of the results is multiplied by the sampling interval T_s of waveform to approximately obtain the width of echo pulses, that is, $\tau = T_s \cdot (n_1 + n_2)$.

(iv) Calculating the position of the echoes within the sampling wave gate. The voltage being certain proportions smaller than the peak (such as 0.5 V_m) is selected as the floating threshold to calculate the number k of sampling points from the selected target sampling point to the front of sampling wave gate.

The number is the data source of generating directly images of target distance through the sampling results of waveform.

(v) Calculating the echo power and the backscattering rate. The power of laser echoes reflects the value of the backscattering rate of a target observed along a certain perspective at this wavelength. In the surrounding of the above L_m, all the point data including the pulse of the entire laser echo are selected to make the quadratic sum operation, then the result of which is divided by pulse width τ. The final result is the relative echo power. According to the power equation, in such circumstance where the transmitting power, optical parameters, working parameters of the system and atmospheric environment are basically constant, the receiving power is inversely proportional to the square of the target distance. Accordingly, the relative value of the backscattering rate of each point can be obtained through the normalization after multiplying the power above by the square of the target distance.

After the above pre-processing, various data files with the laser-triggered cycle as a variable are expected to be generated. These files include the waveform of filtered echo pulses, the distributions of time-delay, power, and amplitude of echoes, width distribution of echo pulses, and the distribution of backscattering rate of ground targets. These data are used to generate laser images of the targets as the data sources for generating 3D images of the targets.

3. **The determination of the distribution of laser sampling points**

The sampling methods through transmitting lasers include laser scanning with a single beam, laser scanning in a linear array, laser staring in an area array and sampling through varying the resolution ratio. The distribution rules of laser spots formed on the targets through the irradiation of laser beams are different by employing different sampling ways. For example, the distribution rules of laser spots in target sampling are shown in Fig. 5.3(a), (b) [1]. In Fig. 5.3(a), the ground target is scanned in an arc form by single beams emitted from the airborne platform. In Fig. 5.3(b), the ground target is scanned through translating by the laser array transmitted from the airborne platform.

When laser beams are emitted to scan the targets, the distribution rules of laser spots formed on the targets by the irradiation of the laser depend on the optical structures for scanning, and directly affect the properties of scanning and ranging, such as the resolution. Generally speaking, the distribution rules of sampling points can be characterized by the track composed by the sampling points in a transverse scanning line. The scanning track is in different shapes, such as straight lines and circular arcs. The so-called straight and circular arc scanning tracks refer to that the connection lines of ground sampling points in each scanning line are a straight line or a circular arc, separately.

The determination of the distribution of target sampling points is to describe the position relation of different sampling points, calculate the coordinates of each sampling point, and analyze their densities, intervals and so on. To realize this, it is necessary to apply various data including platform position, attitude, beam pointing

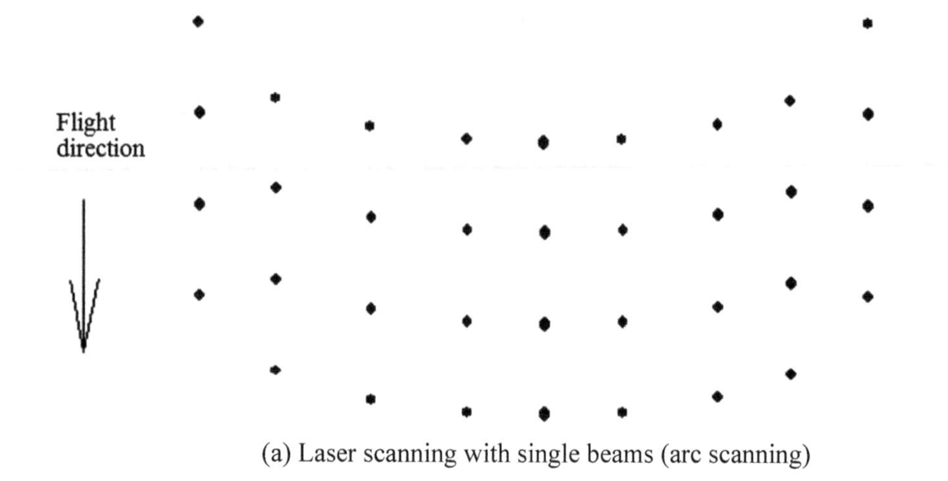

(a) Laser scanning with single beams (arc scanning)

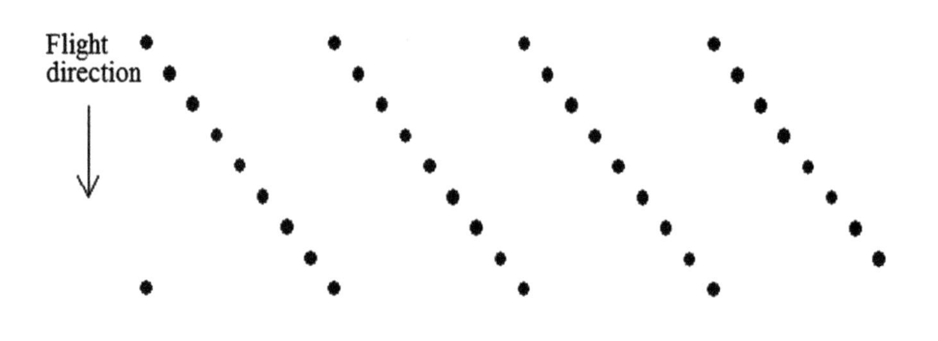

(b) Laser scanning in a linear array

Fig. 5.3 Two distribution types of the laser spots in target sampling

and range in the calculation so as to obtain the center position of each spot, and then conduct statistical analysis.

A typical process flow for obtaining the center position of the laser spot is as follows:

(i) The coordinates of laser footprints in the aircraft coordinate system are determined by using the ranging data.
(ii) The coordinates of laser footprints are transformed from the aircraft coordinate system to the geographic coordinate system by employing the INS data.
(iii) The coordinates of laser footprints are converted from the geographical coordinate system to the world geodetic system 1984 (WGS-84) coordinate system by utilizing the GPS positioning data modified by an eccentricity vector.

(iv) The coordinates of laser footprints are transformed from WGS-84 coordinates to Gauss plane coordinate.

4. **Interpolation processing of the GPS and INS data**

To obtain the center position of a laser spot, it needs to acquire the center positions of the optical system in the transmitting of each laser pulse. However, due to the high repetition frequency of laser, data updating rate of GPS and INS is not so high. In order to obtain the corresponding GPS and INS data for each laser sampling, the GPS and INS data need to undergo the interpolation processing. The procedure for processing GPS data is shown in Fig. 5.4. Its processing flow is as follows [2]:

(i) Differential processing of the GPS data [3]. In order to get the positioning result with high precision, the differential technique is generally needed to obtain 3D positions at meter and even centimeter levels. If the real-time differential technology is used to obtain the 3D position, it is essential to equip a

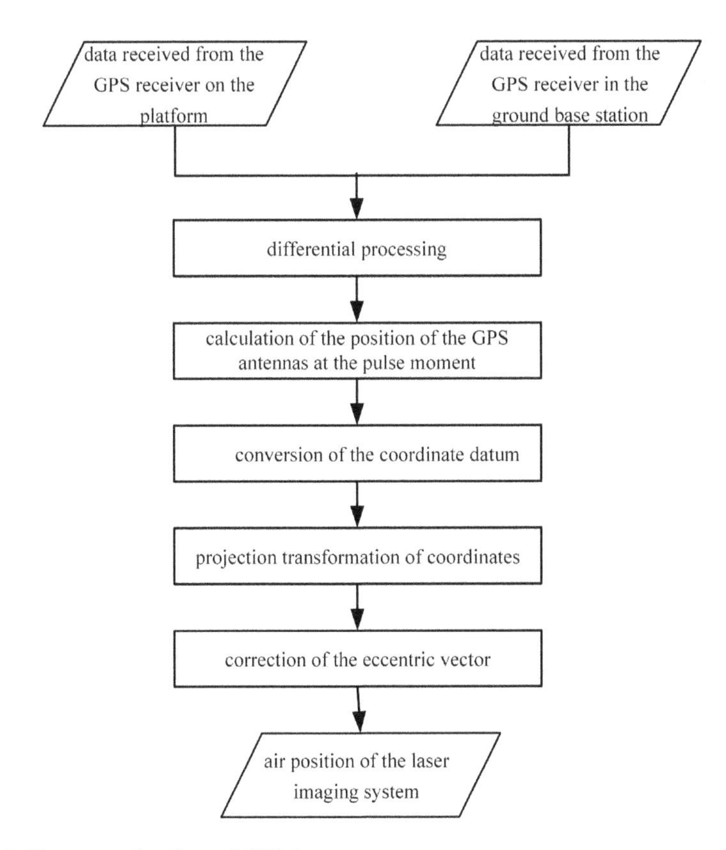

Fig. 5.4 The processing flow of GPS data

data radio transceiver and record the positioning results. The original measurement data received by the GPS receiver in the ground base station and on the platform can be processed with high accuracy afterwards, and the post-processing accuracy is usually higher than that of real-time positioning. In the post-differential processing, the GPS carrier phase data are generally used to perform the dynamic computation. Meanwhile, the carrier phase ambiguity can be calculated by using some mature algorithms including the search algorithm, Kalman filtering algorithm, and so on. At present, the post-processing can be conducted at a high speed, and the accuracy is higher than a few centimeters.

(ii) The calculation of synchronous position of the GPS antennas at the pulse moment. GPS positioning can be used to measure and solve the position of GPS antennas with high precision merely at a fixed sampling rate. While the pulse moment of laser echoes received from the system for acquiring data of the laser imaging is arbitrary, that is, not strictly synchronous with the result provided by the GPS positioning results in time. Therefore, the data of the laser imaging have to be acquired through the fitting and interpolation of the normalized GPS positioning results according to the received pulse moment t. Generally, the 3D position (X, Y, Z) at the pulse moment can be obtained through the precise interpolation by using the polynomial least-squares fitting algorithm. Then the fitting precision can achieve centimeter level through three times of curve-fitting processing using the GPS positioning results 5 s before and after the pulse moment. The interpolation model is

$$\begin{cases} X(t) = a_0 + a_1 t + a_2 t^2 + a_3 t^3 \\ Y(t) = b_0 + b_1 t + b_2 t^2 + b_3 t^3 \\ Z(t) = c_0 + c_1 t + c_2 t^2 + c_3 t^3 \end{cases} \tag{5.1}$$

where a_0, a_1, a_2, a_3, b_0, b_1, b_2, b_3, c_0, c_1, c_2 and c_3 are coefficients, which can be reversed by using the least square method based on the data 5 s before and after the time t.

(iii) The conversion of the coordinate datum. In all the observations and calculations of GPS positioning data, the WGS-84 ellipsoid is used as the coordinate reference datum. While in China, Beijing 54 ellipsoid is widely used as the coordinate reference datum. Therefore, it is essential to convert the coordinate datum according to the formula of Bursa-Wolf conversion model

$$\begin{bmatrix} X \\ Y \\ Z \end{bmatrix}_{BJ-54} = \begin{bmatrix} \Delta X \\ \Delta Y \\ \Delta Z \end{bmatrix} + \begin{bmatrix} 1+\Delta k & \varepsilon_Z & -\varepsilon_Y \\ -\varepsilon_Z & 1+\Delta k & \varepsilon_X \\ \varepsilon_Y & -\varepsilon_X & 1+\Delta k \end{bmatrix} \times \begin{bmatrix} X \\ Y \\ Z \end{bmatrix}_{WGS-84} \tag{5.2}$$

where ΔX,ΔY and ΔZ are three translation parameters of origin coordinates, and ε_X, ε_Y and ε_Z are rotation parameters of the three coordinates. Δk is the scale parameter.

The above seven conversion parameters can be obtained from the National Departments of Surveying and Mapping or can be reversed based on the common points with known coordinates distributed uniformly in the two reference coordinate systems. For the local region without large areas, ε_X, ε_Y, ε_Z and Δk all can be considered to be 0. In such circumstance, the formula (5.2) can be used as the conversion formula of three parameters.

(iv) The projection transformation of coordinates [4]. The geodetic coordinates (η, l) experiencing datum conversion has to undergo projection transformation:

$$\begin{cases} x = s + \frac{\chi^2 N}{2} \sin\eta \cos\eta + \frac{\chi^4 N}{24} \sin\eta \cos^3\eta(5 - tg^2\eta + 9v^2 + 4v^4) \\ \quad + \frac{\chi^6 N}{720} \sin\eta \cos^5\eta(61 - 58\,tg^2\eta + tg^4\eta) \\ y = \chi N \cos\eta + \frac{\chi^3 N}{6} \sin\eta \cos^3\eta(1 - tg^2\eta + v^2) \\ \quad + \frac{\chi^5 N}{120} \cos^5\eta(5 - 18\,tg^2\eta + tg^4\eta + 14v^2 - 58\,tg^2\eta v^2) \end{cases} \tag{5.3}$$

Thereinto:

$$N = \frac{a}{\sqrt{1 - e'^2 \sin^2\eta}}$$
$$s = \frac{N\lambda'' \cos\eta}{\rho''}$$
$$v^2 = e'^2 \cos^2\eta$$
$$\chi = l - l_0,\; a = 6378245.0,\; e'^2 = 0.006738525414684$$

The conversion precision by using this formula is 0.001 m. l_0 is the longitude of the center meridian.

(iv) The correction of eccentric vector [5]. The phase center of GPS antenna is not consistent with the center of the optical system in data acquisition system of the laser imaging. The two centers have a certain distance, namely, the existence of the eccentricity vector (u, w, s) which can be determined precisely through measurement. Meanwhile, the attitude parameters at three axes at this moment can be obtained from the attitude measurement unit, and therefore the position of the optical center can be calculated according to the following conversion formula of coordinates.

$$\begin{bmatrix} X \\ Y \\ Z \end{bmatrix}_{laser} = \begin{bmatrix} X \\ Y \\ Z \end{bmatrix}_{GPS} - \begin{bmatrix} A_1 & B_1 & C_1 \\ A_2 & B_2 & C_2 \\ A_3 & B_3 & C_3 \end{bmatrix} \times \begin{bmatrix} u \\ w \\ s \end{bmatrix} \tag{5.4}$$

Thereinto:

$$\begin{aligned}
A_1 &= \cos\varphi\,\cos\kappa - \sin\varphi\,\sin\omega\,\sin\kappa \\
A_2 &= \cos\varphi\,\sin\kappa - \sin\varphi\,\sin\omega\,\cos\kappa \\
A_3 &= -\sin\varphi\,\cos\omega \\
B_1 &= \cos\omega\,\sin\kappa \\
B_2 &= \cos\omega\,\cos\kappa \\
C_1 &= \sin\varphi\,\cos\kappa + \cos\varphi\,\sin\omega\,\sin\kappa \\
C_2 &= -\sin\varphi\,\cos\kappa + \cos\varphi\,\sin\omega\,\cos\kappa \\
C_3 &= -\cos\varphi\,\cos\omega
\end{aligned}$$

The three attitude parameters, namely, pitching φ, rolling ω and navigation κ are provided by synchronous measurement of the attitude measurement unit (INS).

After the above processing of GPS data, the 3D position of the optical center of the data acquisition system of laser imaging can be obtained at scanning moment. The data of the 3D position of the optical center are expected to be stored so as to calculate the 3D coordinates of the laser sampling points by combining with other data. These 3D coordinates of the laser sampling points can be applied as the basis for calculating the positioning data for generating laser images of targets.

5.2 Generation of Range Images and Gray Images of Targets

The range images and gray images of targets intuitively reflect the entire position characteristics and distribution of the back-reflection of targets. It is established based on the 3D position and the echo intensity data of multiple sampling targets. Each pixel of generated range images and gray images represents the range information and echo intensity corresponding to each target sampling point.

5.2.1 Methods for Generating Images

The range images of targets are generated based on the range data of the target sampling points. The range images are generally expressed by the elevation used in maps to indicate the elevation information of a place in the space. When ground surface is detected as a target, its range image is the digital elevation model (DEM) of the ground. When a target is an object with a certain structure and shape, the range image of this target can be expressed as the height image of each sampling point relative to the reference plane which is selected as the equivalent ground surface. Each pixel of the range images represents the range or height of the corresponding sampling point and all the pixels form a 3D image. According to the corresponding relation of the range information with the shape and spatial distribution of targets, basic characteristics of targets including target structure, shape and size can be inversed on the basis of the range images of targets.

According to the principle of the target detection based on a laser imaging system, the value of energy of laser echoes is proportional to the reflectivity. Meanwhile, the target reflectivity is related to laser wavelength, target material, and smooth degree and brightness of the target surface. Due to the difference in target material as well as smooth degree and brightness of the target surface, different targets present varied reflectivities. On the basis of extracting the energy of laser echoes, the gray images of targets can be generated by retrieving the back-reflection of the illuminated regions of targets employing the mapping relation between the echoes energy and the back-reflection. The gray images of targets demonstrate the back-reflection information of a position in the space. According to the difference in the reflection of different targets in the same detection condition, we can distinguish the targets from the background based on the gray images of targets so as to obtain target characteristics including approximate contour, shape, and spatial distribution.

To generate the range images and gray images of targets, the positions of target sampling points have to be matched with these images. According to the above processing of the data, the relation between the range and gray of each laser sampling point is determined so as to further generate relevant laser images by generating the range images and the gray images.

Generally, the range images and the gray images of targets can be generated using two methods, namely, the pure two-dimensional interpolation and the surface fitting. The former requires the mathematics surface to contain all known points within the interpolation range, that is, the height difference between all reference points and corresponding points on the mathematics surface is 0. The latter does not require the mathematics surface to contain all reference points but has to satisfy a certain condition. For example, the sum of square of the height differences between all reference points and corresponding points on the mathematics surface requires to be minimized, that is, to comply with the least square criterion. The interpolation operation of multi-layer surfaces in the block is an improvement of the above two methods and can be used to generate the range images and gray images of targets effectively. The basis of this method is that a square wave can be synthesized by superimposing multiple periodic harmonics in the series expansion. If this method is used to generate target images, any regular or irregular continuous surface can be approximated through the superposition of multi-layer simple mathematics surfaces. The multi-layer surface is used in two steps. The first step is to spread the multi-layer surfaces with different fluctuant shapes in the extended area of the entire block. The second step is to form an integral continuous surface through the vertical superposition of the surfaces from the perspective of pure geometry so as to constitute the laser images [2].

5.2.2 *Process of Image Generation*

The process for generating the range images and gray images of targets using a target detection system using laser imaging technique is as follows:

(i) Combining range data and laser energy data. The laser energy files segmented based on the GPS event count and the range files are combined into one file to facilitate the subsequent processing.
(ii) Finding the lost parts of the laser energy and range data. The position of the lost parts of the data relating to laser energy and the laser ranging is found out and represented by 0.
(iii) Data formatting. The raw data and the corresponding laser energy as well as the range data are transformed from the matching state of the laser sampling points to the arrangement of the laser images in the grid coordinates.
(iv) GPS data formatting. The original order of GPS data is: event count, X, Y, H event count, X, Y, H event count, X, Y, H … among them, commas are used as decollators, and Hs of each group are not separated from the event count. Here, GPS data are expected to be formatted into the specified order.
(v) Removing the data of the starting and ending parts. The image data and their auxiliary data are defined in the range containing same lines with the GPS positioning data, followed by removing the incomplete semi-cycle data of the starting and ending parts.
(vi) The GPS data and range data as well as gray data are segmented according to the length of three GPS events.
(vii) Formatting and interpolation processing of INS data. The attitude data in degrees, minutes and seconds are converted to those in seconds, and the lost data are generated through the interpolation.
(viii) Processing GPS data and transforming coordinates of the GPS data. The obtained GPS data in the Gauss projection coordinate are converted into those in WGS-84 rectangular coordinate for the subsequent location processing.
(ix) The 3D positioning of laser sampling points. According to the above calculation methods for positioning laser sampling points, the position coordinates of laser sampling points are calculated.
(x) Mesh generation and interpolation processing of coordinates. Figure 5.5 presents a generated range image of the target with surface fluctuation and shows the 3D spatial distribution and fluctuant condition of one side of the target. In this figure, 0–30 in the right coordinate axis demonstrate the number of target sampling points in one direction (width is 150 m), while

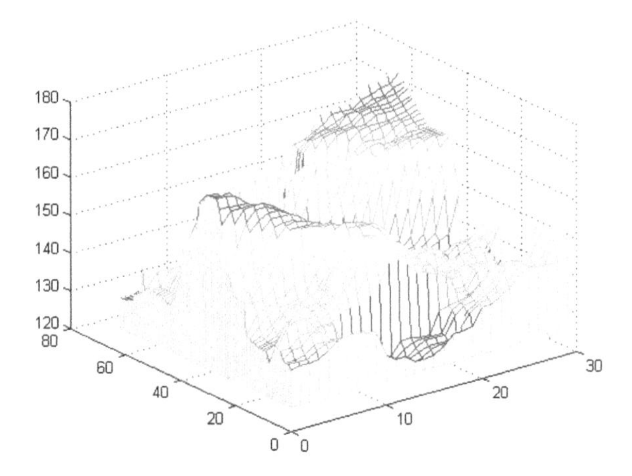

Fig. 5.5 The generated range image of a typical target

the left coordinate axis signifies 80 scanning lines in the other direction. The corresponding target area is 150 m × 200 m and the target fluctuates within 130–170 m.

(xi) Geometric correction of images. The conventional geometric correction method is used to roughly correct the image (without considering the influence of the projection difference). Then the geometric precision correction is performed to the image only considering the correction of projection difference.

(xii) Generating the range images and gray images of targets. Figure 5.6(b) shows the generated range image of a partial region of a building group, as the target, by using the above image generation method. Figure 5.7 illustrates the gray image of a ground target area.

The range images of targets show the information of the spatial distribution of targets and the spatial relation between the targets and surrounding objects, and reflect the shape features and the surrounding environment of targets. The information can be used as an important basis to retrieve target features so as to obtain the information of target features including the target structure, shape and size. In the above images, targets 1–18 in the image grid are three low-rise buildings. The fluctuation of them can be observed obviously from the range image: suddenly changing from 300 m to about 350 m and then declining to about 260 m. In several front channels of laser emission, because the first received laser echo, influenced by trees and electric wires, is not scattered from the buildings but reflected from the trees and electric wires, the range data are abnormal. In the image grid, targets 19–24 basically reflect the information of the high-rise buildings in the distance including the obvious structure and shape of the buildings. The more close relation between the targets features and position characteristics of laser imaging can be acquired by analyzing the range images further so as to further obtain more features of targets.

(a) building group target

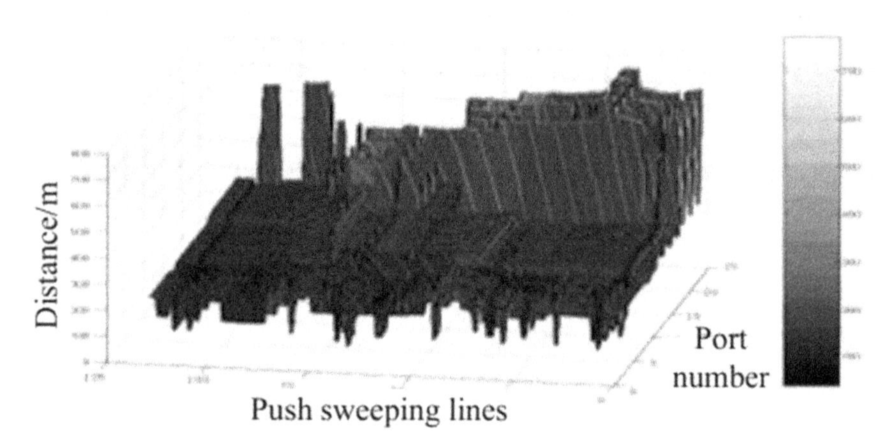

(b) the range image of the region in the red frame of the building group

Fig. 5.6 The building group target and the range image of a part of the region

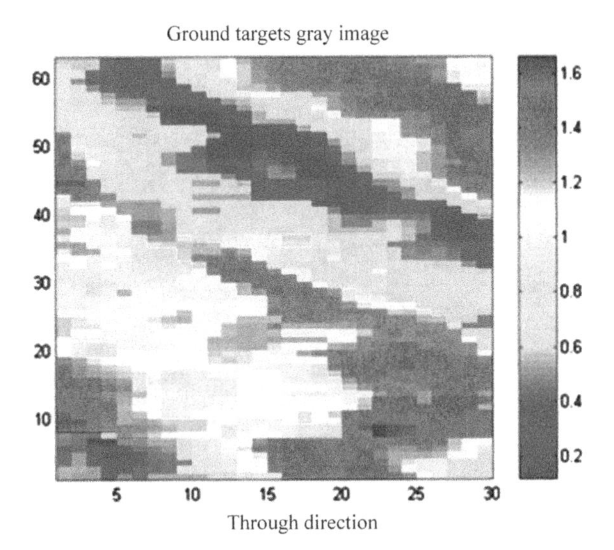

Fig. 5.7 The gray image of the region

The range equation of the target detection based on a laser imaging system reveals that the signal intensity of laser echoes is inversely proportional to the target range while is directly proportional to the target reflectivity. With a certain range, the higher the target reflectivity, the stronger the signal intensity of laser echoes is; and vice versa. The target reflectivity depends on the laser wavelength, target material, and the smooth degree and brightness of targets. The reflectivity of different targets varies with the target material, and smooth degree and brightness of the target surface. According to Fig. 5.6, the place in top right corner with high elevations and large back-reflection is probably hillsides or woods. The stripe stretching from the upper left corner to the lower right corner represents bald hills or roads, which show small back-reflections and low elevations. The place in the left-central part is perhaps slopes at certain heights with modest back-reflections.

On the basis of the energy of laser echoes, targets can be distinguished from the complex background in the gray images of targets according to the reflectivity difference of different objects in the same detection condition. In this way, the outline and shape of targets are identified intuitively so as to provide the information needed for targets determination. However, merely using gray images to extract and recognize targets also shows a certain limitation. Because when the attributes of a target exhibit slight differences from those of a surrounding object, the target and the object show almost similar reflectivity. In this case, it is hard to identify the target based on the reflectivity, let alone the extraction of the general outline and shape of the target. In addition, the back-reflection of targets is not only related to the attributes of targets themselves but also closely associated with the detection conditions. Moreover, the gray images of targets only reflect the information of two-dimensional features of targets. All these result in that the target information obtained based on the gray images is likely to be less comprehensive and therefore difficult to precisely describe targets. Therefore, it is essential to combine the range images to precisely describe targets.

5.3 Generation of Waveform Images of Targets

5.3.1 *Expression of the Waveform of Laser Echoes*

In the target detection based on a laser imaging system, due to the existence of the divergence angle of laser beams, the pulse width, and the size of targets, the echo signals of the laser beam reflected from any a target sampling point in time domain can be expressed as [6]:

$$f_i(t) = a_i \exp[-k(t_i - t)^2/\tau^2] \cdot g(\psi) \cdot \cos(\zeta) \cdot h/s_i^2 \tag{5.5}$$

where, a_i represents the surface reflectivity, which is used to characterize the attributes of targets, while t_i indicates the echo time-delay of the corresponding target sampling points and reflects the spatial distribution of targets. τ denotes the width of the laser pulses and $g(\psi)$ is the energy distribution function of the laser spots, which can be considered to comply with the Gaussian distribution. ζ signifies the angle between the incident angle of the lasers and the normal line of the target surface. h is a coefficient relating to the response of the detector, effective receiving area and atmospheric attenuation. s_i represents the range between the target sampling point and the laser imaging based detection system.

It can be seen from formula (5.5) that the waveform characteristic of the target echo is a combined modulation function relating to various factors of each part in the targets [7]. The factors include time delay (echo time delay), location (the position in laser spots) and attributes (back-reflection and structural fluctuation). Meanwhile, different target types lead to the different echo waveforms. Therefore, waveforms of the target echoes also can be regarded as an important way for reflecting target features.

The waveform images of targets which reflect the amplitude and width of the echo waveforms and distribution law of fission of the echo waveforms with the position of the target sampling points intuitively describe the modulation result of targets for laser pulses. Meanwhile, the waveform images of targets reflect not only the temporal and spatial distribution of the received energy but also the structure characteristics of targets. The most important point is that the characteristics of structure and distribution of targets can be obtained through the waveform images of targets. Moreover, parameters for describing the information of target characteristics including the target range and the back-reflection also can be acquired. Figure 5.8 presents the waveform images of three different targets obtained from the scene simulation and the features extracted based on the waveform images and wavelet transform. Thereinto, the first lines demonstrate the waveform images of different targets. The second, third and fourth lines represent the results of wavelet

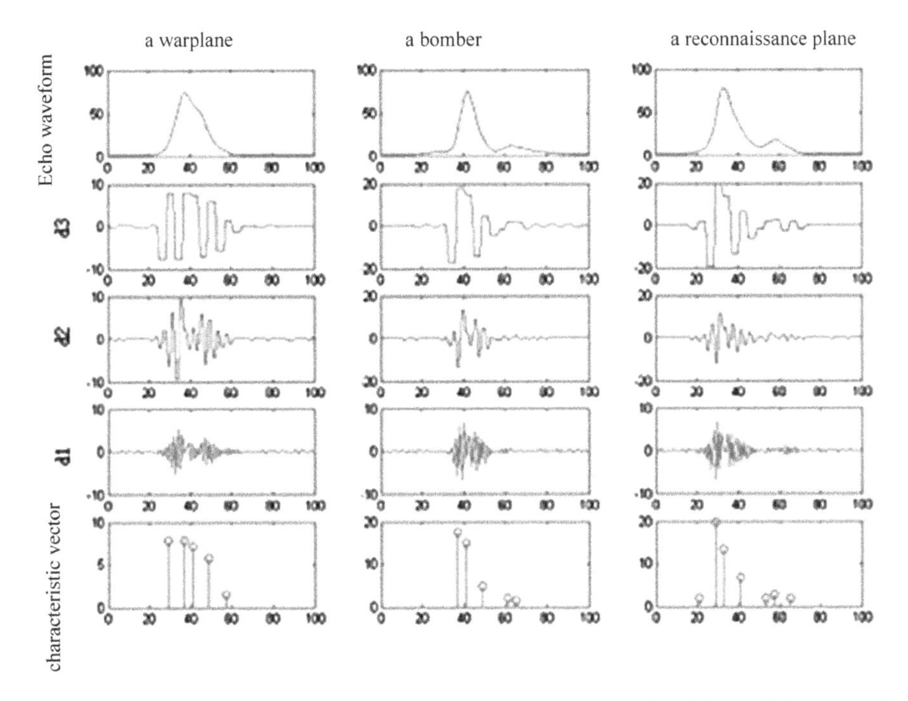

Fig. 5.8 The waveforms of three different targets based on the scene simulation and the extracted features

transform in different scales. The fifth lines illustrate the characteristic vector of the extracted waveforms of echo pulses.

In the information acquisition and processing of the target detection based on a laser imaging system, the laser footprints irradiating the targets form the laser spots with certain areas. If other objects exist in the spot region apart from the detected targets, it is necessary to process the waveform data of the laser echoes reflected from the discrete targets collected after digitizing the waveforms. The data processing includes the decomposition of the echo waveforms and the extraction of the echo waveforms of individual targets to generate real waveform images of the targets so as to reflect the relevant features of targets.

From the perspective of the signal processing, the echo data are equivalent to the convolution of the emission pulses and the system transfer function in time domain. According to the properties of the convolution function, the echo data also can be described using Gaussian function model. Because each echo datum is described using Gaussian function model, the synthesized echo signal can be expressed based on multiple weighted Gaussian functions. Each component in Gaussian function model can be described using the weight, mean value and covariance which represent the intensity factor, center position and width factor of a peak, respectively. The data extraction of target waveforms refers to extracting the waveform data of detected targets from the synthesized waveforms of reflected signals of multiple reflection points in the laser spot area. To realize this, the waveform data of targets can be extracted by using the decomposition method of waveform data, and the essence of this method is to estimate the weight, mean value and variance of each Gaussian function in the fitting model of waveform data. The decomposition methods of waveform data mainly include the nonlinear least square method (LSM) and maximum likelihood estimation algorithm based on expectation maximization (EM) algorithm and improved EM algorithm [8–10].

5.3.2 *The Data Extraction of Target Waveforms Based on LSM*

The data of target waveforms are extracted by using the decomposition method of echo waveforms. Suppose that the number of the sub-echoes is n in the target sampling area, and each sub-echo signal represents a specific sub-target or surrounding object. Then the equation for synthesizing echo signals can be expressed as follows:

$$y = f(t) = \sum_{i=1}^{n} f_i(t) = \sum_{i=1}^{n} a_i \mathrm{e}^{-k(t-t_i)^2/2\sigma_i^2} \tag{5.6}$$

where, k is a constant coefficient, and a_i denotes the echo reflectivity in each sub-target region. t_i represents the time of echo peaks in each sub-target region and σ_i is the half-power width of echoes in each sub-target region.

The decomposition method of echo pulses aims to solve Eq. (5.6) so as to obtain the echo signals in each sub-target region. Suppose that

$$g(t) = \sum_{i=1}^{n} a_i \mathrm{e}^{-k(t_k-t_i)^2/2\sigma_i^2} - f(t) \tag{5.7}$$

The waveform decomposition is to minimize $g(t)$ for each echo in the corresponding synthesized echoes, that is, to satisfy the following formula:

$$\sqrt{\frac{1}{m}\sum_{t=1}^{m} [g(t)]^2} < \varepsilon \tag{5.8}$$

where, ε is the required calculation precision. The selection of ε depends on many factors, such as the precision required by the system and average noise level. For example, the calculation precision can be selected as the triple of average noise level. In formula (5.8), there are totally $3n$ unknown quantities and m known quantities. The iterative equation can be calculated based on the LSM as follows:

$$A(k) = \begin{pmatrix} g_{11}(x(k)) & \cdots & g_{1N}(x(k)) \\ \vdots & \ddots & \vdots \\ g_{m1}(x(k)) & \cdots & g_{mN}(x(k)) \end{pmatrix} \tag{5.9}$$

$$x(k) = -[A(k)^T A(k)]^{-1} A(k)^T f(x(k)) \tag{5.10}$$

$$x(k+1) = x(k) + \Delta x(k) \tag{5.11}$$

In formulas (5.9)–(5.11), $N = 3n$, and k is the iteration number, while $g_{ij}(x) = \partial g(i)/\partial x_j$, and $x(a_1, \ldots a_n, t_1, \ldots t_n, \sigma_1, \ldots \sigma_n)$ represents $3n$ solution quantities.

By using the above algorithms, the waveform data of targets in the local area obtained from the target detection based on a laser imaging system are extracted and processed. The main parameters of the system including the laser pulse energy, half-power width of pulse, the laser divergence angle, and echo sampling resolution are set as 2 mJ, 5 ns, 2 mrad and 0.2 ns, respectively. Then the target which is 60 m away from the detection system is detected. In order to make the size of the laser spot in the detection be equivalent to that of the laser beam in the long distance detection, the lens is added in the initial optical path of the laser beams to enlarge the divergence angle of the output laser to 60 mrad. Under such circumstance, the size of the laser spot on the target is equivalent to that 1.8 km away. The target in front of the experimental system is detected and the optical image of the detection area is shown in Fig. 5.9.

Fig. 5.9 The target area located in the laser spot

In Fig. 5.9, the imaging spot not only covers the clay area, cement pavement, and lawn, respectively, from top to bottom, but also includes two small floor lamps in the upper left and lower right. The laser echo signals received by the detection system and the positions of the calculated local peak points are shown in Fig. 5.10.

In Fig. 5.10, the dotted lines represent the signals of decomposed sub-echoes. According to the positions of peak points of sub-echoes, these lines correspond to sub-echoes 1–4 from left to right. The echo waveforms of the floor lamps in the upper left can be obtained by extracting sub-echoes 1–4, as shown in Fig. 5.11.

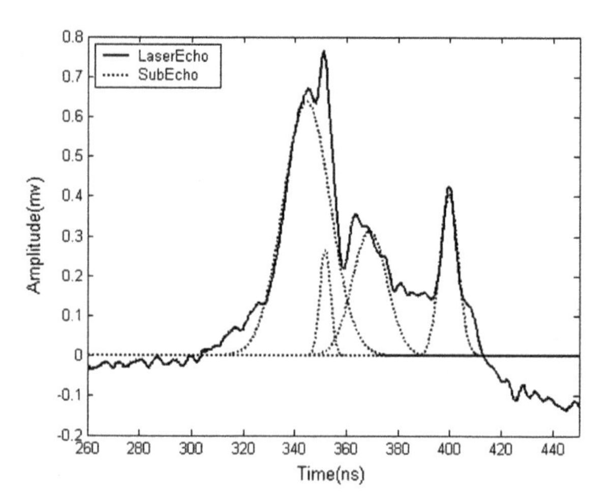

Fig. 5.10 The echo signals of sub-targets obtained by decomposing the waveforms

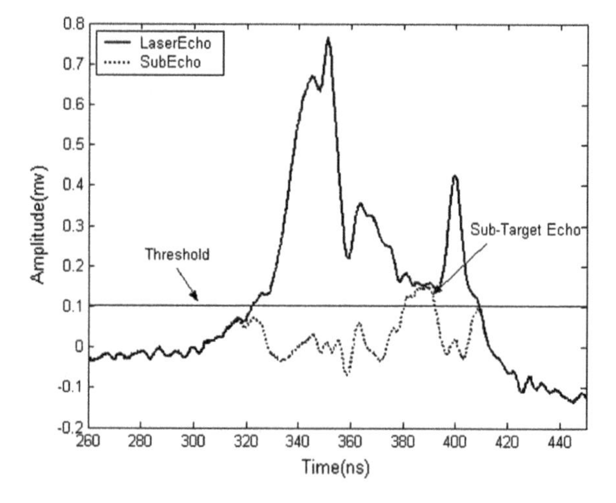

Fig. 5.11 The covered echo signals of sub-targets obtained through the extraction

5.3.3 *Extraction of Waveform Data Based on EM Algorithm*

EM algorithm regards the waveform data as a Gaussian mixture model. Its main purpose is to find a set of Gaussian parameters which conform to the expectations of the original waveform, and calculate the estimations of these expectations by using the maximum likelihood method.

It is assumed that the waveform is composed of a series of simple Gauss distributions, and then the mathematical expression of mixture distribution of Gaussian model (sampled waveform of laser echoes) is

$$f(x) = \sum_{j=1}^{k} p_j \times f_j(x) \quad f_j(x) \in N(\mu_j, \sigma_j^2) \tag{5.12}$$

where, k is the number of Gaussian functions (the number of peaks). $f_j(x)$ is the probability density function of Gaussian distribution. While p_j is the weight of $f_j(x)$, represents the proportion of the distribution in the mixture distribution and satisfies $0<p_j<1, \sum_{j=1}^{k} p_j = 1$. μ_j and σ_j denote the expectation and the standard deviation of Gaussian functions. For each decomposed waveform, the calculated μ_j and σ_j represent the position and width of each echo waveform, respectively.

In the formula (5.12), parameters p_j, μ_j and σ_j of each waveform can be estimated using EM algorithm. The formula for calculating the maximum likelihood estimates of each parameter using EM algorithm can be expressed as:

$$Q_{ij} = \frac{p_j \times f_j(x)}{\sum_{j=1}^{k} p_j \times f_j(x)} \tag{5.13}$$

$$p_j = \frac{\sum_{i=1}^{n} N_i Q_{ij}}{n \times \sum_{i=1}^{n} N_i} \tag{5.14}$$

$$\mu_j = \frac{\sum_{i=1}^{n} N_i Q_{ij}\, i}{n \times p_j \times \sum_{i=1}^{n} N_i} \tag{5.15}$$

$$\sigma_j = \sqrt{\frac{\sum_{i=1}^{n} N_i Q_{ij} (i - \mu_j)^2}{n \times p_j \times \sum_{i=1}^{n} N_i}} \tag{5.16}$$

where, n is the number of the sampled waveforms, N_i represents the amplitude of each sampled i, and p_j is the weight.

What is worth noting is that it is essential to estimate the number of Gaussian functions in Gaussian mixture model and the initial values of parameters of each Gaussian function before using EM algorithm. The initial value of μ_j can be set as the local maximum value of the waveform. The first-order derivatives of the waveforms were calculated using gradient operators. The local maximum can be

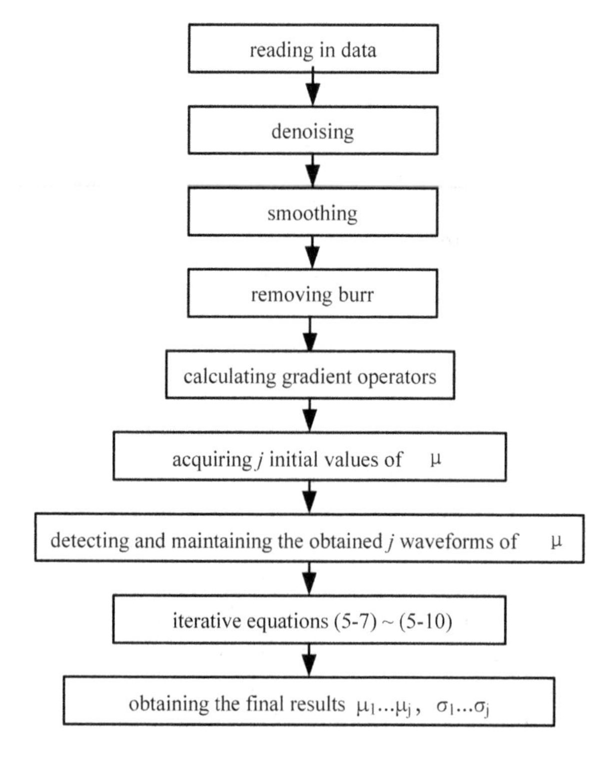

Fig. 5.12 The flow of decomposition of echo waveforms based on EM algorithm

obtained from the sampling points with waveforms whose first-order derivatives are larger than 0 in the left while less than 0 in the right. The number of local maximum values can be used to determine the number of the initial and total Gaussian functions. The decomposition of the waveforms through Gaussian functions based on EM algorithm is shown in Fig. 5.12.

The results of decomposition of waveforms using EM algorithm are displayed in Fig. 5.13.

Because waveforms of target echoes are closely related to the irradiation angle and time, the generated waveform image is a function of irradiation angle and time. Considering this fact, applying the waveform images to reflect target features shows certain shortcomings. Therefore, the target features obtained based on the waveform image are not so complete and accurate. To overcome this disadvantage, there is an effective method which obtains the waveforms of target echoes at different angles and moments by irradiating a target with different angles. Subsequently, a set of waveform images of the target is generated based on the waveforms of target echoes at different angles and moments to eliminate the influences of the uncertainty of time, space and attributes of the target.

Fig. 5.13 The results of decomposition of waveforms based on EM algorithm

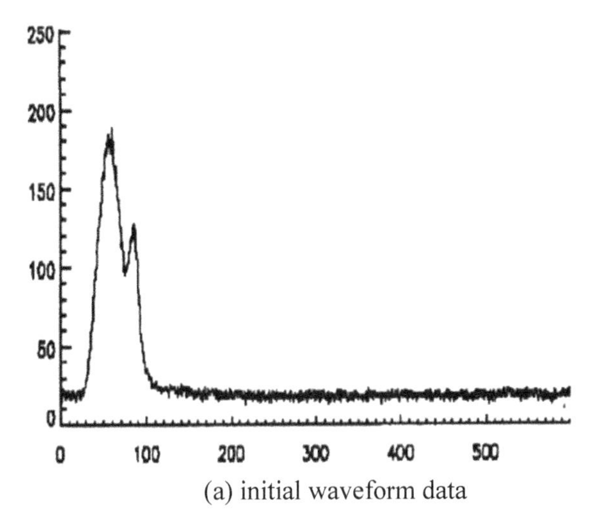

(a) initial waveform data

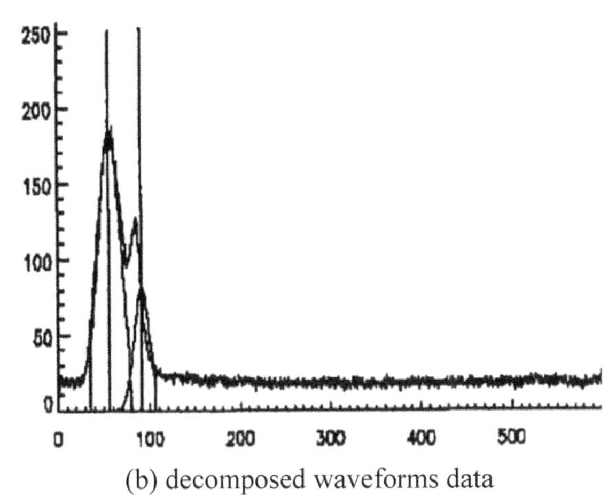

(b) decomposed waveforms data

5.4 Generation of Hierarchy Images of Targets

The hierarchy image of targets which reflects the characteristics of spatial structure of targets is unique to the target detection based on a laser imaging system. Based on the reflection of radiated laser beams from multiple surfaces of targets, multiple echoes along the laser beam can be obtained. Then, hierarchy images of targets are generated by mining the characteristics of multiple echoes including waveform, time delay and energy. The generated hierarchy images of targets are utilized to reflect target information including level and fluctuation, as well as targets covered by some objects in the depth direction. The hierarchy images of targets are mainly suitable for targets with surface fluctuation and obscure targets.

5.4.1 Principle of Level Imaging of Targets

Although the laser beam used in the target detection based on a laser imaging system is narrow, the lasers are expected to form the laser spot with a certain area by irradiating the target in the laser footprint points. If multiple objects in the laser spot show certain spatial relations (for instance, tanks and other targets are located beneath trees or camouflages) or fluctuate, the laser beam can penetrate the interspaces of objects to be separately reflected from the surfaces of different objects. As the surface of targets varies violently, the broadening of the synthesized echoes is expected to be enlarged with the enlargement of the difference in time delay of echoes at different points. Under such circumstance, when the difference of time delay enlarges to some extent, deformations and even fissions are likely to occur in echo waveforms to form multiple sub-echoes, as shown in Fig. 5.14. In the forest area, for example, the echo signals of a same laser in the laser footprint points probably include the echoes reflected from the leaves, branches, and trunks of trees, as well as the ground. The fission, as an important feature of the echo waveform, reflects the characteristics of structure distribution of a target. Therefore, by extracting fission information, it is convenient to acquire the distribution of projection distances of each scattering center of the target in the detection direction.

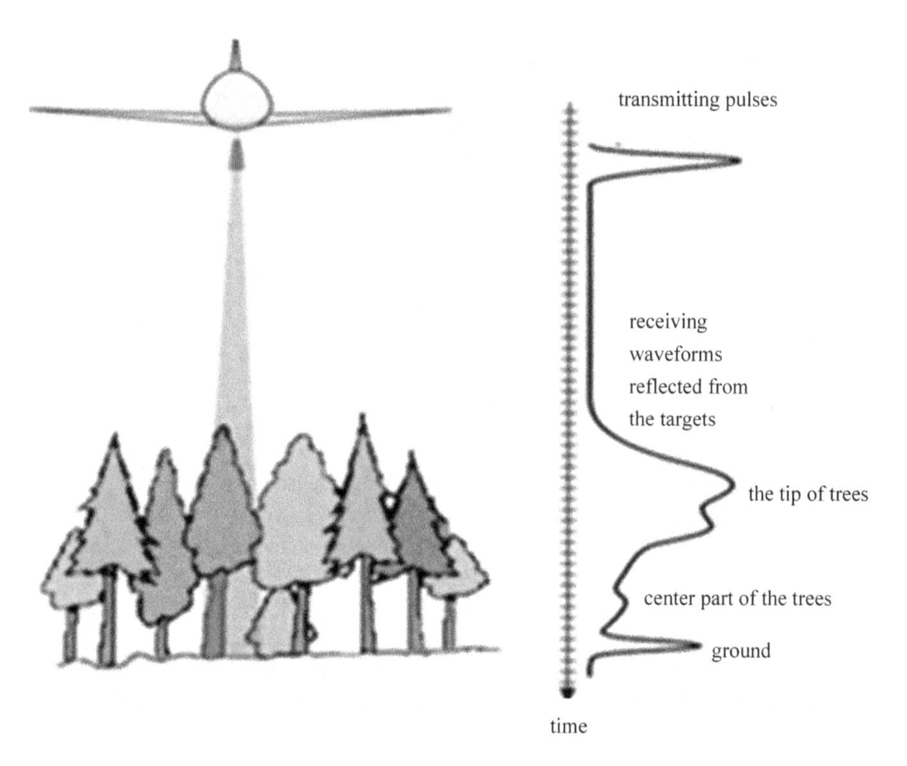

Fig. 5.14 The theory of the level imaging of trees

The differences of time delay of echoes are caused by the spatial distribution of multiple targets in the depth direction. In the precise detection of multiple echoes, these differences are presented as different ranging results in the same laser emission channel. Therefore, the laser echo pulses produced chronologically by the surfaces of different target ranges are segmented into the certain isolated echo pulses by using the decomposition of multi-layer echoes. Then, the above experiment of echo pulses is measured and recorded separately by using the circuit for peak searching and range measuring based on the multiple echo pulses to obtain the multi-layer ranging data of targets. Finally, these ranging data together with the positioning and attitude information of the platform and the detection angle of lasers can be used to obtain the actual spatial relation of targets. The application of this technique can not only acquire ranging information of superficial objects in the detection direction, but also can discover the obscure targets covered by camouflages.

Generally, background (e.g., ground and water) and the camouflages and coverings (e.g., trees and camouflage nets) are always located at the farthest end and the forefront in the depth direction. Therefore, the background and camouflages can be separated from the hierarchy images through analyzing the first and the last ranging information in the multi-echoes ranging. While ranging information of the middle echoes are likely to represent the targets needed to be detected. It is possible to obtain the spatial distribution of detected targets through integrating the results of separate ranging results of the corresponding echoes in the multiple channels of laser emission.

5.4.2 Processing Multiple Echoes

In the target detection based on a laser imaging system, the synthesized waveform of echoes shows different characteristics due to the different fluctuation and surface characteristics of targets in the detection spots. The synthesized waveform formed by the waveforms of multiple laser echoes are processed with the purpose of finding the corresponding relation between echo data and spatial distribution of targets and detecting multi-layer targets. The acquisition of the features of multi-layer targets in the detection spots depends on many factors. On the whole, these factors include angle of laser detection, fluctuation of surface height of multiple targets, pulse width of laser emission and recording precision of echo signals. In general, various processing algorithms for echo data require that the broadening, deformation and fission, especially the effect of waveform fission of echo waveforms need to be changed significantly due to the characteristics of the target surface. In such circumstance, the target regions at different heights in the spots can be easily distinguished.

The premise of generating fractures in echo waveforms is that the difference in time delay of the echoes of each object in the laser spots reaches a certain value, which is related to the pulse width of the emitted laser. According to the simulation, only when the difference in time delay of the sub-echo centers is more than twice as large as the pulse width of laser emission, fractures are expected to be generated in the synthesized echo waveform. Similarly, to record the fractures, the recording precision of echoes needs to be high enough, that is, in the pulse width of 10 ns or so, the recording precision is expected to be less than 1 ns. Suppose that the pulse width of the laser is 7 ns. Under such condition, only when the time delay difference of sub-echoes exceeds 14 ns, that is, the height difference of multi-layer targets in light spots is more than 2.1 m, the targets can be distinguished. In such circumstance, the least height difference of multi-layer targets can be identified is 2.1 m within the light spots. For discovering obscure targets in the woods, the accuracy meets requirements.

Due to the complex corresponding relations between the echo data and the distribution of multi-layer targets in the light spots, the echo data (mainly the data of waveform characteristics) need to be processed separately according to the types of the data. As mentioned above, changes in broadening, deformation and fission are likely to occur to echo waveforms. For the detection of multiple targets and obscure targets, only the latter two changes of echo waveforms are studied concretely.

1. **Processing multiple echoes with known background characteristics**

In that case of detection with known background characteristics, for example, a single background (water or desert) with a large area, multiple echoes can be processed by using the subtraction method.

The waveform of detected targets in water is shown in Fig. 5.15. Figure 5.15a presents the laser echo waveform in the case that there are obscure targets in water. Figure 5.15b shows the laser echoes reflected by the water surface. Figure 5.15c is the waveform difference between the waveforms illustrated in Fig. 5.15a, b and clearly shows the waveforms of the obscure targets and the water. Therefore, the algorithm similar to subtraction method can be used to process the multi-layer echoes so as to find the obscure targets. The processes of the algorithm are as follows: (1) obtaining the curve of backscattering waveforms of the background; (2) acquiring the curve of backscattering waveforms of the targets; and (3) distinguishing the targets from the multi-layer images by subtracting the curve of backscattering waveforms of background from those of targets.

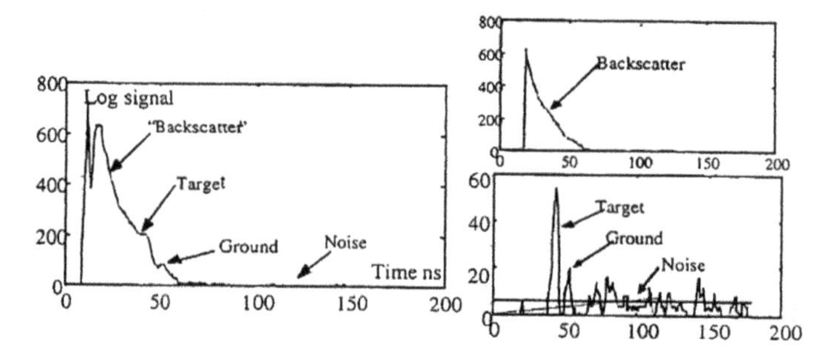

Fig. 5.15 The laser echo detection of obscure targets using the subtraction method

2. Multiple processing for laser echoes with unknown background

For targets with unknown background characteristics, it is essential to use different methods to process echo data for several times, that is, obtaining the spatial distribution relation of multiple objects by analyzing the deformation and the fracture degree of echo waveforms. Moreover, the shape and height of targets can be obtained based on the geometry relationship between the detection angle and time delay of waveforms. The specific processes are as follows:

(i) The obtained echo data are filtered firstly to eliminate the impact of background noise.
(ii) The synthesized waveforms undergo minimum error fitting and sub-echoes decomposition to acquire the features of each sub-echo.
(iii) According to the characteristics (center point position, waveform width, etc.) of sub-echoes, the spatial distribution relationship of objects is obtained. Furthermore, specific shape features of targets are acquired according to the relationship of parameters (detection angle, position, width of detection beam) of laser detection with spatial distribution and sub-echoes synthesis.

Without loss of generality, a one-dimensional target (for example, a cargo truck which is 3 and 2.5 m in the height and width) in approximately rectangular shape is taken as the detection target in the detection spots. While the scanning height of the laser is 1200 m and the pulse width is 4 ns. Meanwhile, the total divergence angle of the laser is 3 mrad and the angle between the scanning laser beams and the nadir is 40°. The relationship between the laser detection beams and the target is shown in Fig. 5.16, and the echo waveform is illustrated in Fig. 5.17.

It can be seen from the echo waveform that the echo is split into two sub-echoes: one is located around 17 ns and the other is around 27 ns. In the figure, the two sub-echoes fail to be separated completely. The main reason is that the height of the target does not suddenly change in the spots. Although the height difference exists between the target top and the ground, the side surface of the target also reflects the laser echo, and therefore, height change of the target is not observed in the figure.

The two sub-echoes can be fitted by using LSM. Moreover, the two sub-echoes are also considered to present Gaussian distribution similar to the waveforms of the emitted laser. The initial quantities of the two sub-echoes calculated using LSM can

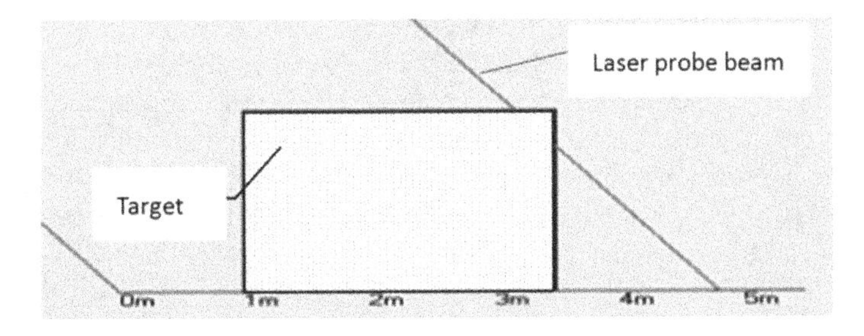

Fig. 5.16 The laser detection beams and the location of the target

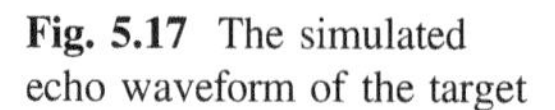
Fig. 5.17 The simulated echo waveform of the target

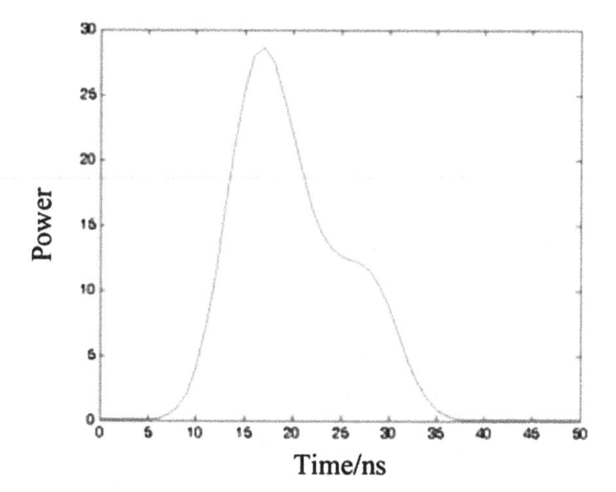

be roughly obtained according to the echo waveform. According to the initial calculation, the two sub-echoes present amplitudes of 27 xxx and 12 xxx, center points of 16 and 27 ns, and widths of 4 ns and 6 ns. The final calculated quantities are as follows: amplitudes of 27.16 and 10.225, center points of 17.137 ns and 27.815 ns, and widths of 6.2882 ns and 4.4511 ns.

It is observed from the center points of the two sub-echoes that the center points of the sub-echoes corresponding to the target top are located at 17.137 ns (where it is nearer to the laser than to the ground). Because the width of the sub-echo is 6.3 ns and then the broadening is 2.3 ns, the distance difference between two ends of the target top, and the center point, can be calculated as: $2.3 \times 0.3 = 0.69$ m. The distance between the corresponding center point of the sub-echoes and the laser and the laser are separately adopted as the radius and the center of a circle to draw a circle. In such circumstance, the corresponding arc is the possible position of the target top. Similarly, two arcs drawn on the basis that the distance difference between two ends of the target top and the center point is 0.69 m correspond to the possible position of the target edge. According to the fact that the laser echo is split to merely two sub-echoes, it can be inferred that the back area of the target is completely blocked, otherwise the sub-echo reflected from the back region is supposed to form another fractured sub-echo (because of the sudden change of its height). Therefore, the height of the intersection points between the two arcs corresponding to the target edge and the laser beams can be considered as that of the target top. Meanwhile, the target shape can be obtained by connecting three arcs (buildings are regarded as models with flat tops) along the horizontal direction through the intersection points. Then according to the ground height obtained from the time delay of the ground, the complete distribution of the height of targets in the laser spots can be acquired.

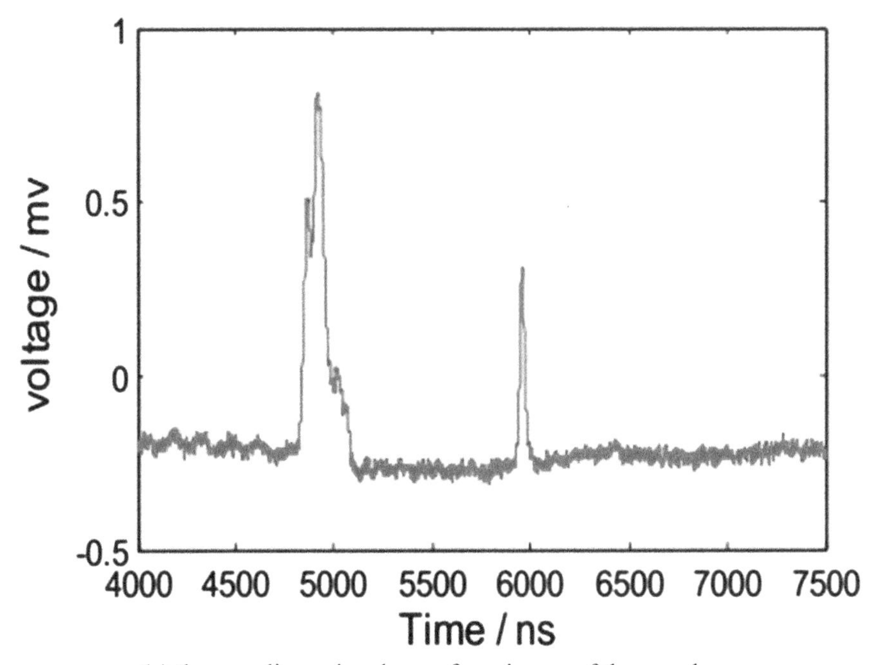

(a) the one-dimensional waveform image of the woods

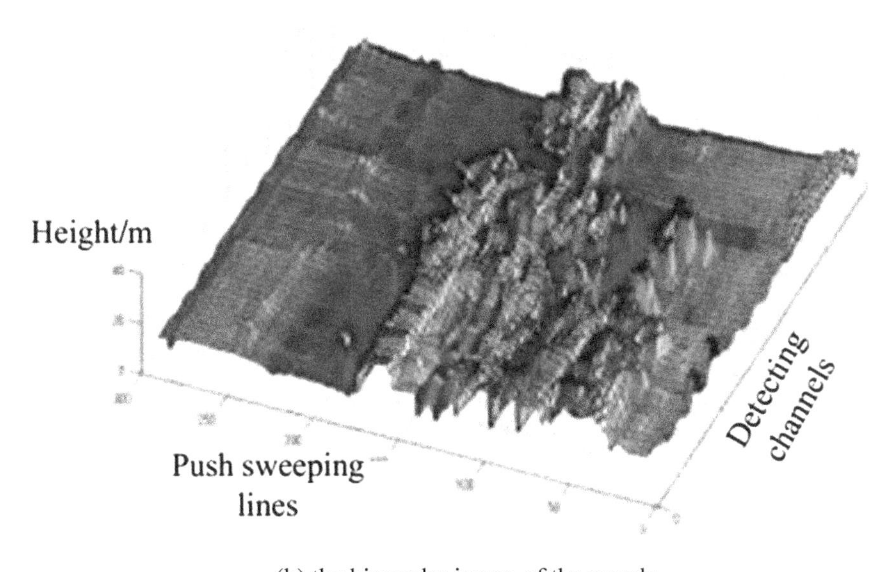

(b) the hierarchy image of the woods

Fig. 5.18 The waveform image and the hierarchy image of the woods

5.4.3 Generation of Hierarchy Images of Targets

The information including waveform, energy and time delay of laser echo at different hierarchies is extracted after decomposing the echoes at each moment so as to generate the waveform images, range images and gray images of multi-layer targets. The extraction method is the same as that for generating the waveform images, range images and gray images of a single-layer target. Figure 5.18 demonstrates the waveform image and the hierarchy image of woods obtained from the laser imaging based detection.

The waveform images, gray images and range images of targets at each layer are generated directly according to the waveform, energy and time delay of laser echoes. Therefore, these images can show the variations in the waveform, energy and time delay change of laser echoes once they change. Therefore, these images can be used to obtain the spatial and structural distribution of targets so as to identify the structural characteristics of targets and discover obscure targets.

References

1. Hu Y, Wei Q, Zhang L (1998) An analysis of the ground resolution of the airborne scanning laser rangefind. Infrared Laser Eng 27(2):37–39
2. Li S, Xue Y (2001) High-efficiency technology integration in three-dimensional remote sensing system. Science Press, Beijing, pp 164–165
3. You H, Li S (2000) Kinematic GPS positioning technique and data processing used in airborne 3D remote sensing system. J Remote Sen 4(1):22–26
4. Zhu H (1900) Common geodetic datum and transformation. The Chinese People's Liberation Army Press, Beijing
5. Liu J, Li Z, Wang Y et al (1993) The principle and application of global positioning system. Surveying and Mapping Press, Beijing
6. Krawczyr R, Goretta O, Kassighian A (1993) Temporal pulse spreading of a return lidar singal. Appl Opt 32(33):6784–6788
7. Zhao N, Hu Y (2006) Relation of laser remote imaging signal and object detail Identities. Infrared Laser Eng 35(2):226–229
8. Hofton MA, Minster JB, Blair JB (2000) Decomposition of laser altimeter waveforms. IEEE Trans Geosci Remote Sens 38(4):1989–1996
9. Persson A, Söderman S, Töpel UT et al (2005) Visualization and analysis of full-waveform airborne laser scanner data. ISPRS workshop 'Laser scanning 2005', Enschede, The Netherlands, 12–14 Sept 2005, pp 103–108
10. Ma H, Li Q (2006) Modified EM algorithm and its application to the decomposition of laser scanning waveform data. J Remote Sens 13:35–41

Chapter 6
Feature Extraction and Processing of Detection Images

Laser images obtained through the target detection using a laser imaging system contain distribution characteristics of target information such as time delay, gray and waveform. Therefore, extracting and processing these features are bases of the extraction and identification of target features. The feature extraction of target images mainly includes extracting features of range, intensity, waveform and level images of targets. Thereinto, feature extraction of range images has been studied extensively by researchers at home and abroad [1–3], and methods for extracting features of intensity images which are similar to the passive image processing approaches have been well developed as well [4, 5]. This chapter mainly discusses the feature extraction and processing of waveform and level images of targets.

6.1 Feature Analysis of Laser Images

As mentioned in the above chapters, there is a complicated modulation relationship between target features and laser echoes. That is, changes of the waveform, time delay and energy of laser echoes reflect the differences of target features. In this section, typical features of laser images, particularly the waveform images, of targets are analyzed.

6.1.1 Width Features of Target Echoes

As mentioned in the above chapters, there is a complicated modulation relationship between target features and laser echoes. That is, changes of the waveform, time delay and energy of laser echoes reflect the differences of target features. In this section, typical features of laser images, particularly the waveform images, of targets are analyzed.

Y. Hu, *Theory and Technology of Laser Imaging Based Target Detection*,
DOI 10.1007/978-981-10-3497-8_6

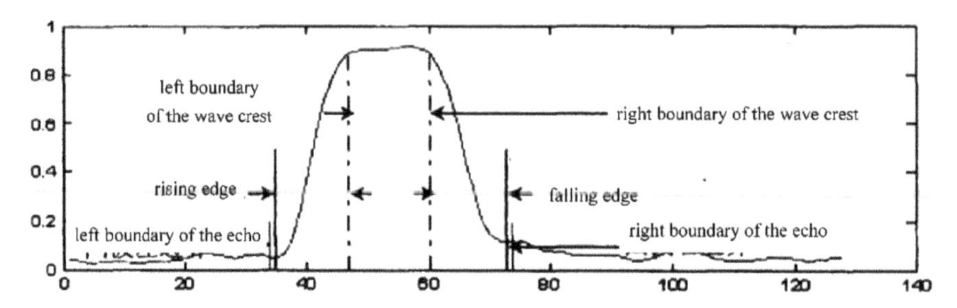

Fig. 6.1 Division of the width of an echo waveform

According to the characteristics of target modulation of lasers, the spatial distribution of targets is expected to broaden the width of echo pulses to some extent. Under conditions with normal incidence or small-angle incidence, pulse broadening is mainly determined by the smoothness and fluctuation of target surfaces [6]. Therefore, the broadening of target echoes in the waveform images of targets contains the dimension information of the targets. In terms of the amplitude and varying gradient, an effective waveform can be divided into a wave crest, a rising edge and a falling edge, as shown in Fig. 6.1. The interval between the left and right boundaries of the echo is defined as the broadened echo width, and the left and right boundaries are determined based on the threshold of the echo signal. The threshold of the normalized echo signal is defined as η. The echo signal is deemed to reach the left boundary when its waveform amplitude is higher than η in the left, and accordingly, it reaches the right boundary when its waveform amplitude is lower than η. The selection of the threshold η is related to noise, and generally, η is set as 3–5 times of the noise amplitude.

6.1.2 Energy Features of Echoes

Energy features of echoes are closely related to the back-reflection of targets. On account of the difference in the texture, smoothness, and brightness, different objects present varied reflectivities to a laser with a certain wavelength. In general, the reflectivity of metals is larger than other objects. Among metals, polishing metals show the largest reflectivity, while oxidized metals have the lowest reflectivity. As for black bituminous pavements and tile roofs, owing to their surfaces can absorb laser signals, they exhibit a small reflectivity so that the echo signals are weak. Experiments show that the reflectivity of natural medium surfaces such as sandy soil generally varies from 10 to 20%, and those of vegetation surface and snow are in ranges of 30–50% and 50–80%, respectively.

Table 6.1 shows the reflectivities of common media to the laser in a wavelength of 0.9 μm. For different targets, there is a great difference in the energy of laser echoes because the energy intensity of laser echo signals is proportional to the

Table 6.1 The reflectivities of different media to the laser [16]

Texture	Reflectivity (%)
White paper	Approximately 100
Timbers in regular shapes (dry pine trees)	94
Snow	80–90
Foam	88
White stone blocks	85
Limestone and clay	Approximately 75
Newspaper with prints	69
Tissue paper	60
Deciduous trees	Typical value 60
Coniferous evergreen trees	Typical value 30
Carbonate sand (dry)	57
Carbonate sand (wet)	41
Coastal beaches and bare deserts	Typical value 50
Rough timbers	25
Smooth concrete	24
Asphalt containing pebbles	17
Volcanic rocks	8
Black chloroprene rubber	5
Black rubber tyres	2

reflective section area and the reflectivity which varies among different targets. In essence, the difference in the energy of target echoes is caused by that in the target features. Thus, echo energy is one of the significant parameters that reflect features including the dimension and the reflectivity information of targets. According to the equation of the laser detection range, echo energy is not only related to the reflective section area and reflectivity of targets, but also closely related to the imaging range, detection angle and system parameters. Therefore, in the extraction of the energy features of echoes, factors including imaging range, detection angle, system parameters, etc. require to be normalized to obtain characteristic values which can reflect features of targets. The energy features of echoes can be obtained by adding up and normalizing the intensities of echo signals within the left and right boundaries of echoes.

6.1.3 Distribution Characteristics of Waveforms

The distribution of target echoes is the combined modulation of echoes in each sub-region of targets. The amplitude distribution of target echoes is affected by the distributions of surface features of targets (distributions of shape and reflectivity). Therefore, the distribution of target echoes is an important character reflecting target features, and also one of significant bases for the classification and identification of targets [7]. However, due to the large data size of echoes and the high distribution dimension of waveforms, target echoes are hard to be processed and identified in a high dimensional space, thus the dimension needs to be reduced. Besides, owing to being polluted by noise, echo signals need to be de-noised so as to extract characteristic quantities that can reflect the nature of signals.

The distribution characteristics of waveforms can be analyzed through wavelet theory. Assume that the function $\psi(t) \in L^2(R)$ meets the permissibility condition:

$$C_\psi = \int_{-\infty}^{\infty} \frac{\left|\widehat{\psi}(\omega)\right|^2}{|\omega|} d\omega < \infty \tag{6.1}$$

Or

$$\int_{-\infty}^{\infty} \psi(t)dt = 0 \tag{6.2}$$

where $\psi(t)$ is a basic wavelet. Through translation and stretching, a family of functions is obtained. The set composed of the family of functions forms a group of wavelet bases in the space $L^2(R)$, that is

$$\psi_{a,b}(t) = |a|^{-1/2}\psi(\frac{t-b}{a}) \quad \{a \in R, b \in R\} \tag{6.3}$$

The wavelet transform of the signal $f(t)$ is the inner product of $f(t)$ and $\psi_{a,b}(t)$ in the space $L^2(R)$:

$$(W_\psi f)(a,b) = \langle f, \psi_{a,b} \rangle \tag{6.4}$$

In order to make it easy to perform, continuous wavelets are discretized. In the typical binary wavelet transform

$$a = 2^{-j}, \ b = k2^{-j}, \quad \{j,k \in z\} \tag{6.5}$$

where j is the scale of binary wavelet transform.

Suppose that the centre and radius of the wavelet $\psi(t)$ are t_0 and Δt respectively; while the centre and radius of $\psi(\omega)$, Fourier transform of $\psi(t)$, are ω_0 and $\Delta\omega$ respectively Then, the wavelet transform is expected to provide partial information of the signal $f(t)$ in a time-frequency rectangular window $[a + at_0 - a\Delta t, b + at_0 + a\Delta t] \times \left[\frac{\omega_0}{a} - \frac{1}{a}\Delta\omega, \frac{\omega_0}{a} + \frac{1}{a}\Delta\omega\right]$ of the time-frequency plane.

With the gradual increase of the wavelet transform scale, the wavelet transform modulus maxima (WTMM) caused by noise decreases rapidly, while that caused by the actual signals is relatively stable. When a signal containing noise is decomposed to the one with low resolution, the noise is reduced greatly due to low pass filtering of the wavelet transform. Therefore, echo signals of lasers are expected to be de-noised effectively through the wavelet transform.

The WTMM points of echo signals in laser detections are effective to detect peak values of echo signals and reflect the structural features of target echoes. The vector $\lambda[\boldsymbol{d}_1, \boldsymbol{t}_1, \boldsymbol{d}_2, \boldsymbol{t}_2, \ldots]$ composed of the values and point locations of WTMMs of laser echo signals is considered as a characteristic value of waveform distribution. In the vector, d_n and t_n are the nth WTMM and the location of the nth WTMM.

The eigenvector function for the waveform distribution of target echoes is defined as:

$$D(t) = \begin{cases} d_i, & t_i = t \\ 0, & others \end{cases} \tag{6.6}$$

The function $D(t)$ is a discrete function, in which the maximum t is the width of the echo waveform. Suppose that k eigenvectors for waveform distribution are obtained by performing wavelet transform for the waveform signals of the echoes for k times, then the average value of points in the vector function for waveform distribution characteristics is:

$$\xi_D(t) = \sum_{h=1}^{k} D_h(t)/k \tag{6.7}$$

The ensemble average of the eigenvector function for the waveform distribution is:

$$\xi_D = \sum_{t=0}^{T} \xi_D(t) \tag{6.8}$$

where T is the width of the waveform of the target echo.

The variance and the standard deviation of the eigenvector function for the waveform distribution are

$$\sigma_D^2 = \sum_{t=0}^{T} \sum_{h=1}^{k} [D_h(t) - \xi_D(t)]^2/k \tag{6.9}$$

$$\sigma_D = \sqrt{\sigma_D^2} \tag{6.10}$$

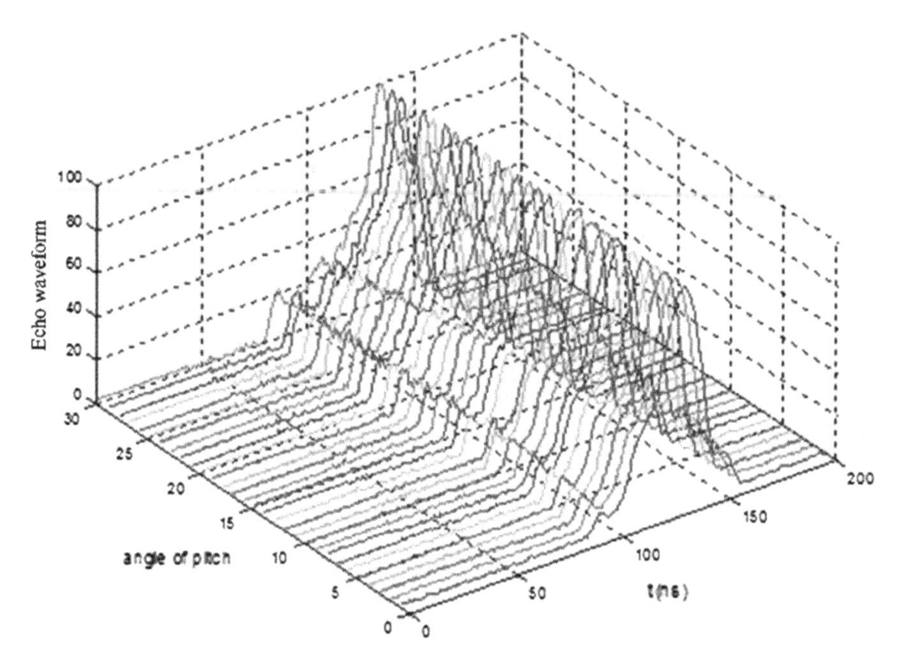

Fig. 6.2 Simulated images of echoes reflected from the F-15 fighter in different angles of pitch

As shown in Eqs. (6.6)–(6.10), the variance and the standard deviation of the vector function for the waveform distribution characteristics present the similarity of vectors of waveform distribution characteristics. Therefore, they can be used as bases for selecting stable features.

For example, by carrying out stable transform for the simulated image of 30 echoes reflected from the digital model of an F-15 fighter in attitude angles of 0°–30°, the waveform distribution of each echo is obtained, as shown in Fig. 6.2.

By applying db1 wavelets of Daubechies wavelet series, the echoes are decomposed into five levels, the wavelet transform results of which are shown in Fig. 6.3.

Then, the averages and the standard deviations of eigenvector functions $\mathbf{D}_1$, $\mathbf{D}_2$, $\mathbf{D}_3$, $\mathbf{D}_4$ and $\mathbf{D}_5$ presenting the waveform distribution are calculated, as shown in Table 6.2.

As shown in Table 6.2, owing to there are large noise components in the wavelet transform of low levels, the computed standard deviations of vector functions for waveform distribution characteristics are large. Thus the eigenvectors for the waveform distribution in the wavelet transform results of the third, fourth and fifth levels are selected as the eigenvectors of waveform distribution of target echoes. Considering the computation complexity, 4, 3 and 2 maximums are selected from the WTMMs in the third, fourth and fifth levels respectively to constitute an eigenvector $\mathbf{D}$ [d_1, t_1, d_2, t_2, d_3, t_3, d_4, t_4, d_5, t_5, d_6, t_6, d_7, t_7, d_8, t_8, d_9, t_9].

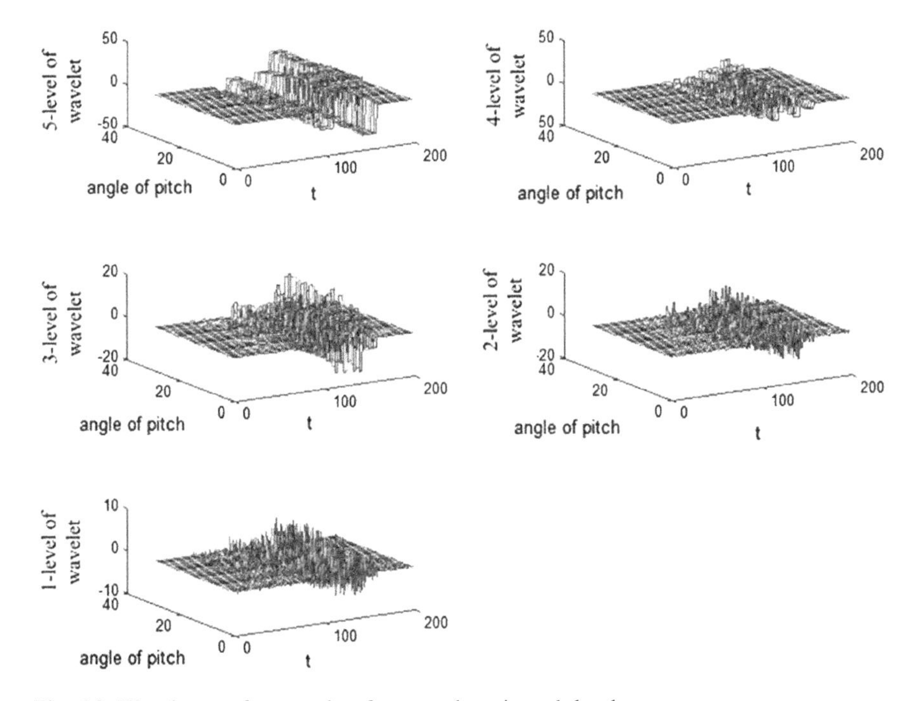

Fig. 6.3 Wavelet transform results of target echoes in each level

Table 6.2 The averages and the standard deviations of eigenvector functions for waveform distribution in each level

	D_1	D_2	D_3	D_4	D_5
Average	64.843	55.647	47.196	42.317	35.387
Standard deviation	3.7431	1.6786	0.71613	0.3652	0.069427

6.1.4 Statistical Characteristics of Waveform Images

Waveform images, as the two-dimensional distribution of intensive sampling results of the waveforms of target echoes, can also serve as the binary images for the waveform functions of echoes. Therefore, waveform images have statistical characteristics of general images as well. For an image, there are various statistical characteristics, among which the moment invariant feature can be used as the statistical characteristics of waveform images of echoes due to its invariance under the conditions of rotation, translation, scaling, etc. Based on the feature, the stability of feature extraction under the range translation and amplitude change of echoes can be realized.

Suppose that the target echoes are expressed by the function $f(t)$, then the corresponding binary image is

$$F(x,y) = \begin{cases} 1 & y \leq f(x) \\ 0 & others \end{cases} \tag{6.11}$$

For a binary image $F(x,y)$, its corresponding geometric moment can be expressed as

$$m_{pq} = \sum_{i=0}^{N1} \sum_{j=0}^{N2} x^p y^q F(x,y), \quad p,q = 0,1,2,\ldots \tag{6.12}$$

where N_1 and N_2 are the upper bounds of the abscissa and the ordinate of the binary image $F(x,y)$, respectively.

The corresponding $(p+q)$ order central moment can be defined as

$$\mu_{pq} = \sum_{i=0}^{N1} \sum_{j=1}^{N2} (x-\bar{x})(y-\bar{y})F(x,y) \tag{6.13}$$

where $\bar{x} = m_{10}/m_{00}$ and $\bar{y} = m_{01}/m_{00}$.

Besides, μ_{00} represents the area of the image and is employed to measure the size of areas. By normalizing the central moment according to μ_{00}, the normalized central moment is

$$\eta_{pq} = \frac{\mu_{pq}}{(\mu_{00})^{\frac{p+q+2}{2}}} \tag{6.14}$$

Based on the second and third moments of the normalized central moment, seven invariant moments are obtained:

$$\phi_1 = \eta_{02} + \eta_{20} \tag{6.15}$$

$$\phi_2 = (\eta_{02} - \eta_{20})^2 + 4\eta_{12}^2 \tag{6.16}$$

$$\phi_3 = (\eta_{30} - 3\eta_{12})^2 + (3\eta_{21} - \eta_{03})^2 \tag{6.17}$$

$$\phi_4 = (\eta_{30} + \eta_{12})^2 + (\eta_{21} + \eta_{03})^2 \tag{6.18}$$

$$\begin{aligned} \phi_5 = {} & (\eta_{30} - 3\eta_{12})(\eta_{30} + \eta_{12})\left[(\eta_{30} + \eta_{12})^2 - 3(\eta_{21} + \eta_{03})^2\right] \\ & + (3\eta_{21} - \eta_{03})(\eta_{21} + \eta_{03})\left[3(\eta_{30} + \eta_{12})^2 - (\eta_{21} + \eta_{03})^2\right] \end{aligned} \tag{6.19}$$

$$\phi_6 = (\eta_{20} - \eta_{02})\left[(\eta_{30} + \eta_{12})^2 - (\eta_{21} + \eta_{03})^2\right] + 4\eta_{11}(\eta_{30} + \eta_{12})(\eta_{21} + \eta_{03}) \tag{6.20}$$

Table 6.3 Statistical data of invariant moments of echo images of the F-15 and F-16 fighters

	ϕ_1	ϕ_2	ϕ_3	ϕ_4	ϕ_5	ϕ_6	ϕ_7
Average	817.1	233,645.5	6537.4	22,093.0	−3.402e+08	2.172e+07	2.794e+09
Standard deviation	12.62	13,232	1026.4	2013.2	9.808e+08	1.616e+06	1.756e+08
Variable coefficient	0.015	0.057	0.157	0.091	−2.883	0.0744	0.0628
Average	743.7	146,647.9	7363.7	18,470.2	−1.112e+08	1.384e+07	2.004e+09
Standard deviation	17.82	8213	626.2	1763.3	3.659e+08	0.943e+06	1.598e+08
Variable coefficient	0.024	0.056	0.085	0.095	−3.291	0.0681	0.0799

$$\begin{aligned}\phi_7 = {} & (3\eta_{21} - 3\eta_{03})(\eta_{30} + \eta_{12})\left[(\eta_{30} + \eta_{12})^2 - 3(\eta_{21} + \eta_{03})^2\right] \\ & + (3\eta_{21} - \eta_{03})(\eta_{21} + \eta_{03})\left[3(\eta_{30} + \eta_{12})^2 - (\eta_{21} + \eta_{03})^2\right]\end{aligned} \tag{6.21}$$

Stable transform is performed for the simulated images of 200 echoes from the digital models of an F-15 and an F-16 fighter in attitude angles of 0°–10°. Then, the invariant moments of the images are calculated according to the above method for computing the invariant moment of waveform images. The averages, standard deviations and variable coefficients of the seven invariant moments are obtained, as illustrated in Table 6.3.

According to Tables 6.2 and 6.3, among the seven invariant moments of echo images of the target, ϕ_1, ϕ_2, ϕ_6 and ϕ_7 present smaller standard deviations, thus they are selected as the statistical characteristics for the waveform images of targets.

6.2 Feature Extraction of Laser Images

Laser echoes are closely related to the spatial characteristics of targets. The time delay characteristics of laser echoes reflect the characteristics of position and structural fluctuation of targets. The distribution characteristics of laser target echoes describe the structural characteristics of targets in the direction of laser beams. Besides, the energy characteristics of echoes illustrate the distribution of the reflectivity of target surfaces. The feature extraction of laser images means extracting typical features that can show the geometrical shape and physical structure of targets from images. Among the characteristics mentioned above, characteristics of energy and time delay can be easily extracted, while the extraction for those of target echoes is relatively complicated.

The waveform characteristics of target laser echoes can be regarded as an identifying feature to classify and identify targets [8]. However, waveforms of target echoes are unstable and varied with the distance and orientation of targets as well as the position in the optical spot. Therefore, it is likely to complexify the detection system if the target echoes are utilized directly to identify targets. Besides, due to the large data size of target echoes, while target echoes used as identifying features directly, a high feature dimension is likely to be generated. This increases the calculation amount and complexity of the system in the classification and identification of targets. Therefore, in the target detection based on a laser imaging system, the following is very significant for the automatic classification and identification of targets: extracting signal features which can represent the essential attributes of targets steadily, reducing the feature dimension of targets, and improving the efficiency and accuracy of classification and identification.

At present, there are many methods for extracting waveform characteristics of echoes. Among which, Fourier transform, though is a traditional signal analyzing method, it can merely obtain the overall features of target owing to it adopts trigonometric functions, which are infinitely extensible in the time domain, as orthogonal basis functions. Fourier analysis is rather effective for the feature extraction of stationary signals. However, for the non-stationary properties of artificial and natural signals, it shows obvious shortcomings in describing local features. In view of this, wavelet transform is an effective method for processing non-stationary signals.

6.2.1 Extraction of Energy Characteristics Based on Multi-scale Wavelet Transform

The coefficient of wavelet transform which reflects the essential characteristics of target echo signals, can be used as the eigenvector for the classification and identification of targets. However, if the coefficient of wavelet decomposition is directly used as the classification feature, then the feature set is oversize and therefore the features are indistinct in the classification and identification. Therefore, it is important to decrease the amount of features and highlight major features before the classification and identification of targets.

1. **Principles of extraction of energy characteristics**

Wavelet transform, in essence, is the filtering for original signals, and the application of different wavelet functions can lead to various decomposition results. However, no matter what wavelet functions are selected, the center frequency and the bandwidth of the filter used in each decomposition scale are in a fixed proportion, namely, present the constant Q. Therefore, smoothing signals and detail signals in each scale space are expected to provide local time-frequency information of original signals, especially the composition information of signals in different frequency bands. After being calculated in different decomposition scales, the signal energies can be arranged in the scale order to form an eigenvector to be used in the identification [9].

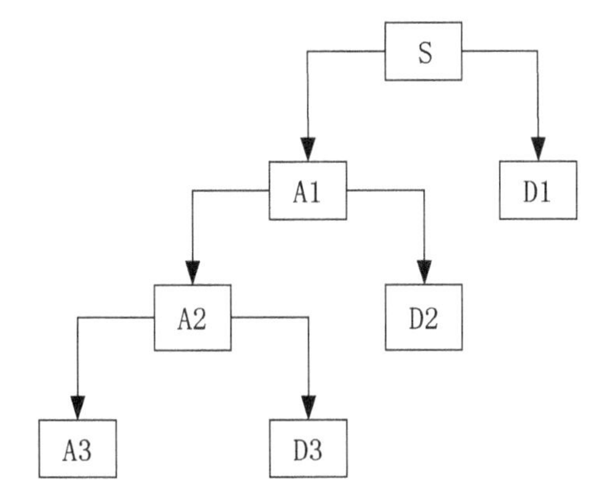

Fig. 6.4 The signal decomposition

Taking the three-level decomposition of binary wavelets for an instance, according to the pyramid decomposition algorithm, signals can be decomposed into low- and high-frequency signals of multiple levels, as shown in Fig. 6.4 [10].

In the figure, A represents low frequencies, D is high frequencies, and the number following A and D denotes the number of decomposition levels. According to Fig. 6.4, the relationship between original signals and decomposed signals is

$$S = A_3 + D_3 + D_2 + D_1 \tag{6.22}$$

Suppose that the lowest and highest signal frequencies are 0 and 1 respectively, then the frequency bands corresponding to A_3, D_3, D_2 and D_1 respectively are shown in Table 6.4.

According to Table 6.4, low- and high-frequency signals correspond to different frequency components of signals, respectively. In addition, for different targets, the corresponding components of frequency spectra are various. By solving the energy distributions of signals with high- and low-frequency components in three levels, an eigenvector can be generated to reflect the essential features of targets.

2. **Methods for extracting energy characteristics**

The extraction of energy characteristics based on wavelet transform is demonstrated in Fig. 6.5 [11].

(i) Pre-processing original signals. The amplitudes of laser echoes change for targets in different distances. Besides, signals are mixed with noise and direct current (DC) components. To eliminate these influences, pre-processing for original signals is required. First, original signals are supposed to undergo low-pass filtering, and then the filtered signals are averaged.

Table 6.4 Frequency bands of the decomposition

Signal	A_1	D_3	D_2	D_1
Frequency	0–0.125	0.125–0.25	0.25–0.5	0.5–1

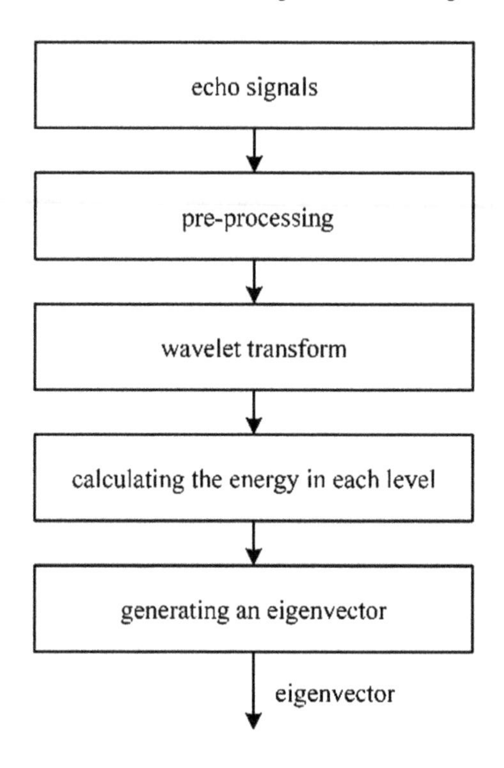

Fig. 6.5 The extraction of energy characteristics based on wavelet transform

(ii) Through wavelet decomposition, signals are decomposed into J levels, the approximation coefficient of the Jth level and detail coefficients from low frequency to high frequency are respectively: $X_{Ja}, X_{Jd}, X_{(J-1)d}, \ldots, X_{1d}$

(iii) Reconstructing wavelet decomposition coefficients and calculating the signal energy in each frequency band. A_J is used to denote the reconstructed signal of X_{Ja}, and accordingly, D_J represents the reconstructed signal of X_{Jd}. Then,

$$S = A_J + D_J + D_{(J-1)} + \cdots + D_1 \tag{6.23}$$

(iv) Calculating the total energy of signals in each frequency band. Suppose that E_{Ja} represents the energy of A_i, then E_{id} $(i = J, J-1, \ldots, 1)$ denotes the energy of $D_i(i = J, J-1, \ldots, 1)$, then,

$$\begin{aligned} E_{Ja} &= \sum_{k=1}^{N} |a_{Jdk}|^2 \\ E_{Jd} &= \sum_{k=1}^{N} |d_{Jdk}|^2 \end{aligned} \tag{6.24}$$

where a_{Jdk} and d_{Jdk} are the amplitudes of the reconstructed low-frequency signal A_J and the reconstructed high-frequency signal D_i, and N is the length of the sampling points of signals.

(v) Constructing eigenvectors by utilizing energies of signals in each level.

$$\boldsymbol{T} = \left[\boldsymbol{E}_{Ja}, \boldsymbol{E}_{Jd}, \boldsymbol{E}_{(J-1)d}, \ldots, \boldsymbol{E}_{1d}\right] \tag{6.25}$$

The following show echo signals extracted from two different ground targets (grassland and the junction of grassland and mud) through experiments. The signals obtained are decomposed into five levels to extract energy characteristics of wavelets in each level, respectively.

Figures 6.6 and 6.7 illustrate the original echo signals and the frequency spectra of echo signals of the two ground targets, respectively.

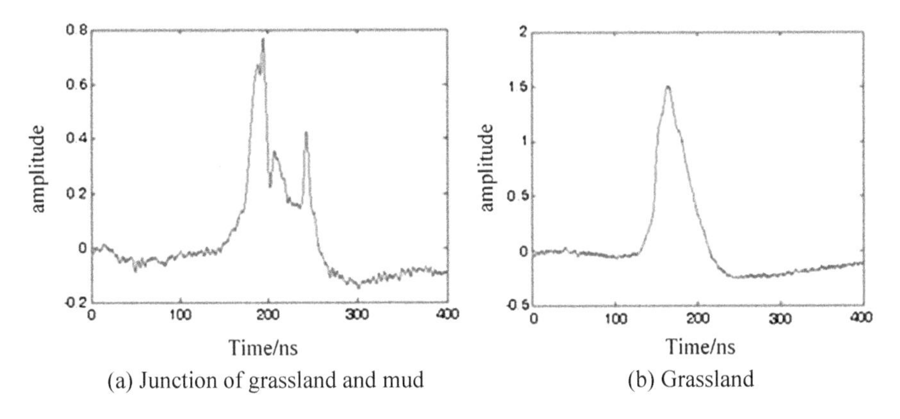

Fig. 6.6 Original echo signals of two types of targets

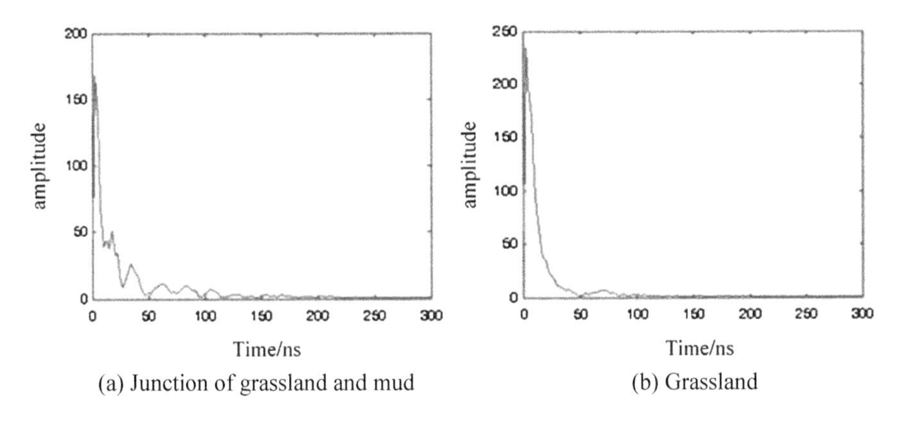

Fig. 6.7 Frequency spectra of two types of target echoes

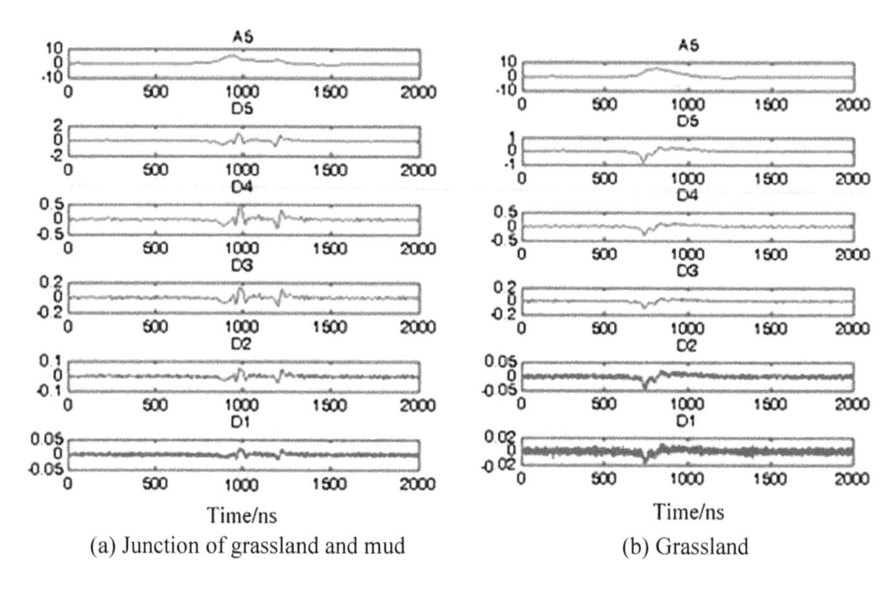

Fig. 6.8 The decomposition and reconstruction of wavelets in each level

The signals obtained are decomposed into five levels. Figure 6.8 shows the approximation and detail signals. Then, the energy characteristics of wavelets in each level are extracted, as displayed in Fig. 6.9.

The echo signals of the two types of targets are decomposed into five levels so as to extract the energy characteristics of wavelets in each level. According to Fig. 6.9, for the two types of targets, the energy proportion varies for the eigenvectors of different dimensions, which presents the dissimilarity of characteristic information among different targets. In the meantime, it is proved that wavelet decomposition is effective to extract energy characteristics.

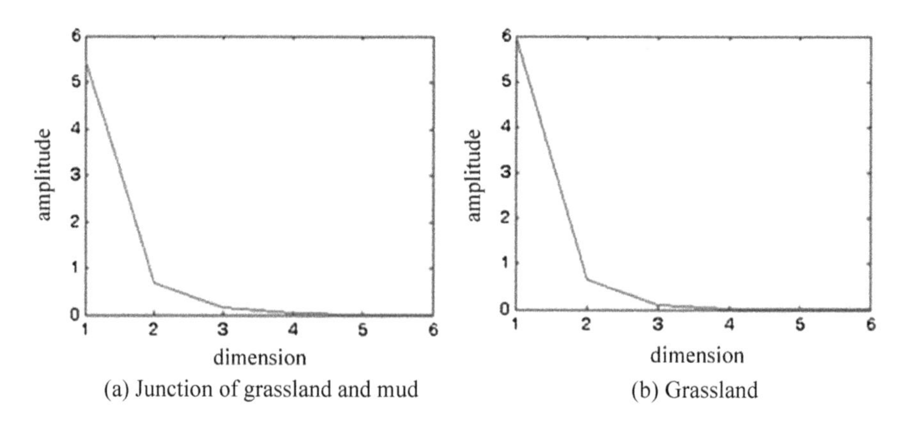

Fig. 6.9 Wavelet-based energy features of two types of target echoes

3. Simulation Analysis

In this section, the extraction of energy characteristics based on wavelet transform is simulated. Besides, the extraction performances of the wavelet-based energy characteristics are contrasted and analyzed for different targets. In the simulation experiment, the targets are 800 m away from the detection system, and the laser beams are emitted towards the targets with laser pulse width, sampling rate and divergence angle of lasers being 5 ns, 1 GS/s, and 10 mrad respectively. Two types of airliners (B747 and DC10) and two fighters (F-15 and F-16) are selected, respectively, as the simulation objects to build four digital 3D models for the four targets, as shown in Fig. 6.10. Assume that the targets surfaces are diffusing surfaces with the same reflectivity, 60 groups of target echoes are obtained, whose SNR, azimuth angle and angle range of pitch are 20 db, 0° and 0°–30° respectively, as shown in Fig. 6.11.

As shown in Fig. 6.11, under different angles of pitch, the target echoes vary for the four airplanes. This is because these airplanes have different structures, and

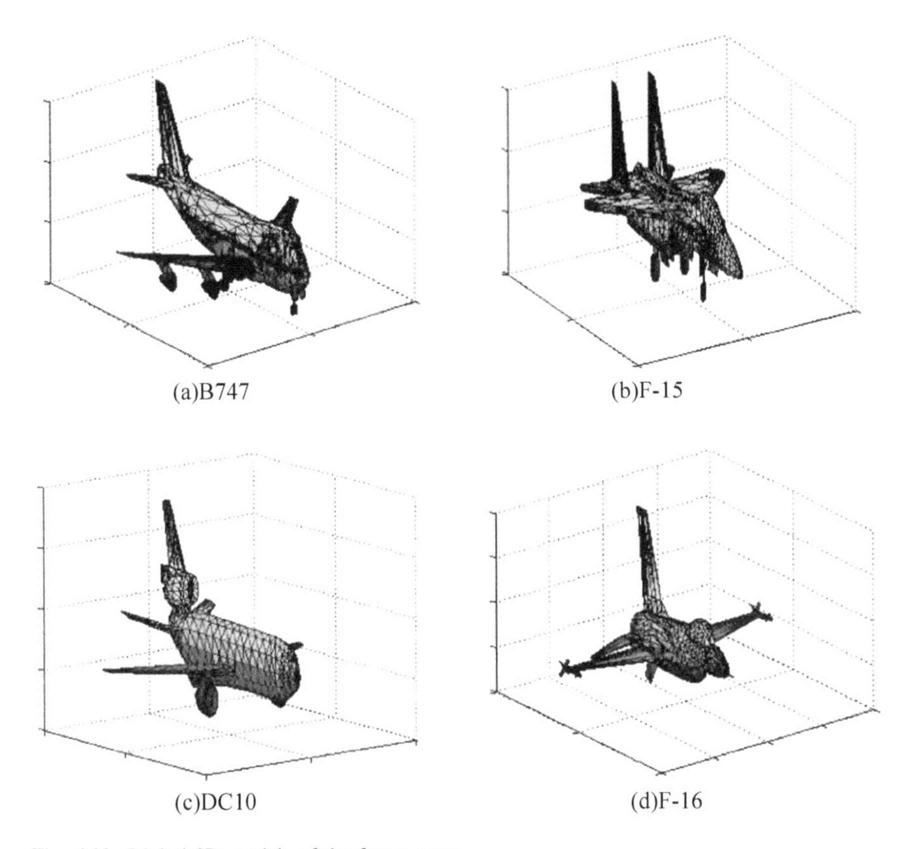

Fig. 6.10 Digital 3D models of the four targets

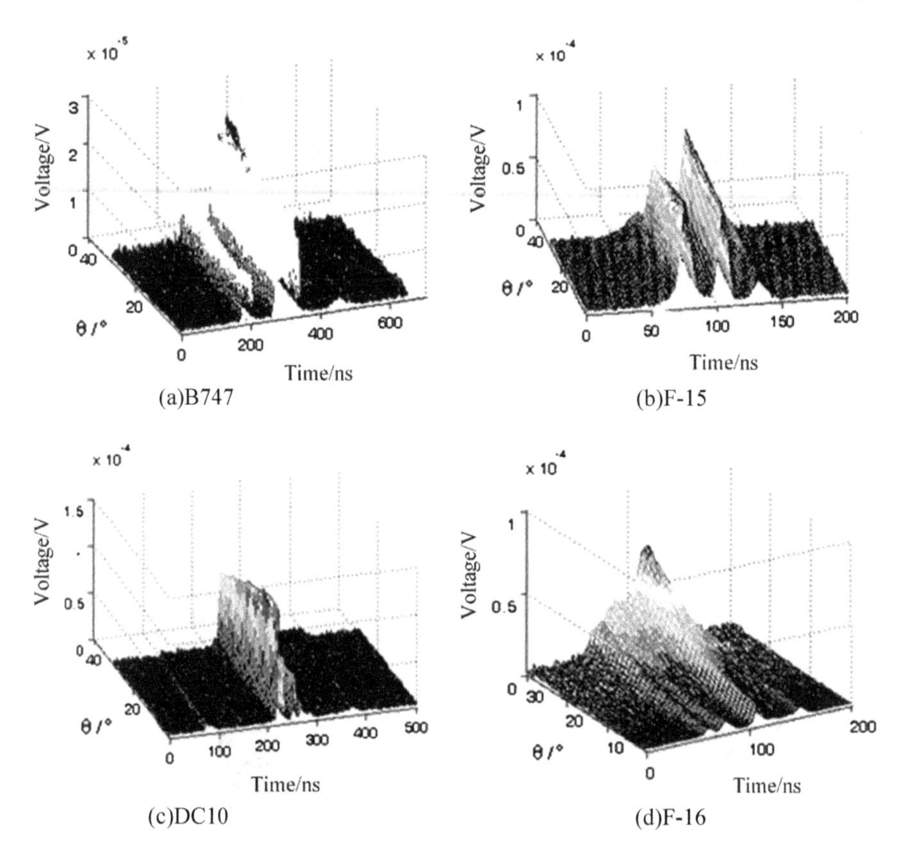

Fig. 6.11 The laser target echoes of the four targets

therefore their sensitivities to attitude variation vary from each other. The results above further testify the sensitivities of laser echoes to targets attitudes, which is likely to cause the instability of the eigenvectors of target echoes obtained. By using wavelet function of db1, the target echoes of the four types of airplanes are decomposed into five levels, so as to extract the energy characteristics under each angle in each level, as shown in Fig. 6.12.

As to the variation trends of wavelet-based energy characteristics of 60 groups of echoes of the four targets with the pitch angle varying from 0° to 30°, B747 shows a slow change, while DC10, F-15 and F-16 present a significant change with the pitch angle. In addition, the variations of the energy characteristics of DC10 and F-16 are more apparent than the other two types of airplanes. In order to explore more details about the variation of energy characteristics of different targets, statistical characteristics of the eigenvectors of wavelet energies of the four targets above are listed in Table 6.5.

As shown in Table 6.5, as the pitch angle of targets ranges from 0° to 30°, the ratio of the standard deviations to the averages are relatively large for the

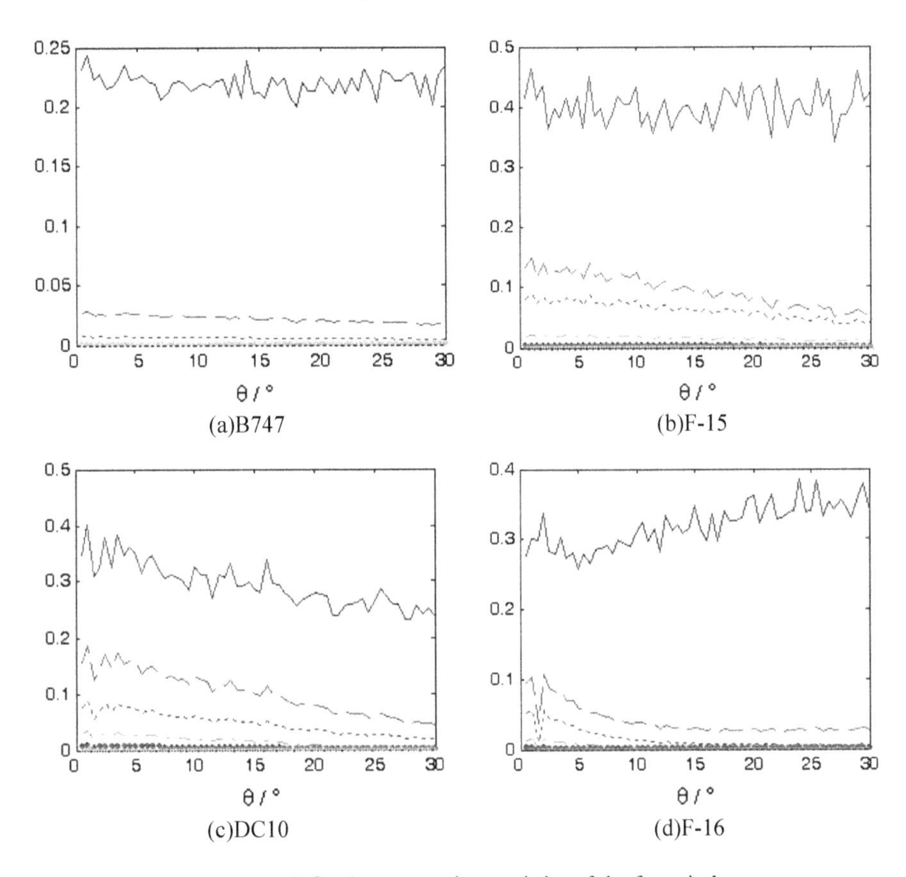

Fig. 6.12 The variation trends for the energy characteristics of the four airplanes

Table 6.5 The wavelet-based energy characteristics of the four types of airplanes

		A5	D5	D4	D3	D2	D1
B747	Average	0.2196	0.0218	0.0058	0.0013	0.00021	0.000032
	Standard deviation	0.0090	0.0030	0.0010	0.0003	0.000057	0.000014
F-15	Average	0.40013	0.0945	0.0616	0.0151	0.00235	0.000319
	Standard deviation	0.02798	0.02645	0.01398	0.0033	0.000527	0.000087
DC10	Average	0.29059	0.10072	0.04633	0.0151	0.00327	0.000480
	Standard deviation	0.042591	0.03922	0.01983	0.00842	0.002088	0.000320
F-16	Average	0.31933	0.03973	0.01442	0.0033	0.00051	0.000080
	Standard deviation	0.03058	0.02089	0.01366	0.0037	0.00059	0.000080

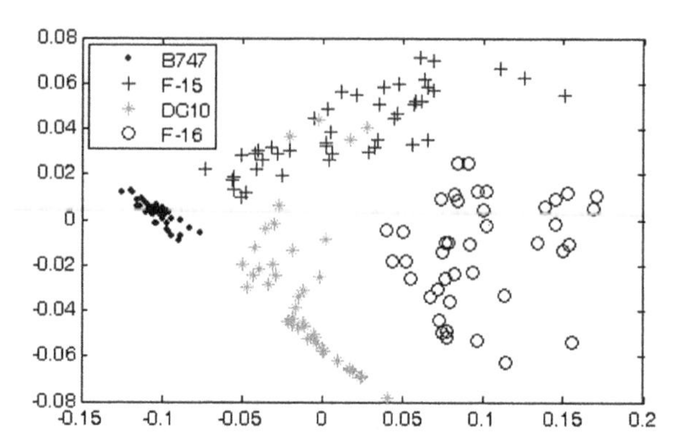

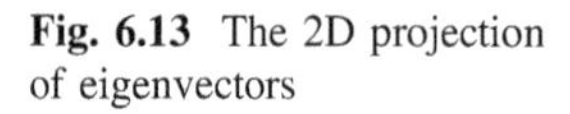
Fig. 6.13 The 2D projection of eigenvectors

approximation signals in the fifth level and detail signals in the fifth, fourth and third levels. While the ratios of those of detail signals obtained by wavelet decomposition in the first and second levels are relatively small. The reasons are that main components of detail signals in the first and second levels obtained by wavelet decomposition are noise, which leads to obvious fluctuation of the signals; while the main components in other levels are echo signals of targets. Likewise, the four airplanes show significantly different averages and standard deviations of wavelet-based energy characteristics in different levels. As shown in Table 6.5, the standard deviations of B747 are relatively small, while those of other targets are larger. Apart from these, the eigenvectors of DC10 and F-16 are overlapped, which adversely affects the classification of the three types of airplanes. Figure 6.13 shows the 2D projection distributions of wavelet-based energy characteristics of the four types of targets.

As shown in Fig. 6.13, in the 2D projection characteristic space, the four types of targets show distinct boundaries. By using different wavelet functions, the obtained coefficients of wavelet decomposition obtained vary, and accordingly, the statistical characteristics extracted are different. In view of that, for different echo signals, it is significant to select proper wavelet functions so as to extract eigenvectors with high stability and good classification performance.

According to Table 6.6, the eigenvectors extracted using different wavelet functions have various statistical characteristics. By comparing the ratios of the standard deviations to the averages, statistical characteristics obtained using db1 wavelets are optimal in all the four targets. Therefore, db1 wavelets are used to extract energy characteristics.

Energy characteristics of the four airplanes are extracted by employing db1 wavelets. Figure 6.14 shows the correlations of the waveforms and the eigenvectors of original echoes with the pitch angle of the targets ranging from 0° to 30°. In Fig. 6.14, the correlations obtained under other pitch angles are compared with that under a pitch angle of 0°.

According to the experimental results shown in Fig. 6.14, the correlations among the echoes at different angles of the original echoes in targets are smaller than 0.9, and decrease with the increase of the pitch angle. Among these targets,

Table 6.6 The statistical characteristics of the four types of targets when different wavelet functions are used

		B747	F-15	DC10	F-16
db1	Average	0.2460	0.5740	0.4565	0.3817
	Standard deviation	0.0112	0.0411	0.0618	0.0394
db2	Average	0.2538	0.5912	0.4799	0.3851
	Standard deviation	0.0124	0.0516	0.0660	0.0502
db3	Average	0.2526	0.5913	0.4847	0.3933
	Standard deviation	0.0114	0.0462	0.0698	0.0544
db4	Average	0.2520	0.5838	0.4859	0.3909
	Standard deviation	0.0137	0.0494	0.0745	0.0514
dmey	Average	0.2543	0.5865	0.4866	0.3937
	Standard deviation	0.0129	0.0522	0.0698	0.0588

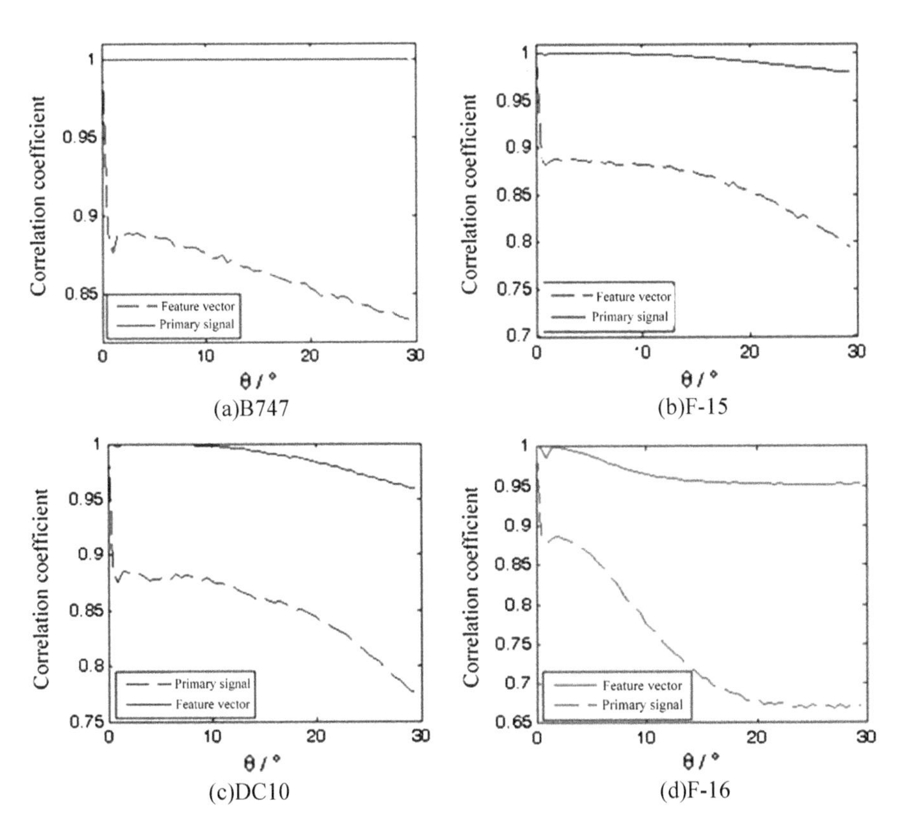

Fig. 6.14 The correlations of the waveforms and the eigenvectors of original echoes with the pitch angle varying from 0° to 30°

B747 shows the smallest change in the correlation, whose minimum value is more than 0.8; while F-16 presents the most apparent change and the minimum correlation is approximate to 0.65. Moreover, all the correlation coefficients of eigenvectors of the echoes reflected from the four types of targets are above 0.95, among which that of the eigenvectors of B747 is highest and approximate to 1 in each pitch angle.

6.2.2 The Extraction of Waveform Characteristics Through Wavelet Transform Based on WTMMs

In the process of extracting characteristics of laser images, the distribution characteristics of waveforms are mainly represented by the distribution of peak points of target echoes. In the target detection based on a laser imaging system, the peaks of target echoes reflect the distribution of projection distances in the detecting direction and the laser scattering abilities of each scattering center of targets. Therefore, the peaks of target echoes are a significant index for describing the structural characteristics of targets using echo signals and therefore important for target identification [12].

1. **WTMMs and signal singularity**

Signal singularity refers that signals are discontinuous at some points, as well as their derivatives of certain orders. Obviously, infinite differential functions are smooth, that is, non-singularity. In the target detection based on a laser imaging system, the hierarchy and obvious fluctuation of target surfaces lead to the division of target echoes, which thereby present multiple peak points and singular points of wavelets in the time domain. By extracting the peak points and singular points of wavelets, structural information of targets can be obtained according to the mapping relation between the division of target echoes and the structures of targets. The WTMM based feature extraction can be conducted to extract the characteristics of peak values and singular points of echo signals.

Concretely, the regularity of a signal $S(t)$ at a point ω can be represented by the attenuation of its WT $|WT(t,j)|$ with the decrease of the scale j in the neighborhoods of υ. The attenuation of $|WT(t,j)|$ cannot be directly obtained from the time-scale plane (t,j), but can be obtained through the local WTMM. WTMM refers that, for any point (t,j) in the time-scale plane, WT $|WT(t,j_0)|$ obtains the local WTMM when $t = t_0$, that is, the derivative of wavelet transform at this point equals to 0 [13].

$$\left.\frac{\partial |WT(t,j)|}{\partial t}\right|_{t_0,s_0} = 0 \tag{6.26}$$

It is noteworthy that the WTMM point (t_0, j_0) is the local maximum in both the left and right neighborhoods of t_0. Therefore, it can be proved that, if $WT(t,j)$ does not have a WTMM in small scales, then the function shows local regularity. That is, when j_p tends to 0, $WT(t,j)$ has a group of maximum points (t_p,j_p), $p \in N$ which are converged to the point υ, then the function is singular, that is, $\lim_{p\to+\infty} t_p = \upsilon$ and $\lim_{p\to+\infty} j_p = \upsilon$. For different types of singular points, the WTMM points exhibit varied change tendencies with the scale Hence, WTMMs favorably show the singularity and characteristics of signals.

2. **WTMM based characteristic extraction method**

The values and point locations of WTMMs of laser echo signals reflected from targets constitute a vector $\lambda[d_1, t_1, d_2, t_2, \ldots]$, which can be used as an eigenvalue of waveform distribution. In the vector, d_n and t_n are the value and location of the nth WTMM, respectively [14].

Target echoes are sensitive to the ranges and attitudes of targets at a certain degree. In this case, eigenvectors constructed for classifying waveforms vary evidently with the ranges and attitudes due to the involvement of the time location of WTMMs, which is not beneficial to the stable extraction of characteristics. Based on the variations of eigenvectors caused by the changes of ranges, the waveforms need to be calibrated to eliminate the influences of time shifting on characteristics extraction in scale spaces.

Figure 6.15 shows the process of the WTMM based characteristics extraction.

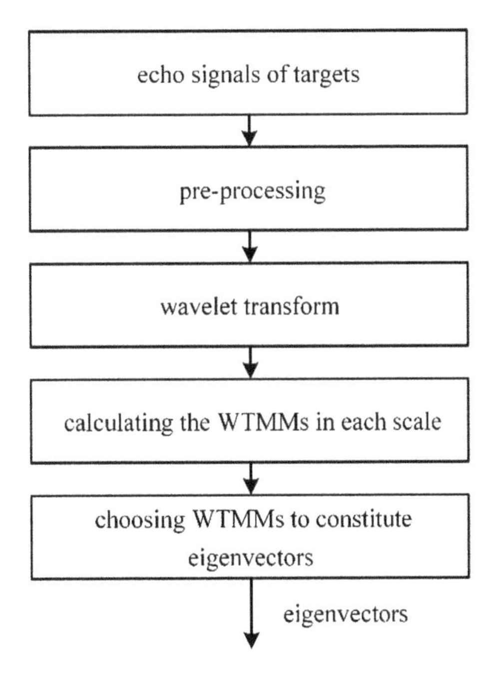

Fig. 6.15 The WTMM based characteristics extraction

The steps for extracting characteristics based on WTMMs are as follows:

(i) Pre-processing laser echo signals. According to the above analysis, the pre-processing mainly includes denoising, normalizing and calibrating the signals. In the transmission and detection, owing to the echoes are mixed with noisc which affects the detection and characteristics extraction, the signals need to be de-noised in advance. In order to remove the influence of range changes, the signals require to be normalized and calibrated. The normalization can be realized by selecting modulus maxima of the echo signals. For the calibration, the sliding correlation method is employed, that is, a position where the maximum correlation coefficient exists is selected as the standard position for calibrating the signals.
(ii) Decomposing the signals by carrying out wavelet transform. By using a proper wavelet function, signals are decomposed into J-level wavelets. Afterwards, the wavelet coefficient of each level is reconstructed.
(iii) Computing WTMMs in each scale. Utilizing the method for calculating the WTMM points in the above section, WTMMs of the decomposed signal in each scale are calculated and their positions are marked.
(iv) Constituting eigenvectors by WTMMs. In characteristic extraction, scale-space decomposition is performed in the same order, time-shift parameters and amplitudes of the WTMM are selected to constitute a 2 m-dimension eigenvector $\vec{\mathbf{\Pi}}$, as shown in Eq. (6.27).

$$\vec{\mathbf{\Pi}} = (T(\Pi_1), V(\Pi_1), T(\Pi_2), V(\Pi_2), \ldots, T(\Pi_m), V(\Pi_m)) \tag{6.27}$$

The eigenvector function for the distribution of target echoes is defined as:

$$D(t) = \begin{cases} d_i, & t_i = t \\ 0, & others \end{cases} \tag{6.28}$$

$D(t)$ is a discrete function, and the maximum value of t is the width of target echoes. Suppose that k eigenvectors for waveform distribution are obtained by performing wavelet transform for the waveform signal of target echoes received at the kth time, the mean value of points in the eigenvector function for the distribution of waveforms can be expressed as:

$$\xi_D(t) = \sum_{h=1}^{k} D_h(t)/k \tag{6.29}$$

The ensemble average of the function is:

$$\xi_D = \sum_{t=0}^{T} \xi_D(t) \tag{6.30}$$

where, T is the width of target echoes.

The variance and the standard deviation of the function are respectively:

$$\sigma_D^2 = \sum_{t=0}^{T} \sum_{h=1}^{k} [D_h(t) - \xi_D(t)]^2 / k \tag{6.32}$$

$$\sigma_D = \sqrt{\sigma_D^2} \tag{6.33}$$

According to Eqs. (6.29) to (6.32), the variance and the standard deviation of the function characterize the similarity of eigenvectors for waveform distribution. Therefore, they can be regarded as references for selecting stable characteristics.

3. **Characteristic extraction based on singular value decomposition**

WTMMs of signals depict the singularity of signals, that is, WTMMs in different scales can be considered as the singularity characteristics of the signals, and hence be used to identify different signals. However, if these WTMMs are merely expressed in graphs, they are not favorable for the automatic identification. Besides, due to different ranging times, sampling results for a same signal are likely to show time shifting, which makes it difficult to accurately reveal the singularity characteristics of signals, extract characteristics of signals, and classify signals. To solve this problem, echo signals of targets need to be calibrated in the pre-processing stage, which has been discussed in the above section. In the calibration, a template and moving average are needed. Ifan improper template is selected or signals are calibrated inaccurately, the positions of WTMM points are likely to shift, which is not beneficial to characteristic extraction.

To solve this problem, based on above methods, the singular value decomposition (SVD) method is employed here to extract characteristics. According to the SVD [15] method, if there is an r-rank matrix X, unit matrixes U and V which meet the following conditions with the size of $N \times N$ are obtained.

$$X = U \begin{bmatrix} \Lambda^{\frac{1}{2}} & O \\ O & 0 \end{bmatrix} V^H \tag{6.34}$$

$$Y \equiv \begin{bmatrix} \Lambda^{\frac{1}{2}} & O \\ O & 0 \end{bmatrix} = U^H X V \tag{6.35}$$

where $\Lambda^{\frac{1}{2}}$ is a $r \times r$ diagonal matrix with $\sqrt{\lambda_i}$ as its element, λ_i represents the non-zero eigenvalue of a correlation matrix $X^H X$, and O refers to a null matrix.

According to Eq. (6.34), it can be obtained that

$$X = \sum_{i=0}^{r-1} \sqrt{\lambda_i} \boldsymbol{u}_i \boldsymbol{v}_i^H \tag{6.36}$$

where $\boldsymbol{u}_i$ and $\boldsymbol{v}_i$ are the first r columns of U and V respectively, as well as the eigenvectors of XX^H and X^HX separately. In addition, the eigenvalue λ_i represents the singular value of X.

As characteristics, singular values have two features. First, singular values of matrixes are stable, that is, they barely change with the small variation of elements in matrixes. Second, singular values are inherent features of matrixes.

The SVD based method for extracting characteristics is conducted in the following steps: Firstly, on the basis of extraction of WTMMs in each level in the above section, a WTMM matrix $M = \{m_{i,j}\}$ is established by the WTMM vectors $m_{i,j}$, where i and j are the time and scale of wavelet transform respectively. Secondly, after being constructed, the WTMM matrix is decompose into non-zero singular values using the SVD method. Thirdly, these non-zero singular values are arranged in a certain order to constitute an eigenvector.

By conducting experiments, echo signals reflected from two types of ground targets (the junction of grassland and mud, and grassland) are extracted. These echo signals are decomposed into five levels by using db1 wavelets to extract eigenvectors for waveform distribution of the wavelets. Figure 6.16 shows the five-level

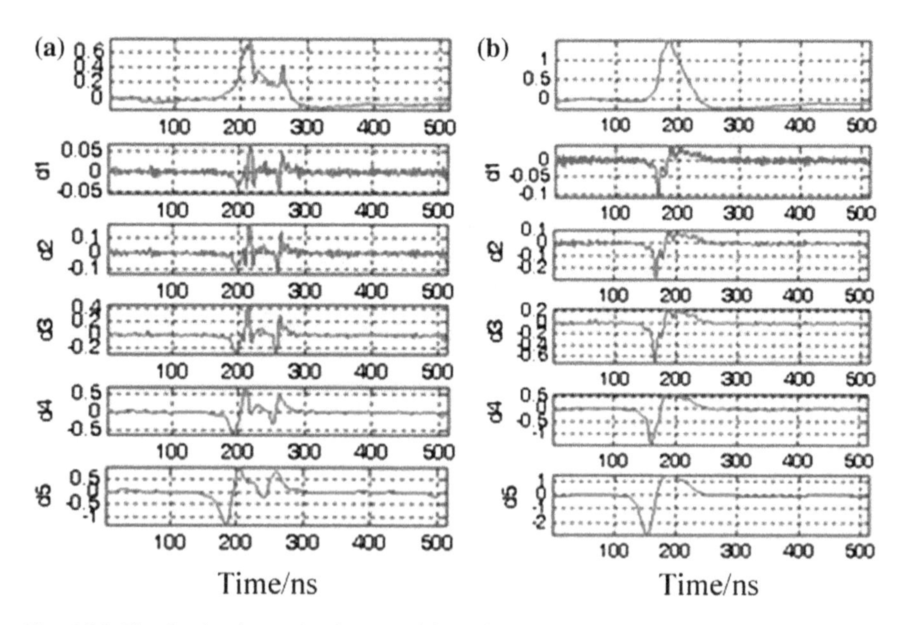

Fig. 6.16 The five-level wavelet decomposition of the echoes reflected from the two types of ground targets

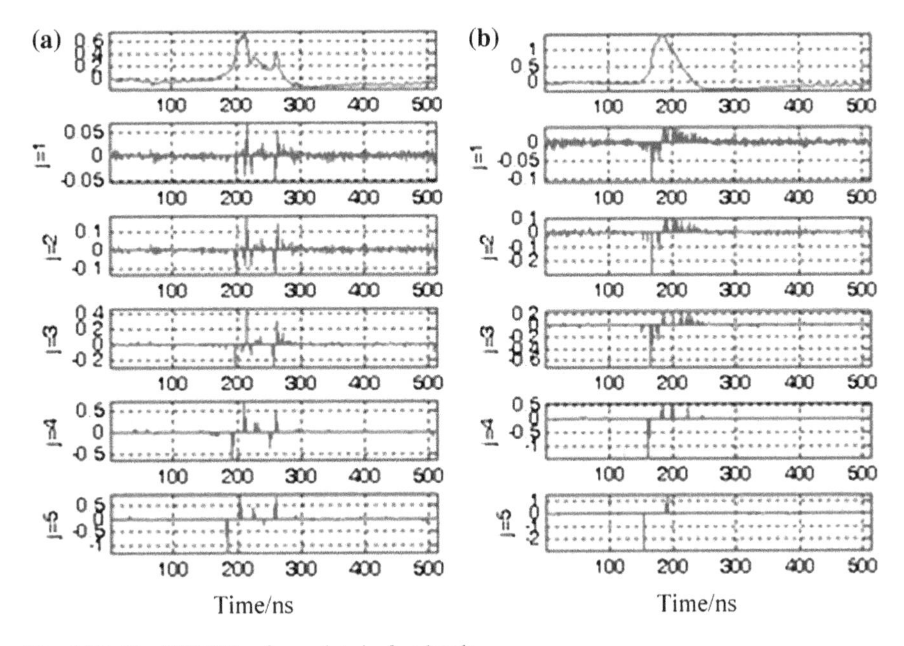

Fig. 6.17 The WTMMs of wavelets in five levels

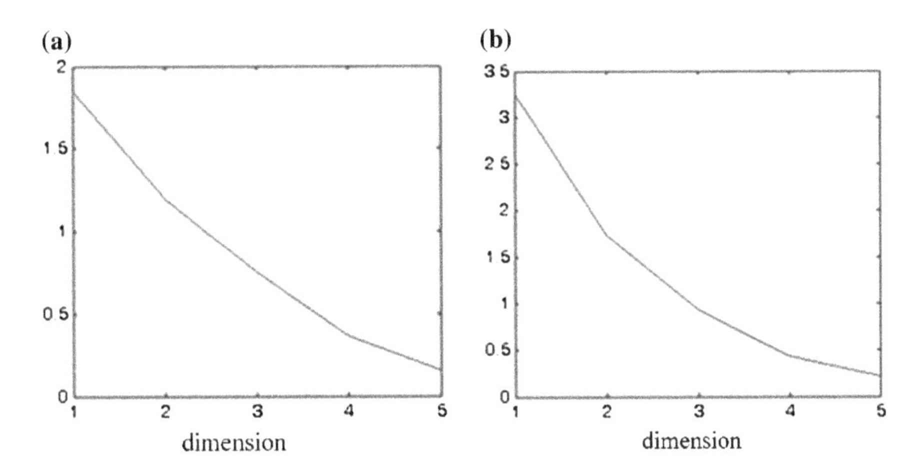

Fig. 6.18 The eigenvectors for waveform distribution of echoes

wavelet decomposition results, and Fig. 6.17 demonstrates the results of WTMM extraction based on five-level wavelet decomposition. Additionally, Fig. 6.18 shows the extracted eigenvectors for waveform distribution.

The two types of target echoes are processed through five-level wavelet decomposition to extract distribution characteristics of waveforms of wavelets in each level. As shown in Fig. 6.18, for the two types of targets, the distribution proportion of energies of eigenvectors in each dimension is greatly different from each other. This result presents the dissimilarity in feature information of different targets and verifies that the wavelet transform is effective to extract distribution characteristics of waveforms.

6.3 Evaluating the Separability of Characteristics

Feature selection directly affects the design and performance of a classifier as well as the final classification result, so it is one of the key issues in characteristic extraction. After some characteristics of laser echoes of targets are extracted, their separabilities are measured and evaluated to confirm whether they are typical enough to achieve an optimal classification result [17].

6.3.1 *The Measurement of Class Separability*

In characteristic extraction, the separability can be measured by various criteria, among which the most commonly used distance criterion is selected in the research. In a characteristic space, an eigenvector of a pattern can be regarded as a point, and the sample set in this pattern is a point set. In general, the greater the average distance between classes is, the smaller the average distance within classes and the better the separability. Therefore, the distance between samples is the most intuitive criterion for studying the distribution of samples and measuring class separability. Suppose that the joint eigenvector set of c pattern classes $\omega_1, \omega_2, \ldots, \omega_c$ is $\{x^{(i,k)},\quad i = 1, 2, \ldots, c,\quad k = 1, 2, \ldots, N_i\}$. Where $x^{(i,k)} = \left(x_1^{(i,k)}, x_2^{(i,k)}, \ldots, x_m^{(i,k)}\right)^t$ is the kth m-dimensional eigenvector of class ω_i, and N_i is the number of the eigenvectors in class ω_i. The average distance S_i among all the eigenvectors in class ω_i is calculated and defined as:

$$S_i = \frac{1}{2}\frac{1}{N_i}\sum_{j=1}^{N_i}\frac{1}{N_i - 1}\sum_{k=1}^{N_i}\left\|x^{(i,j)} - x^{(i,k)}\right\|^2 \tag{6.37}$$

By averaging S_i $(i = 1, 2, \ldots, c)$, the average inner-class distances is

$$S_\omega = \frac{1}{c}\sum_{i=1}^{c} S_i \tag{6.38}$$

By expanding Eq. (6.37), the average inner-class distance is derived as:

$$S_{\omega} = \frac{1}{c}\sum_{i=1}^{c}\sum_{l=1}^{m}\frac{1}{N_i - 1}\sum_{k=1}^{N_i}\left(x_l^{(i,k)} - \mu_l^{(i)}\right)^2 \tag{6.39}$$

Thereinto,

$$\mu_l^{(i)} = \frac{1}{N_i}\sum_{k=1}^{N_i} x_l^{(i,k)} \tag{6.40}$$

where $\mu_l^{(i)}$ is the mean value of the lth components of eigenvectors for samples in class ω_i. Generally, the average vector of a class can be employed to represent the class. Assume that the average vector of samples in class ω_i is $\mu^{(i)}$ and the average vector of the overall samples is μ, then the average distance of c classes can be defined as:

$$S_b = \frac{1}{c}\sum_{i=1}^{c}\left\|\mu^{(i)} - \mu\right\|^2 = \frac{1}{c}\sum_{i=1}^{c}\sum_{l=1}^{m}\left(\mu_l^{(i)} - \mu_l\right)^2 \tag{6.41}$$

where μ_l is the lth component in the average vector of the overall samples.

$$\mu_l = \frac{1}{c}\sum_{i=1}^{c}\frac{1}{N_i}\sum_{k=1}^{N_i} x_l^{(i,k)} \tag{6.42}$$

A preferable separability can be obtained only with small inner-class distances and large distances of classes. Therefore, the distance criterion is defined as the ratio of the average distance of classes to the inner-class distance, that is

$$J_A = \frac{S_b}{S_{\omega}} = \frac{\frac{1}{c}\sum_{i=1}^{c}\sum_{l=1}^{m}\left(\mu_l^{(i)} - \mu_l\right)^2}{\frac{1}{c}\sum_{i=1}^{c}\sum_{l=1}^{m}\frac{1}{N_i-1}\sum_{k=1}^{N_i}\left(x_l^{(i,k)} - \mu_l^{(i)}\right)^2} \tag{6.43}$$

Equation (6.43) is the ratio of the variance of average distance of classes to that of the average inner-class distance of each feature component.

6.3.2 Evaluation of Wavelet Characteristics

According to the above analysis, owing to target echoes are sensitive to the attitude, the characteristics of wavelet energies and WTMMs extracted are expected to fluctuate at a certain range of attitude angles. In order to reduce the sensitivity of target echoes, attitude angles at a certain range are considered while extracting features of targets. This is because templates are not likely to be established for a certain angle while establishing a template library and performing classifier training.

Three types of targets are selected including two airliners and a fighter. By employing simulation technology, forty groups of target echoes are obtained for each target with a SNR of 20 db and attitude angle varying from 0° to 10°. Energy characteristics of wavelets and characteristics of singular values of WTMMs are extracted from these data. Figure 6.19 demonstrates the distributions of energy

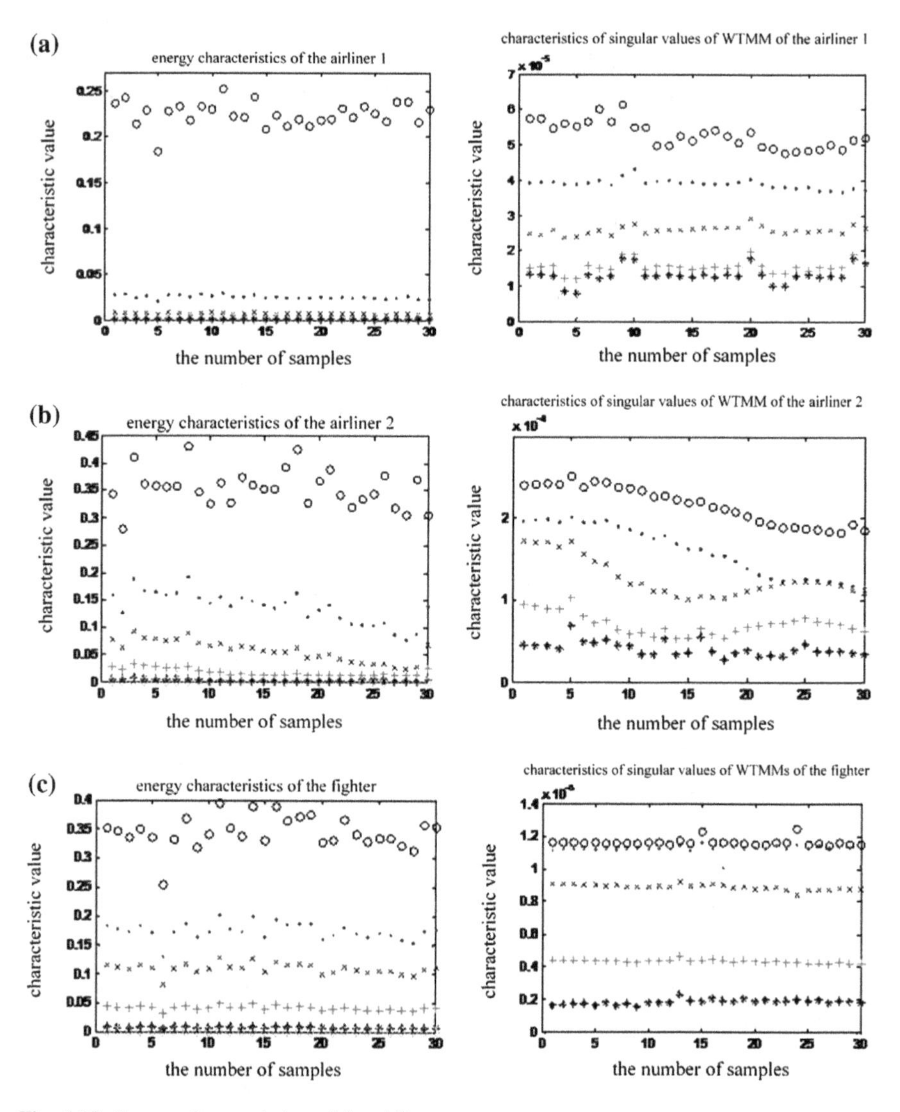

Fig. 6.19 Energy characteristics of the airliner

Table 6.7 The ratios of the distance between classes to inner-class distance of the two types of characteristics

	Energy characteristics of wavelets	Characteristics of singular values of WTMMs
The ratio of the distance between classes to the inner-class distance	7.7568	20.6183

characteristics of wavelets and characteristics of WTMM singular values for the three targets.

By using the distance criterion for features in Sect. 6.3.1, the ratios of the distance between classes to the inner-class distance of the two types of characteristics are calculated, as shown in Table 6.7.

The simulation results that the ratio of the distance between classes to the inner-class distance of characteristics of singular values of WTMMs is obviously larger than that of energy characteristics of wavelets. Hence, the characteristics of singular values of WTMMs are more favorable for the classification and identification of targets. That is, while using distance criterion to evaluate the eigenvectors extracted, the distribution characteristics of waveforms have better performance than the energy characteristics.

References

1. Chen X (2010) Algorithms for target identification based on laser radar images. National University of Defense Technology, Changsha
2. Holmes QA, Zhang X, Zhao D (1997) Multi-resolution surface feature analysis for automatic target identification based on laser radar images. In: Proceedings of the international conference on image processing, Santa Barbara, CA, USA, pp 468–471
3. Koksal AE, Shapiro JH, Wells WM (1999) Model-based object recognition using laser radar range imagery. Automatic target recognition IX, Orlando, FL, USA, SPIE vol 3718, pp 256–266
4. Pal NR, Cahoon TC, Bezdek JC, Pal K (2001) A new approach to target recognition for LADAR data. IEEE Trans Fuzzy Syst 9(1):44–52
5. Soliday SW, Perona MT, McCauley DG (2001) Hybrid fuzzy-neural classifier for feature level data fusion in LADAR autonomous target recognition. Automatic target recognition XI, Orlando, FL, USA, SPIE vol 4379, pp 66–77
6. Hu Y, Shu R, Xue Y (2001) Extraction of echo characteristics based on laser imaging in earth observation. The 15th national seminar about infrared science and technology and memoir of national academic communication on photoelectric technology, Ningbo, pp 457–461
7. Hu Y (2008) Methods for object inversion based on laser remote imaging-progress report in 2008. Hefei Electronic Engineering College, Hefei
8. van den Heuvel JC, Schoemaker RM, Schleijpen RHMA (2009) Identification of air and sea-surface targets with a laser range profiler. In: Proceedings of SPIE, vol. 7323
9. Zhang J, Zhang B, Jiang X (2000) Analysis of methods for characteristic extraction based on wavelet transform. Sig Process 16(2):156–162

10. Jiang L, Gong S, Hu W (2003) A method for extracting characteristics of signals based on wavelet transform. College J Wuhan Univ Technol (Edit: Transportation science and engineering) 27(3):358–360
11. Wang Q, Ni H, Xu Y (2003) Mining detection based on energy characteristics of wavelets. J Data Acquisition & Proc 18(2):156–160
12. Li L (2010) Extraction of echo characteristics and target classification based on laser remote imaging. Hefei Electronic Engineering College, Hefei
13. Watson JN, Addison PS (2002) Spectral-temporal filtering of NDT data using wavelet transform modulus maxima. Mech Res Commun 29:99–106
14. Ouni K, Ktata S, Ellouze N (2006) Automatic ECG segmentation based on wavelet transform modulus maxima. In: IMACS multi conference on computational engineering in systems applications (CESA), pp 140–144
15. Theodoridis S, Koutroumbas K (2006) Pattern recognition, Third edn. Publishing House of Electronics Industry
16. Lai X (2010) The rudiments and application of airborne laser radar. Electronic Industry Press, Beijing
17. Chen X, Hu Y, Zhang J (2005) Characteristic selection of remote sensing images based on minimum entropy and genetic algorithm. Remote Sens Inf (5):3–5

Chapter 7
The Extraction of the Features of Targets Detected Using a Laser Imaging System

Target features can be divided into three types, namely, structure features, physical features and statistical features, all of which describe the basic attributes of targets. Targets can be found by detecting the target features and classified and recognized according to characteristic differences of different targets. Feature extraction is not only one of key points, but also the difficulty in the research of target detections based on a laser imaging system. The extraction of target features means extracting the following features including structures and reflectivity of targets, and range, displacement and attitude of moving targets by processing the time delay images, the gray images and the waveform images of targets. The purpose of the extraction of target features is to detect, classify and recognize targets. This chapter mainly introduces the basic principles and specific methods of feature extraction in the target detection using a laser imaging system and discusses a universal method for extracting target features by taking ground and aerial targets as examples.

7.1 The Principle of the Extraction of Target Features

It can be known from the above-mentioned chapters that after the target modulations for laser echoes, the target features are mainly shown in the differences of time delay, energy and pulse waveform of laser echoes. The laser images of targets detected using a laser imaging system, comprehensively reflect the essential features of targets. The target features such as the structures, physical properties and the statistic features of the distribution of the targets were extracted using a laser imaging technique based on the range, gray, waveform and hierarchy of the targets image.

Y. Hu, *Theory and Technology of Laser Imaging Based Target Detection*,
DOI 10.1007/978-981-10-3497-8_7

7.1.1 Extracting Structure Features of Targets Based on the Time Delay and Level of Echoes

In the target detection based on a laser imaging system, the structure features of targets is mainly extracted on the basis of the time delay and level information of laser echoes. The range images and level images of targets can be acquired according to the time delay and level information of laser echoes, respectively. On this basis, the structure features of targets can be further extracted by analyzing the spatial relation of targets. The overall process of the extraction of structure features of targets based on the time delay and level of echoes is shown in Fig. 7.1.

1. Edge detection for the depth of targets

For laser range images of targets, the depth mutations frequently occur in the target edges. The edge ordinarily means the end of one region and the start of the other region. Under the circumstance, the edge information extracted by detecting the depth edges of targets is an important basis for the subsequent extraction of target features.

The depth edges of laser images have two features, namely, direction and amplitude. The range information changes gently along the edge direction while dramatically in the direction perpendicular to the edges. The latter is likely to present the step or ramp type changes. The edge information shows the characters of first-order peaks and the second-order zero crossing by taking first-order and

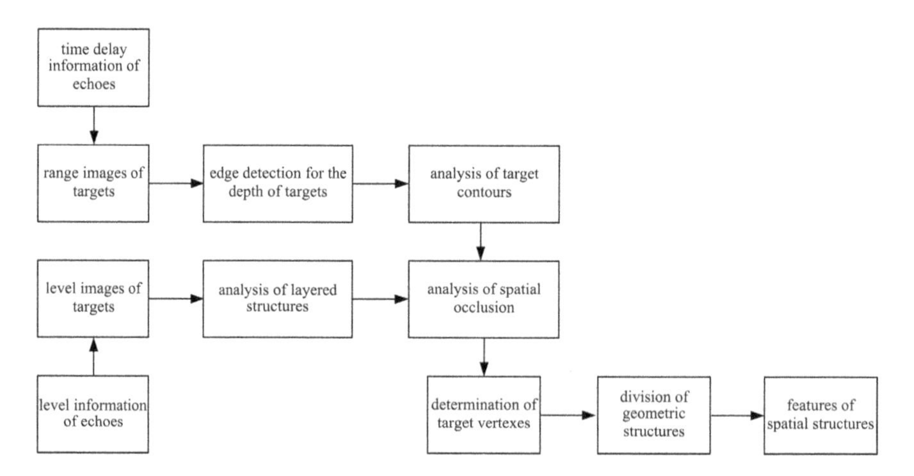

Fig. 7.1 The flow of extracting of structure features targets based on the time delay and level of echoes

second-order derivatives of the range information of all points on the edges. Then, through performing the above operations for the whole laser range images and detecting the corresponding first-order peaks and second-order zero crossing terms, the depth edges of targets can be extracted effectively.

2. Analysis of contours

The extracted depth edges of targets are generally discontinuous and likely to contain the fake edges produced by noise. Therefore, it is essential to connect the right target edges by using certain analysis methods for contours to obtain effective and true target contours, so as to provide the basis for further analyzing the structure features of targets.

In the analysis of contours, the depth edges with large enough first-order and second-order responses can be detected firstly followed by connecting the edges with obvious features of the target contours in the scene. After dividing major regions of the scene based on main contours, the boundary lines with few changes of depth edges are predicted and determined according to prior knowledge. This process mainly aims to estimate the effectiveness and collinearity of each point on the edges. The main principles are as follows: (1) The number of edge points has to exceed a certain threshold; (2) The minimum variance of the boundary and the predicted boundary needs to be smaller than a certain threshold; and (3) The difference between the direction of the line fitted with boundary points and the predicted direction is smaller than a certain threshold. In the analysis, the above processes are repeated until all target contours in the scene are extracted.

3. Analysis of layered relation

After obtaining the level images of targets according to the level information of laser echoes, the spatial geometric relations of the level images are matched, together with the classification of the point cloud data of the levels. The spatial points of each level located in the same geometric region are grouped together through the spatial geometric relations matching. Then, the point cloud data at each level in the same spatial region are analyzed in terms of the statistical features such as the variance of spatial distribution to further classify the point cloud data which possibly belong to a same target at each level. The data of spatial levels in the scene can be divided through the above analysis of layered relations according to different spatial regions and distribution features of targets.

4. Analysis of spatial occlusion

After the analysis of the spatial contours and layered relations of targets, the information of the two aspects can be used to analyze the spatial occlusion of each boundary. The extracted target boundaries are divided into four types according to their physical features, namely, convex, concave, occluding and occluded boundaries. Each extracted boundary is analyzed regarding the spatial geometric constraints and then the spatial layered relations of boundaries to determine the type and spatial occlusion of boundaries, which are used afterwards to predict the trend of occluded boundaries.

5. Determination of target vertexes

The spatial vertexes of targets can be further obtained by extracting boundary lines and spatial occlusion of targets. These vertexes are mainly divided into two categories: one type of vertexes is formed by the intersections of visible boundaries, and the other type is composed of the intersections of occluding and occluded boundaries. In the analysis, all possible intersections of boundary lines are analyzed first to list the set of all possible vertexes. Then, all vertexes are connected and analyzed regarding the space geometry to remove fake vertexes which do not confirm to the logic of space connection. In this way, a vertex set of targets following the rules of the spatial connection is obtained.

6. Division of spatial structures

After obtaining the boundary representation and the vertex set of targets, it needs to further examine whether targets can be further divided into more smaller spatial objects. The trend of target boundaries is analyzed according to the trend of spatial boundaries and the geometrical principle so as to further divide targets into several spatial regular substructures and obtain the features of spatial structures of targets. These features mainly include the boundary features, the features of space vertexes, the spatial occlusion and the sub structural features.

7.1.2 Extracting Features of Physical Properties of Targets Based on the Echo Energy

The target detection based on a laser imaging system can obtain the information of the 3D range and the echo energy of targets simultaneously to generate the 3D range images and the gray images of targets, respectively. Each pixel of the range images and the gray images is generated by different feature information of the same echo. Therefore, each pixel of the range images corresponds to that of the gray images. Because the energy of laser echoes and the square of target range obey the range equation, the intensity information of target echoes can be corrected through fusing data at the pixel level and then considering the atmospheric transmission model. The purpose of this information correction is to extract more accurate information of the scattering cross section or the surface reflectivity of targets, which directly reflects the physical properties of targets.

For a point target, suppose that its echo energy and range are E_i and R_i, respectively. In addition, the laser pulse energy emitted by the system, the divergence angle of the laser beams, distribution function of beams are expressed as E_0, α and K, separately. Meanwhile, the attenuation factor of laser pulse energy in the two-way transmission in the atmosphere and the gain of system transmitting and receiving are T_i and η, respectively. Then, according to the range equation, the scattering cross section of the target is

$$\sigma_i = \frac{E_i}{E_0} \cdot \frac{T_i \eta \alpha^2}{K \pi^2 R_i^4} \tag{7.1}$$

For an area target, it can be obtained from the range equation that its surface reflectivity is

$$\rho_i = \frac{E_i}{E_0} \cdot \frac{T_i \eta}{4 \pi^2 R_i^2} \tag{7.2}$$

For an area target, it is also essential to perform space division for the distribution difference of the reflectivities of each point to obtain the results of space division of targets after acquiring the surface reflectivities of each point based on the distribution of echo intensities. Afterwards, the statistical features of the reflectivity distribution for all parts of targets can be obtained through the statistic analysis of the reflectivity information of targets within the area. In the processing, the reflectivity information of all points can be regarded as the gray images of reflectivity so that extract the statistical features of target reflectivity through the filtering, the edge division and the statistical analysis.

The reflectivity image of each point in the target can be expressed as $I(x, y)$. Because there is no target information at the noisy points of the image, the intensity of the noisy points is obviously lower than the gray values of target points. Therefore, the noisy points can be filtered through median filter.

Characterized by local smoothing, median filter can overcome the fuzziness of image details caused by linear filters, and is the most effective method for filtering pulse interference and grain noise. The principle of median filter is to substitute the gray value of a point with the medium values of its neighborhood points in the window with a fixed size, that is

$$I(x, y) = median[x_1, x_2, \ldots, x_n] \tag{7.3}$$

Median filter can be used to effectively restrain random noise and reserve edge information.

The positions of mutational points of reflectivity are extracted using the edge extraction method in the filtered image of target reflectivity. Then the target boundaries are obtained by connecting edge points to realize the region division. For the divided region, the features of reflectivity distribution can be extracted including mean values and variances of reflectivity distribution and texture distribution of reflectivity.

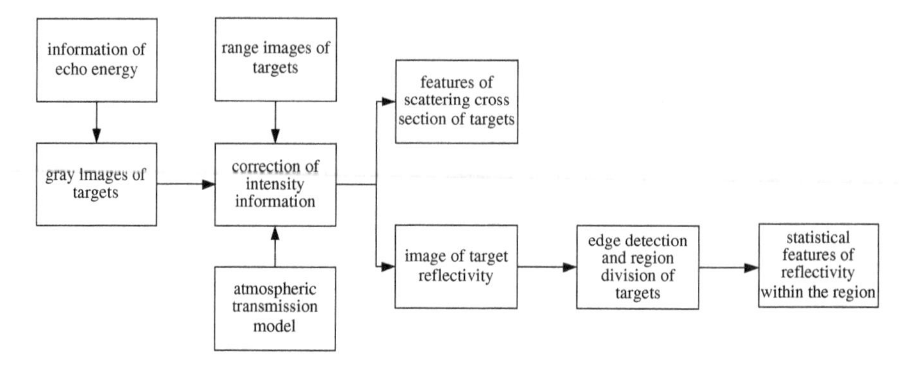

Fig. 7.2 The basic procedure of extracting physical property features of targets based on the echo energy

The basic procedure of extracting physical property features of targets based on the echo energy is illustrated in Fig. 7.2.

7.1.3 The Extraction of Comprehensive Features of Targets Based on Echo Waveforms

The waveform images detected using a laser imaging system contain rich information of targets. The laser pulse signals received by the detector exhibit modulated waveforms in time domain after undergoing spatial and reflectivity modulation of targets. Based on the waveforms, the comprehensive features of targets can be extracted from the waveform images to some extent. These comprehensive features include the structural features, the features of physical properties relating reflectivity and the statistical features.

When aerial targets are detected using a laser imaging system, the detection background is the atmosphere. According to the principle of laser imaging, the sampling values of each pixel of gray images reflect the received echo intensity. When there are targets in the transmission channel of optical waves, the echo intensity is relatively strong, so the target area included in the gray images has relatively higher gray values. While without such characteristic, range images cannot characterize echo intensity. The feature of gray images can be employed to effectively restrain the interference of non-target information such as noise for range images so as to strengthen the target information.

After processing the gray images, the noisy points can be effectively filtered, so that the gray images preferably reserve the edge features. The mean value of obtained gray images is calculated by the following formula.

$$T = \frac{1}{M \times N} \sum_{x,y=1}^{\substack{x=M \\ y=N}} I(x, y) \tag{7.4}$$

The images are subjected to the binarization processing by comparing with this threshold.

$$I'(x, y) = \begin{cases} 1 & I(x, y) \geq T \\ 0 & others \end{cases} \tag{7.5}$$

The range images are filtered by multiplying the range images by the binary images. Suppose that the range image of the target is $R(x, y)$, then the processed range image is

$$R'(x, y) = R(x, y) \cdot I'(x, y) \tag{7.6}$$

The unchanged features of the processed range images are extracted. At present, the commonly used features include Hu invariant moment, which is characterized by the translation invariance and is a preferable image feature.

The division results of spatial region can be obtained by combining level images with range images and gray images to finally realize the precise division of targets. Then, by extracting the comprehensive features from the data of divided target images, the features including the invariant moments of depth images and gray images, as well as the spatial size and the relation of spatial position of targets. The basic procedure of extracting the comprehensive features of targets is demonstrated in Fig. 7.3.

7.2 The Feature Extraction of Ground Targets Based on Echo Waveforms

A remarkable characteristic of ground targets is that they are mixed with the background of ground objects. Therefore, the feature extraction of ground targets is influenced significantly by the distribution, structures and physical properties of ground objects. For ground targets with large areas, the laser images generated by the echo signals of target sampling points distributed according to certain rules are the basis for extracting target features. The target features are extracted by obtaining

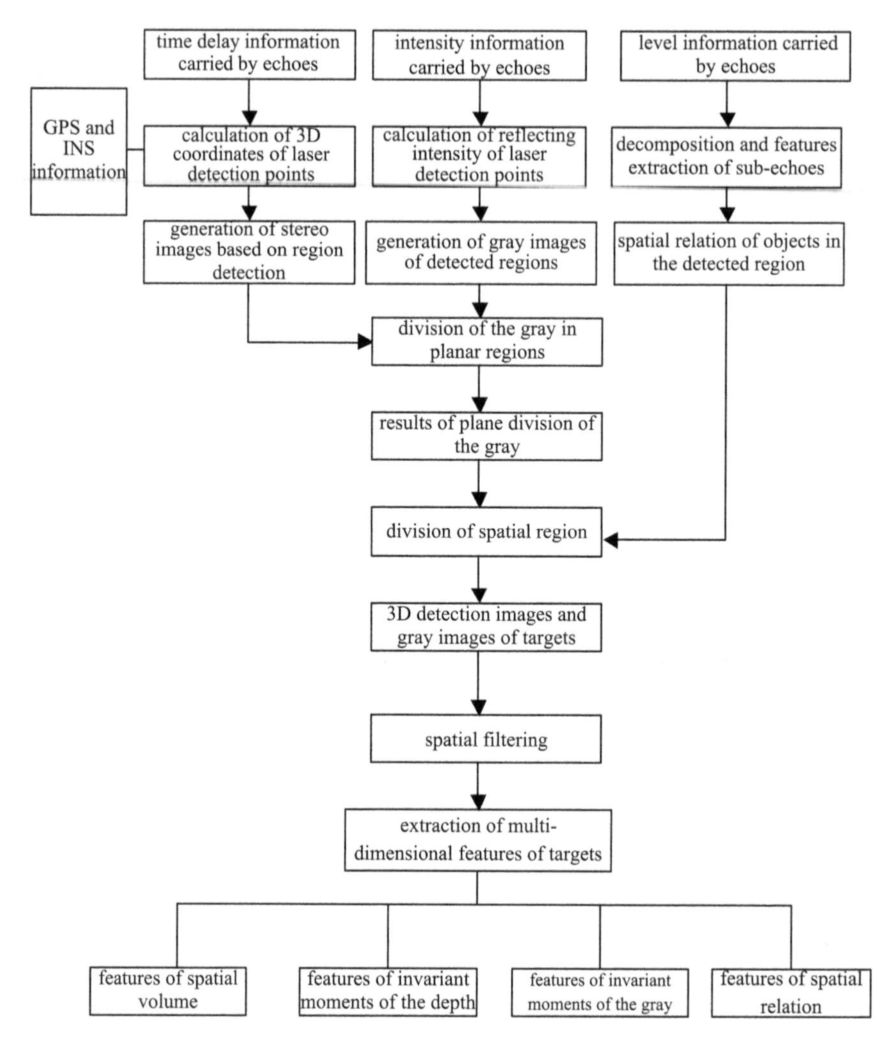

Fig. 7.3 The basic procedure of extracting the comprehensive features of targets

the differences of target features based on those of echo images of each sampling point. The methods of such features extraction of large targets have been discussed extensively in some literatures relating laser remote sensing. For the long-range target detection, targets generally can be illuminated by a beam because of their small areas. Under such circumstance, it needs to study the features of a laser echo containing common modulation of targets and backgrounds of ground objects to extract the features of ground targets [1, 2].

7.2.1 *The Model of Laser Echoes Reflected from Ground Targets*

According to the modulation mechanism of targets for laser pulses given in Sect. 2.4 that the shape, fluctuation and reflectivity of the targets influence the energy and waveform of echo pulses, the modulation features of ground targets for laser pulses can be acquired. In order to facilitate the analysis and not lose the generality, the relationship between the laser spot and ground surface is obtained for a one-dimensional case, as shown in Fig. 7.4.

Suppose that the ground target and the surface are diffuse reflecting targets, which subject to Lambert cosine law, then the echo corresponding to the any angle ψ in the laser spot of Fig. 7.4 can be expressed as [3].

$$f_i(t) = a_i \exp[-\frac{k(t_i - t)^2}{\tau^2}] g(\psi) \cos(\zeta) \frac{h}{s_i^2} \tag{7.7}$$

where, t_i indicates the time delay of the echo of the corresponding target sampling point and is used to characterize the shape distribution of the target. While a_i represents the surface reflectivity. $g(\psi)$ is the energy distribution function of the laser beams relating ψ and is considered to comply with Gaussian distribution. ζ signifies the angle between the incident angle of the lasers and the normal line of the target surface. h is a coefficient relating the response of the detector, the effective receiving area and atmospheric attenuation, and can be considered as a constant in primary detections. s_i represents the distance between the target sampling point and the target detection system.

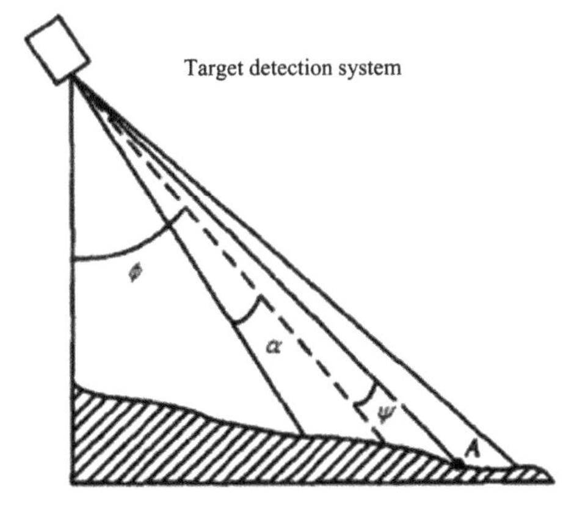

Fig. 7.4 The relationship between the laser spot and the ground surface

Therefore, the pulse power of the compound echo at t moment can be expressed by solving the integral of the power of echo pulses with any angle ψ.

$$
\begin{aligned}
P(t) &= \int_{\psi_1}^{\psi_2} f_i(t)d\psi = \int_{\psi_1}^{\psi_2} a_i \exp[-\frac{k(t_i-t)^2}{\tau^2}]g(\psi)\cos(\zeta)\frac{h}{s_i^2}d\psi \\
&= a_i \cos(\zeta)\frac{h}{s_i^2}\exp[-\frac{k(t_i-t)^2}{\tau^2}]\int_{\psi_1}^{\psi_2} g(\psi)d\psi \\
&= b_i\left\{\int_{-\infty}^{\psi_2} g(\psi)d\psi - \int_{-\infty}^{\psi_1} g(\psi)d\psi\right\}
\end{aligned}
\tag{7.8}
$$

Let $b_i = a_i \cdot g(\psi) \cdot \cos(\zeta) \cdot \frac{h}{s_i^2}$ and suppose that $g(\psi)$ obeys Gaussian distribution whose mean value is (χ, σ^2), then formula (7.8) can be rewritten as

$$
\begin{aligned}
P(t) &= b_i\left\{\int_{-\infty}^{\psi_2} g(\psi)d\psi - \int_{-\infty}^{\psi_1} g(\psi)d\psi\right\} \\
&= \frac{b_i}{\sqrt{2\pi}\sigma}\left[\int_{-\infty}^{\psi_2} e^{-\frac{(\psi-\chi)^2}{2\sigma^2}}d\psi - \int_{-\infty}^{\psi_1} e^{-\frac{(\psi-\chi)^2}{2\sigma^2}}d\psi\right] \\
&= \frac{b_i}{2}\left[erf\left(\frac{\psi_2-\chi}{\sqrt{2}\sigma}\right) - erf\left(\frac{\psi_1-\chi}{\sqrt{2}\sigma}\right)\right]
\end{aligned}
\tag{7.9}
$$

where, $\psi_1 = \phi - \alpha$ and $\psi_2 = \phi + \alpha$.

It can be seen from formulas (7.7) and (7.8) that signals of laser echoes are influenced by the joint modulation of the height and reflectivity distribution of ground targets, and thus affect the one-dimensional time distribution of echo power.

According to the expression form of signals of discretized laser echoes, a series of values $\{y_k : k = 1, \ldots, N\}$ corresponding to the echo signals obtained by sampling are acquired by overlapping multiple echoes of ground targets. These echoes can be considered to comply with Gaussian distribution, then

$$
\begin{aligned}
y = f(t) &= \sum_{i=1}^{n} a_i \cdot g(\psi) \cdot \cos(\zeta) \cdot \frac{h}{s_i^2}\exp[-\frac{k(t_i-t)^2}{\tau^2}] \\
&= \sum_{i=1}^{N} b_i \exp[-\frac{k(t_i-t)^2}{\tau^2}]
\end{aligned}
\tag{7.10}
$$

Because there are N known quantities and $2N$ unknown quantities (b_i, t_i) in this equation, the equation cannot be solved directly. However, these problems can be solved by using least mean square error (LSME) to make the following formula valid.

$$\sqrt{\frac{1}{N}\sum_{k=1}^{N}\left(f(t_k)-y_k\right)^2}<\varepsilon \tag{7.11}$$

where, ε is the required solution accuracy. The selection of ε depends on many factors, such as the number of iteration, precision required by the system and average noise level. For example, the solution precision can be selected as the triple of the average noise level.

The time delay information of echoes of all points on the ground can be acquired through the above calculating operations followed by obtaining the height distribution of the ground through converting coordinate positions.

7.2.2 Parameter Analysis of Ground Targets

1. Distribution of surface reflectivity

The distribution of surface reflectivity directly influences the intensity of echo waveforms. The echo intensity changes suddenly with the variation of reflectivity. By taking the derivative of echo signals, the position of mutational points of reflectivity can be obtained.

Suppose that the position of mutational points of reflectivity is M, where the reflectivity changes at a rate of μ, then it can be obtained from formulas (7.7) and (7.10) that

$$\begin{cases} \sum\limits_{i=1}^{M-1} y_i = \sum\limits_{i=1}^{M-1}\sum\limits_{\psi=1}^{M-1} ag(\psi)\exp\left[-\frac{k(t_\psi-t)^2}{\tau^2}\right] = h\sum\limits_{\psi=1}^{M-1} g(\psi) \\ \sum\limits_{i=M}^{N} y_i = \sum\limits_{i=M}^{N}\sum\limits_{\psi=M}^{N} (1-\mu)ag(\psi)\exp\left[-\frac{k(t_\psi-t)^2}{\tau^2}\right] = h(1-\mu)\sum\limits_{\psi=1}^{M-1} g(\psi) \end{cases} \tag{7.12}$$

To make the above equations valid, the following condition is assumed, that is, if the range of the compound echo of these echoes located at $\psi = 1, \ldots, k$ is still within $\{y_i, i = 1, \ldots, k\}$, then its accumulated value can be substituted with the sum of y_i. This assumption can be summarized as that the echo time delay in different terrain ranges is separated from each other in the time, which is called simply the separation condition of echo time delay.

Suppose that the target is scanned with an angle of 45°, the laser detector is positioned 5 km above the surface, and $\psi_m - \psi_{m-1} = \Delta\psi = 0.1\,\mathrm{mrad}$, then the horizontal range of two ground target points is

$$\Delta l = H\tan(\varphi - \Delta\psi) - H\tan(\varphi) \approx 0.006\,\mathrm{m} \tag{7.13}$$

On horizontal ground, the difference between the distances from the two target points to the laser detector is

$$\Delta s = H/\cos(\varphi + \Delta\psi) - H/\cos(\varphi) \approx 0.71\,\mathrm{m} \tag{7.14}$$

In order to meet the assumption, the height difference between M point and $M-1$ point needs to be smaller than the range difference on horizontal ground to make the time delay of echoes satisfy the separation condition. Under the above given condition, when the horizontal range of terrain is 1 m, the height difference is smaller than 0.71 m. It is thus clear that this assumption can be met in general terrain conditions. Here, the distribution information of surface reflectivity can be acquired from formula (7.11).

2. The analysis of the distribution of surface height

In order to acquire the distribution information of surface height in the laser spots, it is essential to obtain the time delay of all sampling points on the ground by solving formula (7.10) and then calculating the height distribution.

The solution of formula (7.10) based on LSME needs to obtain the estimated initial values of the unknown quantities b_i and t_i firstly and then perform the iterative operation. Because the intensity of the compound echo is proportional to that of the echo intensity of all points, and the time delay of the compound echo meets the separation condition in general terrain conditions, the estimated initial values are

$$b_i = yi; \quad t_i = i\,(i = 1, \ldots, N) \tag{7.15}$$

By substituting formula (7.12) into formula (7.7) to perform the iterative operation, the error of kth iteration is obtained as

$$\eta_i = y_i - f(t) \tag{7.16}$$

The estimated value is corrected according to the iteration error.

$$b_{(k+1)i} = b_k + c\eta_i \quad t_{(k+1)i} = t_k + d\eta_i \tag{7.17}$$

where, c and d represent the correction coefficients of the iteration error, separately. Through the multiple iterations, the estimated values of the unknown quantities meeting formula (7.10) can be acquired.

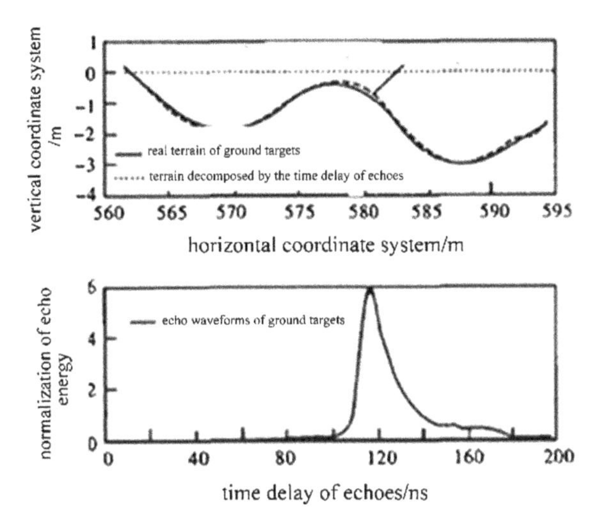

Fig. 7.5 The simulation result of height distribution of the terrain

3. Simulation

The simulation is conducted with the scanning angle, the half divergence angle of lasers, pulse width of lasers, and the elevation of the detector being $\phi = 30°$, $\alpha = 6$ mrad, $\tau = 2$ ns, and H = 1000 m, respectively. The simulation result of height distribution of the terrain calculated through the above operations is illustrated in Fig. 7.5.

This figure shows the distribution of terrain height obtained by analyzing detailed features of targets based on received echo signals, and the average error is 0.08 m. However, the calculated average error of the terrain distribution is supposed to reach 1.56 m while merely selecting the time delay of peak echo signals as the average time delay of terrain in the laser spots but not analyzing echo signals. It can be seen that using above algorithms can effectively improve the solution accuracy of surface height in the case that the terrain changes obviously.

7.3 Feature Extraction of Aerial Targets Based on Echo Waveforms

Compared with ground targets, aerial targets are characterized by smaller detection areas, simpler detection backgrounds, and greater mobility, and their detection is more likely to be influenced by the attitude and the range. When one aerial target is detected using a laser imaging system, especially in a long range, the target is generally illuminated by a beam, so it needs to extract target features according to the waveforms of single-beam laser echoes.

7.3.1 Expression of Laser Echoes of Aerial Targets

According to the analysis results of the modulation of targets for laser beams in the previous chapters, the echo pulses of lasers are modulated by the range, structure and reflection properties of targets [4, 5]. For an aerial target, its echo pulses are also modulated by its attitude and motion. Taking a plane as an example, the relation between the plane and the laser detection beam is shown in Fig. 7.6. As exhibited in the figure, α indicates the half divergence angle of lasers and ψ is the angle between an arbitrary point in the beam and the central axis of the beam.

Considering that the laser beam basically illuminates the entire target or most region of the target, the modulation function of the various parts of the target for the laser echo is not the same due to the different structures and reflectivities. By adopting the method of grid partition, the laser echoes of the target are analyzed and expressed.

The surface of the aerial target is divided into $N \times N$ imaging sub-regions, and each region is considered to present the same attributes (including the imaging range, imaging angle and reflectivity), as shown in Fig. 7.7.

In Fig. 7.7, O represents the central position of the laser beam. While Z_{kj} denotes the distance between the imaging sub-regions and the detection system. ψ_{kj} is the angle between the sub-region and the central axis of the laser beam.

Assume that the angle between the incident angle of lasers and the normal line on the surface of the target is δ_{kj}, then the sub-images of the echoes reflected from the imaging sub-regions can be expressed as

$$p_{kj}(t) = a_{kj} \exp[-k(2z_{kj}/c - t)^2/\tau^2] \cdot g(\psi) \cdot \cos(\zeta) \cdot h/z_{kj}^2 \qquad (7.18)$$

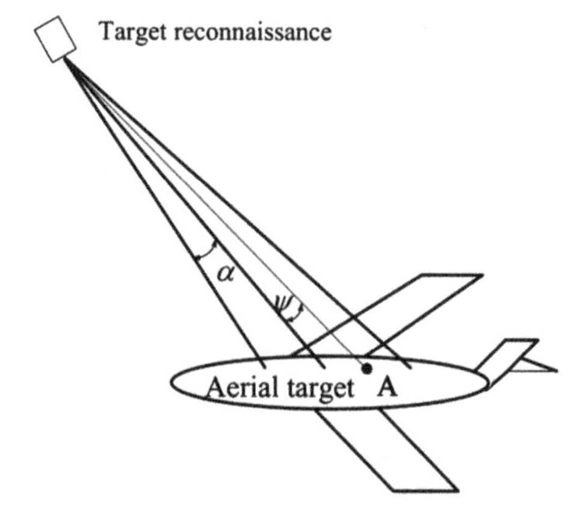

Fig. 7.6 The relationship between the laser spot and the aerial target

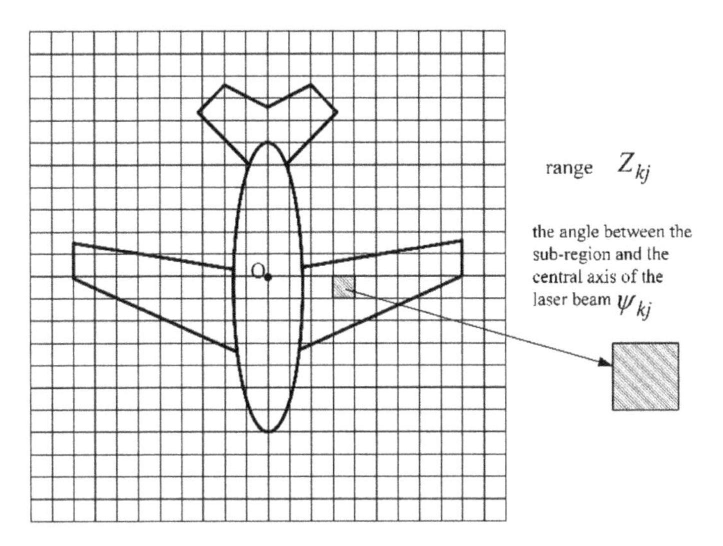

Fig. 7.7 The division of the imaging sub-regions of the aerial target

The result of comprehensive modulation of each sub-region in one laser beam for the laser echoes is

$$p(t) = \sum_{k=1}^{N}\sum_{j=1}^{N} p_{kj}(t) \tag{7.19}$$

7.3.2 *The Model of Extracting the Range Features of Aerial Targets*

If only the distance between the aerial target and the target detection system using a laser imaging technique is changed, then the imaging relation of the aerial target at different ranges is illustrated in Fig. 7.8. In the condition, the target attitude and the relative position between the center point of the target and the laser beam are remained the same.

In Fig. 7.8, the original distance from the intersection point O of the central axis of the laser beam and the target to the laser detection system is Z_1 and the changed range is Z_2. Suppose that $k = Z_2/Z_1$, and the angles between the imaging sub-region i and the central axis of the laser beam before and after the change of the range are ψ_1 and ψ_2, respectively. Then, it can be obtained from the geometrical relationship that

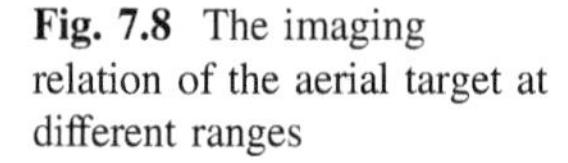

Fig. 7.8 The imaging relation of the aerial target at different ranges

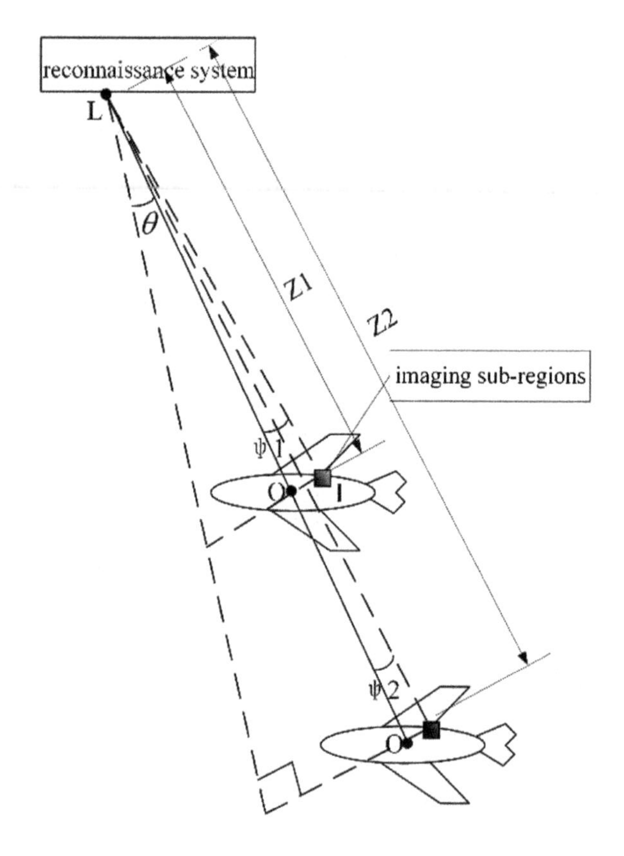

$$Z_1 \cdot \cos\theta \cdot \tan(\theta + \psi_1) - Z_1 \cdot \sin\theta = Z_2 \cdot \cos\theta \cdot \tan(\theta + \psi_2) - Z_2 \cdot \cos\theta \quad (7.20)$$

where, θ is the angle between the vertical line which is perpendicular to the intersecting line between the LOI plane and the plane of the imaging region and the straight line LO.

According to the figure, $0 \leq \theta \leq \phi$, there into, ϕ is the angle between the beam of laser detection and the orientation of the target. Suppose that Z_1 = 1000 m and Z_2 = 2000 m, the ratios of ψ_1 to ψ_2 obtained by formula (7.20) are illustrated in Fig. 7.9.

It can be known from Fig. 7.9 that $\psi_2/\psi_1 < 0.505$, which can be considered that $\psi_2/\psi_1 \approx 0.5 = Z_1/Z_2 = 1/k$. That is to say, the variation of the angle between each imaging sub-region and the central axis of the laser beam is inversely proportional to the imaging range of the beam center.

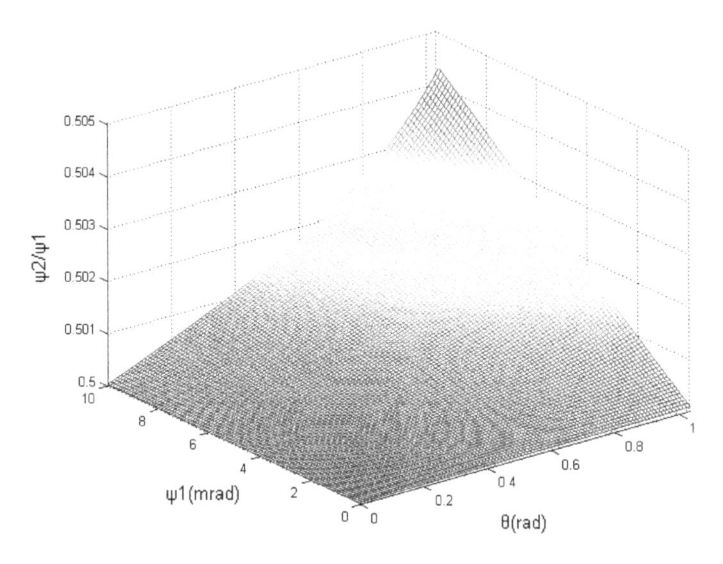

Fig. 7.9 The ratios of ψ_1 to ψ_2 at different imaging ranges

Under two circumstances, the ratio of the imaging ranges in the imaging sub-region i is

$$\begin{aligned}\xi &= \frac{Z_{i2}}{Z_{i1}} = \frac{Z_2 \cdot \cos\theta/\cos(\theta+\psi_2)}{Z_1 \cdot \cos\theta/\cos(\theta+\psi_1)} = \frac{Z_2}{Z_1} \cdot \frac{\cos(\theta+\psi_1)}{\cos(\theta+\psi_2)} \\ &= k \cdot \frac{\cos(\theta+\psi_1)}{\cos(\theta+1/k \cdot \psi_1)}\end{aligned} \tag{7.21}$$

According to formula (7.21), the ratios of imaging ranges in imaging sub-regions can be obtained, as shown in Fig. 7.10.

According to Fig. 7.10, when $1.98 < \xi < 2.002$, it can be considered that $\xi \approx 2 = Z2/Z1 = k$, that is, the imaging range of different imaging sub-regions of the target varies with an equal ratio with that of the beam center.

It can be obtained from formula (7.18) that after the imaging range is changed, the imaging result of the imaging sub-regions is

$$\begin{aligned}p_{i2}(t) &= a_i \exp[-k(2k * z_{i1}/c - t)^2/\tau^2] \cdot g(\psi_{i1}/k) \cdot \cos(\zeta) \cdot h/(z_{i1}^2 * k^2) \\ &= \frac{1}{k^2}\frac{g(\psi_{i1}/k)}{g(\psi_{i1})}\rho_{i1}(t_2)\end{aligned} \tag{7.22}$$

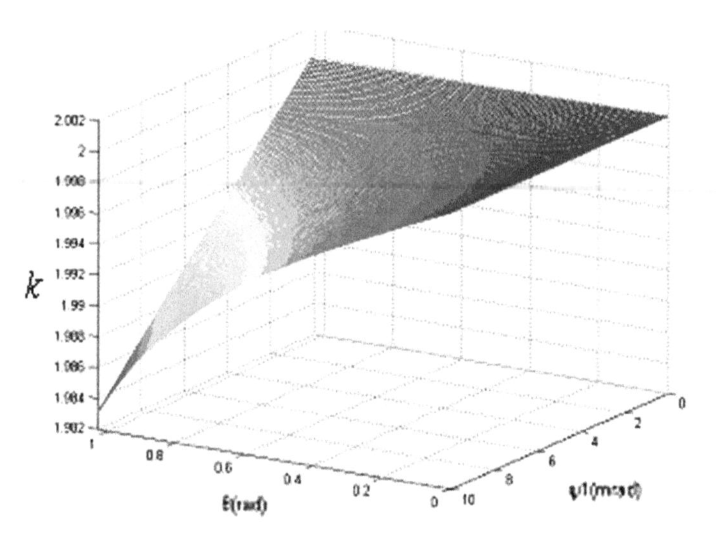

Fig. 7.10 The variation of the ratios of the imaging ranges in imaging sub-regions

where:

$$t_2 = t - 2(k - 1)z_{i1}/c$$

According to formula (7.22), the imaging result of the imaging sub-region i changes in the two aspects. The first is that the echo intensity decreases with the square of range due to the change of the range. The second is the changes of the value of the energy distribution function of the laser beams caused by the variation of the angle between the imaging sub-regions and the central axis of the laser beam.

For the joint imaging results of the whole target, it can be known from formula (7.19) that the change of the results is related to not only imaging detection results of single imaging sub-regions, but also relative distribution of echo time delay of each imaging sub-region. That is to say, the echo waveforms of the joint imaging detection are expected to change with the variation of the distribution of echo time delay of each imaging sub-region. The relative distribution of echo time delay of each imaging sub-region is consistent with the relative changes of the imaging range of each sub-region. Assume that the center imaging range is Z_1, and the angles from the imaging sub-regions i and j to the central axis of the laser beam are ψ_{i1} and ψ_{j1}, respectively, then the difference in the imaging ranges of the sub-regions i and j are

$$\Delta Z_1 = Z_1 \cdot \cos\theta/\cos(\theta + \psi_{i1}) - Z_1 \cdot \cos\theta/\cos(\theta + \psi_{j1}) \qquad (7.23)$$

According to formulas (7.22) and (7.23), when the imaging range of the beam center is Z_2, the imaging range of sub-regions is

$$\begin{aligned}\Delta Z_2 = & \sqrt{[(Z_1 \cdot \cos\theta \cdot \tan(\theta + \psi_{i1}) - Z_1 \cdot \sin\theta) + Z_2 \cdot \sin\theta)]^2 + (Z_2 \cdot \cos\theta)^2} \\ & - \sqrt{[(Z_1 \cdot \cos\theta \cdot \tan(\theta + \psi_{j1}) - Z_1 \cdot \sin\theta) + Z_2 \cdot \sin\theta)]^2 + (Z_2 \cdot \cos\theta)^2}\end{aligned} \tag{7.24}$$

Suppose that ψ_{i1} = 0.007 rad, Z_1 = 1000 m and Z_2 = 2000 m. According to formulas (7.23) and (7.24), the relations of the relative variation $dZ = \Delta Z_2 - \Delta Z_1$ of the imaging range for imaging sub-regions with θ and ψ_{j1} are demonstrated in Fig. 7.11.

It can be seen from Fig. 7.11 that the absolute value of the relative variation of the imaging range for imaging sub-regions *i* and *j* is smaller than 0.015 m, leading to an echo time delay the relative variation of which is smaller than 0.1 ns. Therefore, the waveforms of echo images are slightly changed. The changes of target imaging are mainly caused by the variation of the imaging range of each imaging sub-region and the value of the energy distribution function of the laser beams.

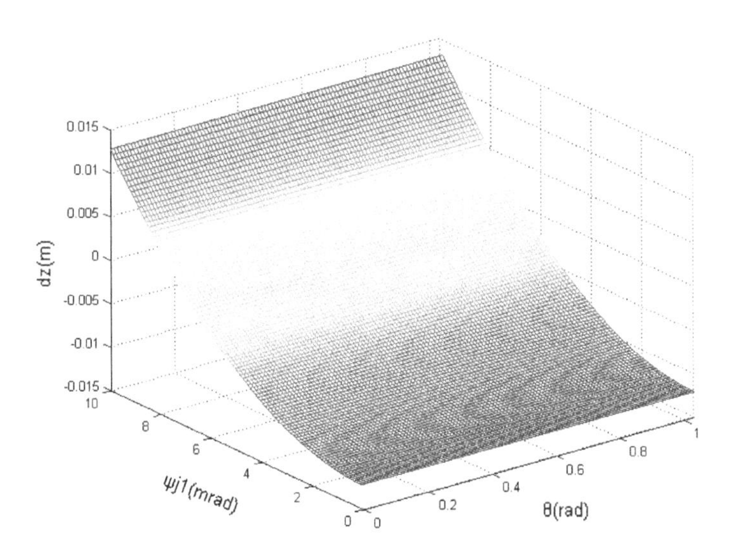

Fig. 7.11 The relations relating the relative variation of the imaging range for imaging sub-regions

7.3.3 The Model of Extracting the Displacement Features of Aerial Targets

When the laser detection beams are emitted to irradiate an aerial target and the relative position between the center of the beams and the target center is changed, the position of the target in the laser detection beams is shifted horizontally. As a result, echo images of the target are changed, see the relation in Fig. 7.12.

In Fig. 7.12, O, ϕ and I represent the central point of the laser beam, the angle between the laser beam and the nadir, and an imaging sub-region, respectively. The range between the central point of the laser beam and the detection system is Z. Meanwhile, x_i and y_i represent the distances from the imaging sub-region I to the x axis and they axis which originate from the point O in the horizontal plane, respectively. While, x_i' and y_i' are the distances from the imaging sub-region I undergoing horizontal displacement to the x axis and y axis that start from the point O, separately.

Suppose that the horizontal displacement quantities of the target along the x axis and y axis are dx and dy, respectively. According to the imaging relation shown in Fig. 7.12, it can be obtained that

$$x_i' = x_i + dx \quad y_i' = y_i + dy \tag{7.25}$$

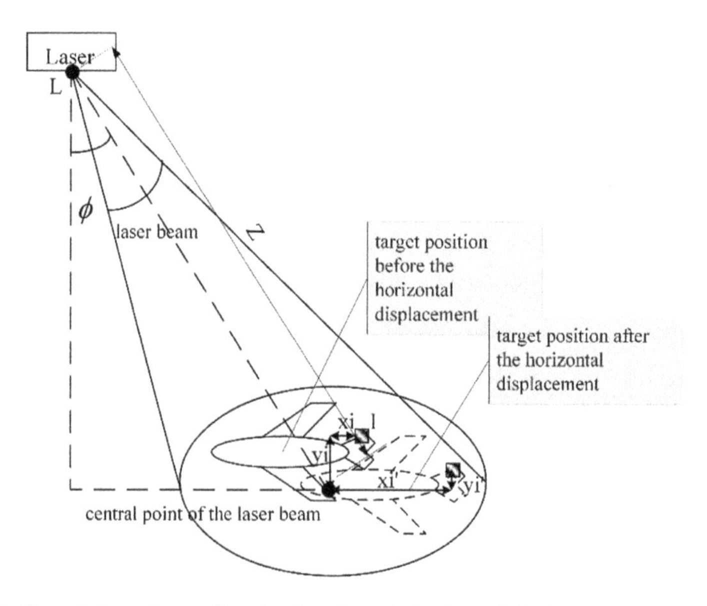

Fig. 7.12 The relation of target imaging based on the horizontal displacement

According to geometrical relationship, the imaging range of the imaging sub-region I before the horizontal displacement is

$$Z_{i1} = \sqrt{(Z \cdot \cos \phi)^2 + (Z \cdot \sin \phi + x_i)^2 + y_i^2} \tag{7.26}$$

The imaging range of the imaging sub-region I after the horizontal displacement is

$$\begin{aligned} Z_{i2} &= \sqrt{(Z \cdot \cos \phi)^2 + (Z \cdot \sin \phi + x_i')^2 + y_i'^2} \\ &= \sqrt{(Z \cdot \cos \phi)^2 + (Z \cdot \sin \phi + x_i + dx)^2 + (y_i + dy)^2} \end{aligned} \tag{7.27}$$

Assume that $Z = 2000$ m and the divergence angle of the laser beam is 10 mrad, then $dx \leq 2000 \times 0.01 = 20\,\text{m}$, $dy \leq 2000 \times 0.01 = 20\,\text{m}$, $x_i \leq 2000 \times 0.005 = 10\,\text{m}$, and $y_i \leq 10\,\text{m}$. Let $\phi = \pi/3$, and dx and dy are set as the maximum values of 20 m, then the value of ratio $\xi = Z_{i2}/Z_{i1}$ of imaging ranges of the imaging sub-region before and after the horizontal displacement is shown in Fig. 7.13.

According to Fig. 7.13, the ratio of imaging ranges of the imaging sub-region before and after the horizontal displacement is approximately equal to 1, which means that the intensity variation of echo images caused by the range changes can be ignored.

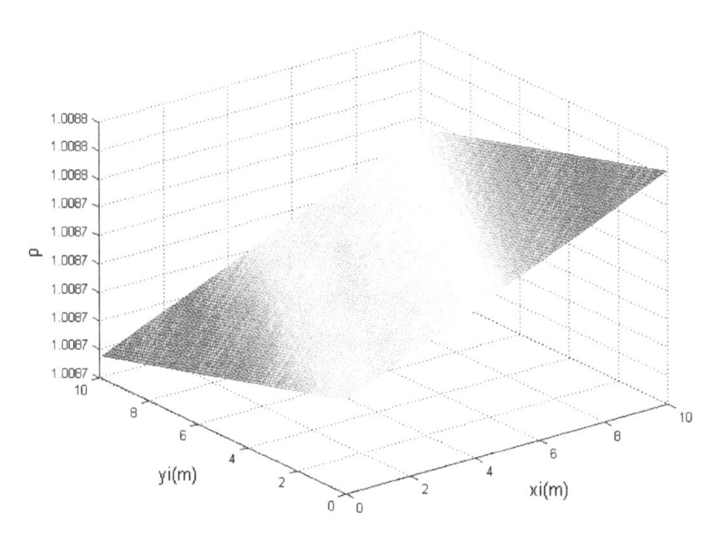

Fig. 7.13 The ratio of imaging ranges of the imaging sub-region before and after the horizontal displacement

For the imaging sub-region I, before the horizontal displacement, the angle between it and the central axis of the laser beam can be expressed as

$$\begin{aligned}\psi_{i1} &= \angle ILO = a\cos[(IL^2 + LO^2 - IO^2)/(2 \cdot IL \cdot LO)] \\ &= a\cos\left(\frac{(Z \cdot \cos\phi)^2 + (Z \cdot \sin\phi + x_i)^2 - x_i^2 + Z^2}{2Z \cdot \sqrt{(Z \cdot \cos\phi)^2 + (Z \cdot \sin\phi + x_i)^2 + y_i^2}}\right)\end{aligned} \tag{7.28}$$

After the horizontal displacement, the angle between the imaging sub-region and the central axis of the laser beam is

$$\psi_{i2} = a\cos\left(\frac{(Z \cdot \cos\phi)^2 + (Z \cdot \sin\phi + x_i + dx)^2 - (x_i + dx)^2 + Z^2}{2Z \cdot \sqrt{(Z \cdot \cos\phi)^2 + (Z \cdot \sin\phi + x_i + dx)^2 + (y_i + dy)^2}}\right) \tag{7.29}$$

Assume that dx = 5 m, dy = 5 m, Z = 2000 m, and $\theta = \pi/3$, then the variation of the angle between each imaging sub-region and the central axis of the laser beam with the changes of x_i and y_i can be acquired, as displayed in Fig. 7.14.

It can be seen from Fig. 7.14 that when the target is displaced horizontally in the laser beam, the angle between each imaging sub-region and the central axis of the laser beam changes non-linearly. That is to say, the value of the energy distribution function of the laser beam in different imaging sub-regions present non-linear changes, affecting the echoes of each imaging sub-region and the joint imaging result.

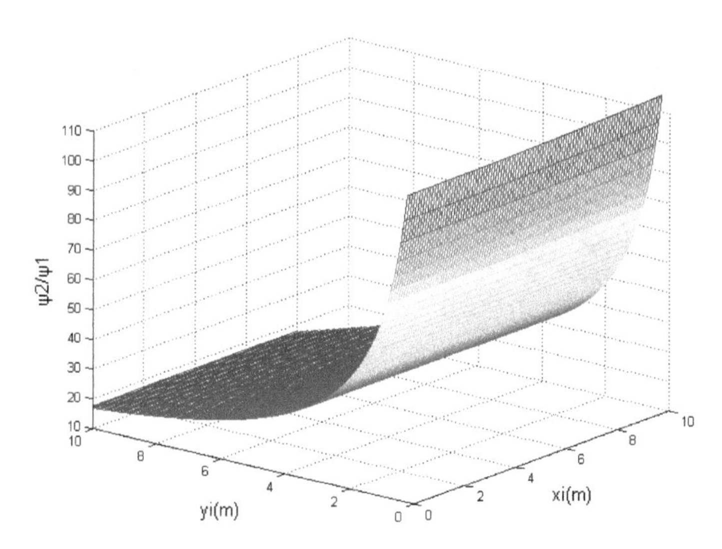

Fig. 7.14 The variation of the angle between each imaging sub-region and the central axis of the laser beam in the horizontal displacement of the target

For two imaging sub-regions I and J, the time delay difference of echoes changes with the horizontal displacement of the target so that affects the joint imaging result of the whole target. Suppose that the distances from the imaging sub-region I to the x axis and y axis that originates from point O are x_i and y_i, separately; likewise, the imaging sub-region J presents distances of x_j and y_j to the axes, respectively. Then the variation quantity of the difference in the imaging range before and after the horizontal displacement is

$$\begin{aligned} dz = & \sqrt{(Z \cdot \cos\theta)^2 + (Z \cdot \sin\theta + x_i)^2 + y_i^2} \\ & + \sqrt{(Z \cdot \cos\theta)^2 + (Z \cdot \sin\theta + x_j + dx)^2 + (y_j + dy)^2} \\ & - \sqrt{(Z \cdot \cos\theta)^2 + (Z \cdot \sin\theta + x_j)^2 + y_j^2} \\ & - \sqrt{(Z \cdot \cos\theta)^2 + (Z \cdot \sin\theta + x_i + dx)^2 + (y_i + dy)^2} \end{aligned} \tag{7.30}$$

Assume that dx = 5 m, dy = 5 m, Z = 2000 m, $\theta = \pi/3$, $x_i = 1$ m, and $y_i = 1$ m, then the variation of dz with the change of x_j and y_j is obtained, as illustrated in Fig. 7.15.

According to Fig. 7.15, $|\mathrm{dz}| \leq 0.025$ m, and the variation quantity of time delay difference of echoes is smaller or equal to 0.17 ns, which is relatively small. This indicates that the waveform variation of echo images is not mainly induced by the change of time delay difference of imaging sub-regions.

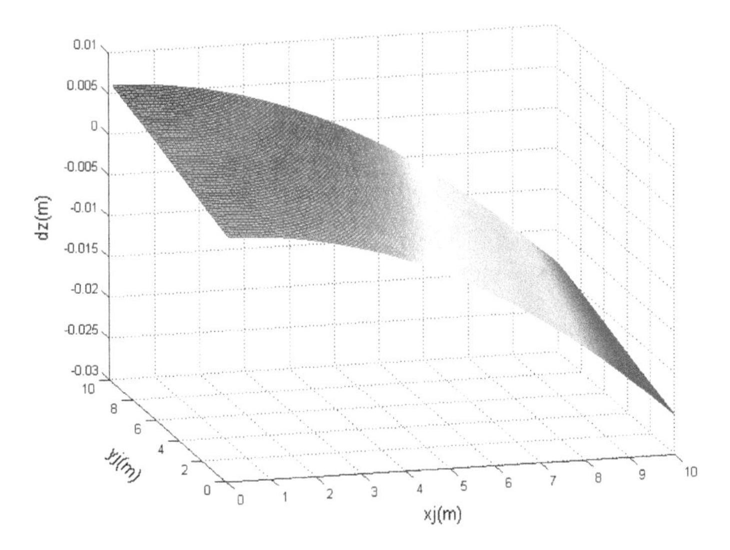

Fig. 7.15 The variation of imaging range difference at the time of horizontal displacement of the target

It can be obtained from formulas (7.20) and (7.24) that when the aerial target is displaced horizontally in the laser detection beam, the imaging result based on the target echoes is

$$
\begin{aligned}
p(t) = \sum_i \{a_i \exp[-k(2\sqrt{(Z \cdot \cos\theta)^2 + (Z \cdot \sin\theta + x_i + dx)^2 + (y_i + dy)^2}/c - t)^2/\tau^2] \\
\cdot g\left\{a\cos\left(\frac{(Z \cdot \cos\theta)^2 + (Z \cdot \sin\theta + x_i + dx)^2 - (x_i + dx)^2 + Z^2}{2Z \cdot \sqrt{(Z \cdot \cos\theta)^2 + (Z \cdot \sin\theta + x_i + dx)^2 + (y_i + dy)^2}}\right)\right\} \cdot \cos(\delta_i) \\
\cdot h/(Z \cdot \cos\theta)^2 + (Z \cdot \sin\theta + x_i + dx)^2 + (y_i + dy)^2\}
\end{aligned}
\tag{7.31}
$$

After acquiring the echoes, the displacement parameters of the aerial target can be obtained by solving formula (7.31).

7.3.4 *The Model of Extracting the Attitude Features of Aerial Targets*

When an aerial target flies in the sky, its echo image is supposed to change accordingly with the variation of the relative attitude between the target and the laser detection system. Considering the relative relationship between targets and laser beams, here, the author mainly discusses the changes of the pitch angle and the flight direction and equivalently analyzes the side-rolling influence. When the target attitude is changed, the echo image changes not only owing to the attitude variation of imaging region but also the mutual occlusion of imaging sub-regions. Here, the author merely discusses the image changes in the slight variation of the target attitude. Suppose that the imaging sub-regions are not occluded by each other, merely the imaging changes of the imaging sub-regions themselves are studied.

When the relative position between the central points of the target and the laser beam is fixed, the corresponding flight direction and attitude of the target change with the rotation of the target around the central point on the horizontal level. For ease of analysis, it is assumed that the rotation center of the target is at the center of the beam. In other conditions when the target rotates around other centers, they can be analyzed based on the previous horizontal displacement with the combination of center rotation. The imaging of the rotating target is shown in Fig. 7.16.

In Fig. 7.16, the rotation angle of the target around O point is υ. The distances from the imaging sub-region I to the x and y axes originating from the point O are x_i

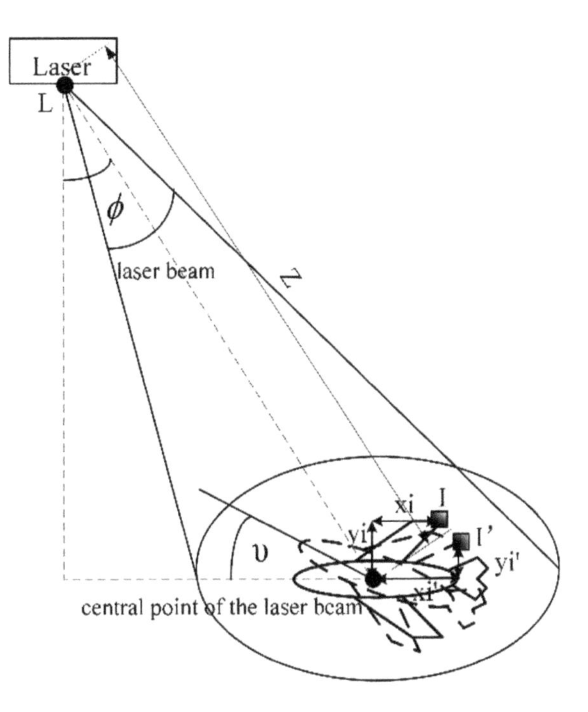

Fig. 7.16 The imaging of the rotating aerial target

and y_i before the rotation, respectively. While after the rotation, the distances change to x_i' and y_i', separately. According to the geometrical relationship, the angles between the sub-region and the central axis of the laser beam before and after the target rotation are equal, that is, $\psi_{i1} = \psi_{i2}$. The imaging range of the sub-region I before the rotation can be expressed by formula (7.26), while that of the sub-region I after the rotation is

$$\begin{aligned} Z_{i2} &= \sqrt{(Z \cdot \cos\phi)^2 + (Z \cdot \sin\phi + x_i')^2 + y_i'^2} \\ &= \sqrt{\begin{array}{l}(Z \cdot \cos\phi)^2 + (Z \cdot \sin\phi + (x_i^2 + y_i^2)^{1/2} \cdot \sin(a\tan(yi/xi) - \upsilon))^2 \\ + (x_i^2 + y_i^2) \cdot \cos^2(a\tan(y_i/x_i) - \upsilon)\end{array}} \end{aligned} \tag{7.32}$$

Suppose that the imaging sub-region I presents distances of x_i and y_i to the x and y axes that originate from the point O, while the distances from the imaging sub-region J to these axes are x_j and y_j before the rotation, respectively. Then the variation of the imaging range before and after the rotation is

$$
\begin{aligned}
dZ = & \sqrt{(Z\cdot\cos\phi)^2 + (Z\cdot\sin\phi + x_j)^2 + y_j^2} \\
& + \sqrt{\begin{aligned}&(Z\cdot\cos\phi)^2 + (Z\cdot\sin\phi + (x_i^2+y_i^2)^{1/2}\cdot\sin(a\tan(y_i/x_i) - \upsilon))^2 \\ &+ (x_i^2+y_i^2)\cdot\cos^2(a\tan(y_i/x_i) - \upsilon)\end{aligned}} \\
& - \sqrt{(Z\cdot\cos\phi)^2 + (Z\cdot\sin\phi + x_i)^2 + y_i^2} \\
& - \sqrt{\begin{aligned}&(Z\cdot\cos\phi)^2 + (Z\cdot\sin\phi + (x_j^2+y_j^2)^{1/2}\cdot\sin(a\ \tan(y_j/x_j) - \upsilon))^2 \\ &+ (x_j^2+y_j^2)\cdot\cos^2(a\tan(y_j/x_j) - \upsilon)\end{aligned}}
\end{aligned} \tag{7.33}
$$

Assume that $Z = 2000\,\text{m}$, $\phi=\pi/3$, and $\upsilon = \pi/4$, then the ratio of the imaging ranges of the imaging sub-region before and after the rotation of the target with the change of the x_i and y_i can be obtained, that is, $\xi = Z_{i2}/Z_{i1}$, as shown in Fig. 7.17.

According to Fig. 7.17, $0.99 < \xi < 1.005$, and the ξ which deviates most significantly from 1 is 0.0992. Let υ change from 0 to π, then the curve of maximum deviation values of ξ is drawn, as shown in Fig. 7.18.

It can be seen from Fig. 7.18 that when the target rotates around the central point, the ratio of the imaging ranges of each sub-region in two times of imaging is approximately equal to 1. In other words, the intensity variation of echo images caused by the range change can be neglected.

Suppose that the distances from the imaging sub-region I to the x and the y axes that start from the point O are x_i and y_i before the rotation of the target, and those from the imaging sub-region J to the x and the y axes are x_j and y_j, respectively.

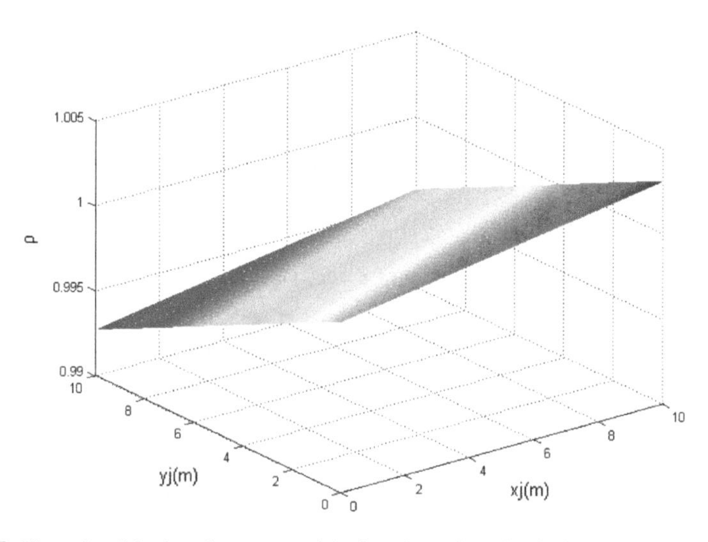

Fig. 7.17 The ratio of the imaging ranges of the imaging sub-region before and after the rotation of the target

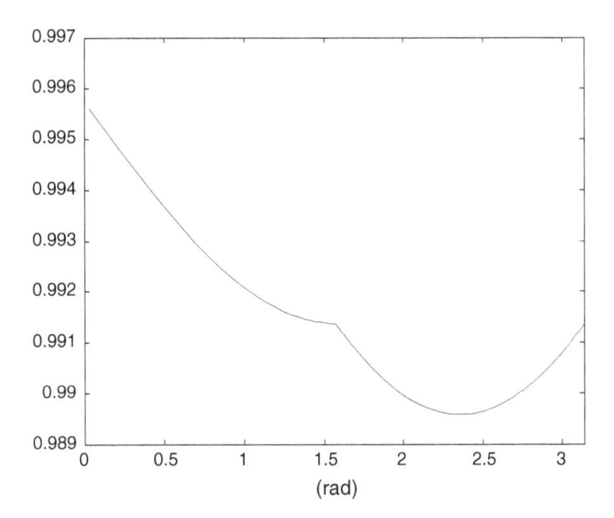

Fig. 7.18 The curve of the maximum deviation values of ξ

After the rotation of the target, the time delay of echoes changes accordingly. Under such condition, the variation quantity of the imaging range difference is

$$
\begin{aligned}
dZ = & \sqrt{(Z \cdot \cos\phi)^2 + (Z \cdot \sin\phi + x_j)^2 + y_j^2} \\
& + \sqrt{\begin{aligned}&(Z \cdot \cos\phi)^2 + (Z \cdot \sin\phi + (x_i^2 + y_i^2)^{1/2} \cdot \sin(a\tan(y_i/x_i) - \upsilon))^2 \\ &+ (x_i^2 + y_i^2) \cdot \cos^2(a\tan(y_j/x_j) - \upsilon)\end{aligned}} \\
& - \sqrt{(Z \cdot \cos\phi)^2 + (Z \cdot \sin\phi + x_i)^2 + y_i^2} \\
& - \sqrt{\begin{aligned}&(Z \cdot \cos\phi)^2 + (Z \cdot \sin\phi + (x_j^2 + y_j^2)^{1/2} \cdot \sin(a\tan(y_j/x_j) - \upsilon))^2 \\ &+ (x_j^2 + y_j^2) \cdot \cos^2(a\tan(y_j/x_j) - \upsilon)\end{aligned}}
\end{aligned}
\tag{7.34}
$$

Suppose that $Z = 2000\,\text{m}$, $\phi = \pi/3$, $x_i = 1\,\text{m}$, and $y_i = 1\,\text{m}$, then the variation quantity of the imaging range difference of the sub-regions I and J with the changes of x_i and y_i can be obtained, as illustrated in Fig. 7.19.

According to Fig. 7.19, before and after the rotation, the imaging range differences of each imaging sub-region change significantly. Thereinto, the maximum imaging range difference reaches 13.83 m, and the corresponding variation of time delay difference of echoes of the two imaging sub-regions is 92 ns. Evidently, due to the significant changes in time delay difference of echoes of all imaging sub-regions, the distribution positions of echo images in the time axis vary correspondingly, causing the waveform changes of the entire echo image of the target. According to the previous analysis, this is the main factor causing the waveform change of echo images of the target when the target rotates.

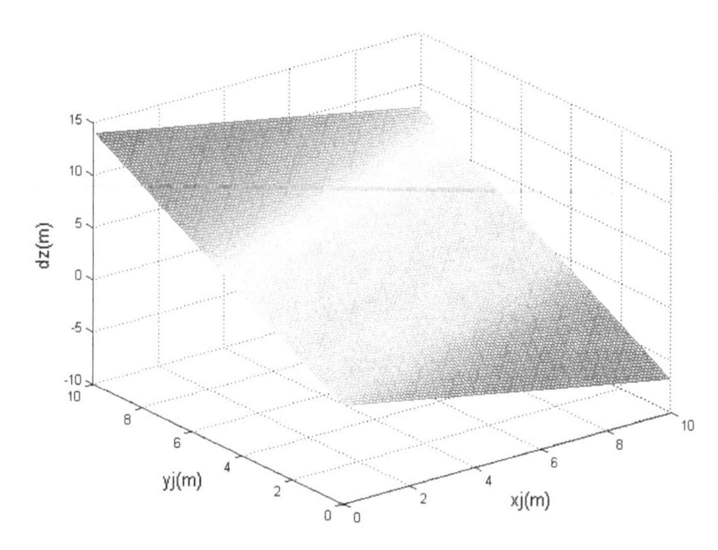

Fig. 7.19 The variation quantity of the imaging range difference of the sub-regions before and after the rotation of the target

Under the circumstance that the relative position between the central point of the target and the center of the laser beam is constant, when the target rotates around the central point in the pitching direction, the corresponding pitch attitude of the target is supposed to change. Here, it is assumed that the rotation center of the target is located at the beam center. For other conditions where the target rotates around other centers, they can be discussed based on the previous horizontal displacement together with pitch rotation. While the pitch attitude of the target changes, the imaging condition is illustrated in Fig. 7.20.

In Fig. 7.20, υ is the variation quantity of the attitude angle of the target in the pitching direction. Suppose that the coordinate of the imaging sub-region I is (x_i, y_i, z_i) with O as the origin, then the coordinate $(x_i^{'}, y_i^{'}, z_i^{'})$ after the rotation of the target in the pitching direction can be represented as

$$\begin{cases} x_i{}' = x_i \cos \upsilon + z_i \sin \upsilon \\ y_i{}' = y_i \\ z_i{}' = z_i \cos \phi - x_i \sin \phi \end{cases} \tag{7.35}$$

The angle between the imaging sub-region I and the central axis of the laserbeam before the rotation can be expressed as

$$\begin{aligned} \psi_{i1} &= \angle ILO = a\cos[(IL^2 + LO^2 - IO^2)/(2 \cdot IL \cdot LO)] \\ &= a\cos\left[\frac{(Z \cdot \sin \phi + x_i)^2 + (Z \cdot \cos \phi - z_i)^2 + Z^2 - x_i^2 - z_i^2}{2Z \cdot \sqrt{(Z \cdot \sin \phi + x_i)^2 + (Z \cdot \cos \phi - z_i)^2}}\right] \end{aligned} \tag{7.36}$$

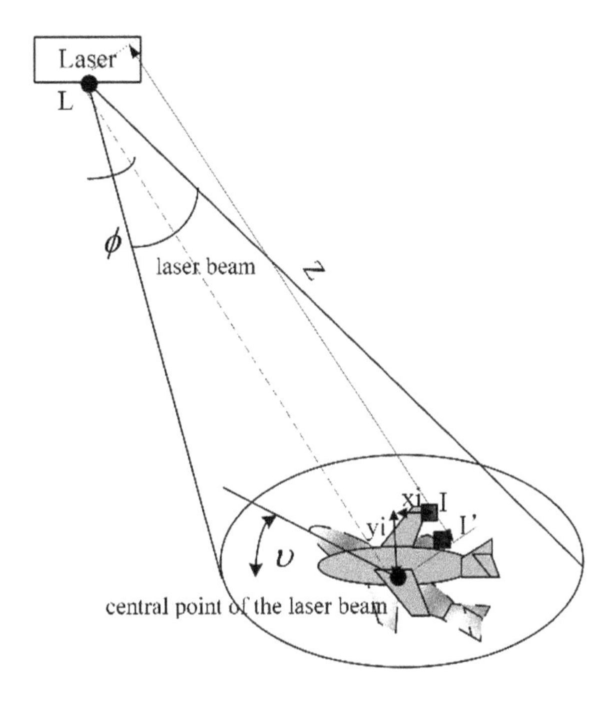

Fig. 7.20 The imaging relation with the variation of the pitch attitude of the target

Then, after the rotation, the angle changes as

$$
\begin{aligned}
\psi_{i2} &= a\cos\left[\frac{(Z\cdot\sin\phi + x_i')^2 + (Z\cdot\cos\phi - z_i')^2 + Z^2 - x_i'^2 - z_i'^2}{2Z\cdot\sqrt{(Z\cdot\sin\phi + x_i')^2 + (Z\cdot\cos\phi - z_i')^2}}\right] \\
&= a\cos\left[\frac{\begin{array}{c}(Z\cdot\sin\phi + x_i\cos\upsilon + z_i\sin\upsilon)^2 + (Z\cdot\cos\phi - z_i\cos\phi + x_i\sin\phi)^2 \\ + Z^2 - (x_i\cos\upsilon + z_i\sin\upsilon)^2 - (z_i\cos\phi - x_i\sin\phi)^2\end{array}}{2Z\cdot\sqrt{\begin{array}{c}(Z\cdot\sin\phi + x_i\cos\upsilon + z_i\sin\upsilon)^2 + \\ (Z\cdot\cos\phi - z_i\cos\phi + x_i\sin\phi)^2\end{array}}}\right]
\end{aligned}
\tag{7.37}
$$

According to formulas (7.36) and (7.37), the angle between the imaging sub-region and the central axis of the laser beam changes with the rotation of the target, and its variation quantity is related to those of the sub-regional coordinates and the attitude angle. Furthermore, the imaging result is influenced by the change of the value of the energy distribution function for the echoes of the sub-regions.

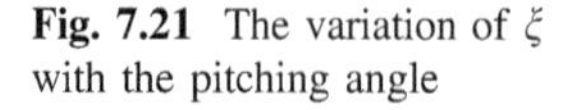

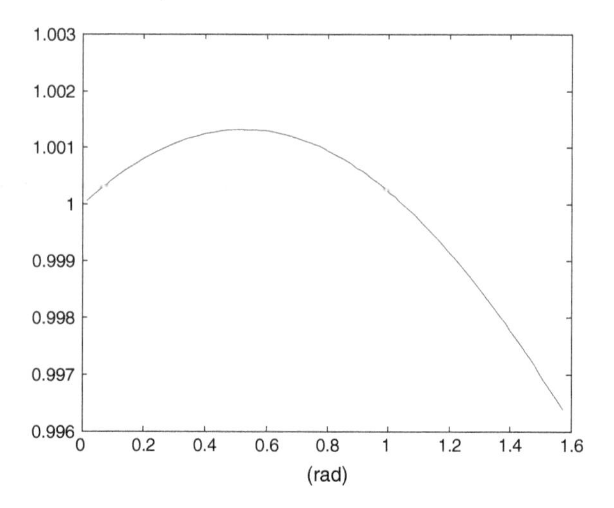

Fig. 7.21 The variation of ξ with the pitching angle

Before and after the changes in the pitching angle of the target, the ratio of the imaging ranges of the imaging sub-region *I* is

$$\xi = \frac{Z_{i2}}{Z_{i1}} = \frac{\sqrt{(Z \sin \phi + x_i')^2 + (Z \cos \phi - z_i')^2 + y_i'^2}}{\sqrt{(Z \sin \phi + x_i)^2 + (Z \cos \phi - z_i)^2 + y_i^2}} \tag{7.38}$$

Suppose that $x_i = 10\,\text{m}$, $y_i = 10\,\text{m}$, $z_i = 0\,\text{m}$, $Z = 1000\,\text{m}$, and $\phi = \pi/3$, then the variation of ξ with the pitching angle is obtained, as illustrated in Fig. 7.21.

It can be seen from Fig. 7.21 that $|\xi| \approx 1$, that is, the intensity change of echo images induced by the range variation can be ignored before and after the change of the pitching angle of the target.

The angle between the incident light and the surface normal line of the target tends to change when the pitch angle of the target varies. According to the formula (7.35), the imaging echoes are supposed to change as well. It can be known from the geometrical relationship that if the angle between the normal line on the surface of the imaging sub-region and the laser beam is δ_1 before changing the target attitude, then after the change of target attitude, the angle can be approximately expressed as

$$\delta_2 = \delta_1 + \upsilon \tag{7.39}$$

Assume that the distances from the imaging sub-regions I and J to the x and y axes that originate from the point O are x_i and y_i, as well as x_j and y_j before the rotation of the target, respectively. Then the time delay difference of echoes changes after the rotation of the target. Under such circumstance, the imaging range difference changes by

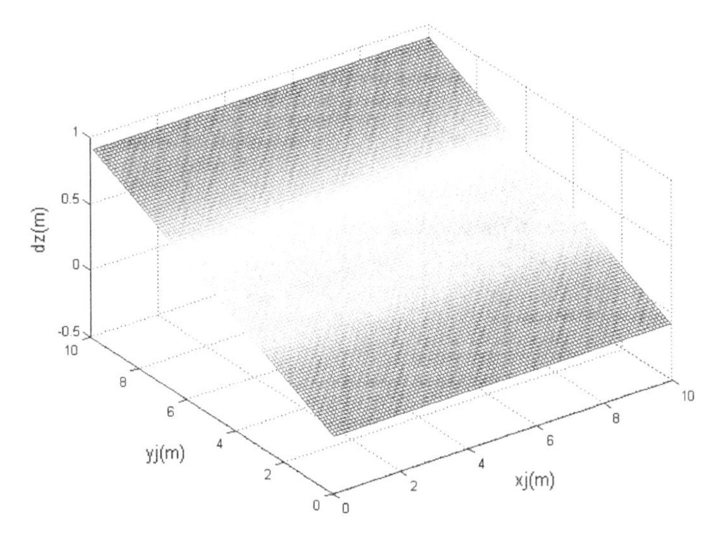

Fig. 7.22 The variation values of imaging range difference of imaging sub-regions before and after the change of the pitching angle

$$\begin{aligned} dz = & \sqrt{(Z\sin\phi + x_j')^2 + (Z\cos\phi - z_j')^2 + y_j'^2} + \sqrt{(Z\sin\phi + x_i)^2 + (Z\cos\phi - z_i)^2 + y_i^2} \\ & - \sqrt{(Z\sin\phi + x_i')^2 + (Z\cos\phi - z_i^2)^2 + y_i'^2} - \sqrt{(Z\sin\phi + x_j)^2 + (Z\cos\phi - z_j)^2 + y_j^2} \end{aligned} \quad (7.40)$$

Suppose that $Z = 2000\,\text{m}$, $\phi = \pi/3$, $\upsilon = \pi/4$, $x_i = 1\,\text{m}$, $y_i = 1\,\text{m}$, $z_i = 0\,\text{m}$, and $z_j = 0\,\text{m}$, then the variation values of the imaging range difference of the sub-regions I and J with the change of x_j and y_j are calculated, as shown in Fig. 7.22.

According to Fig. 7.22, the imaging sub-regions present significant changes in the imaging range difference before and after the rotation of the target. Meanwhile, the maximum imaging range difference reaches 0.89 m, and the corresponding variation value of time delay difference of echoes reflected from the two imaging sub-regions is 5.9 ns. Obviously, when the time delay difference of each sub-region changes, the distribution positions of echo images in the time axis vary as well, leading to the change in the entire waveform images of targets.

7.4 The Extraction of Target Features Based on the Reference Echo Pulses

In the extraction of target features based on echo waveforms, the echo waveform is expected to be modulated only by targets so that the target features can be extracted accurately from the echo waveforms. However, in fact, the echo waveforms are closely related to the features of receiving channels in the laser detection. In general, the influence of receiving channels on echo waveforms is stable and can be eliminated through normalization while processing the laser echoes. Therefore, it is necessary to construct a kind of reference echo pulse, which is the most typical and normalized laser echo pulse. This echo pulse is mainly related to the receiving channels and reflects the effect of system response of the receiving circuit except the target and the environment on the distribution of laser pulses in the time and space [6]. The reference echo pulse can be acquired through experiments after developing the system. Meanwhile, it also can be used to reversely infer the features of system response so as to eliminate the influence of the receiving channels in the extraction of the target features.

7.4.1 Constructing Principle of the Reference Echo Pulse

The system for acquiring laser echo waveforms is a linear system. Suppose that the waveform of laser echoes is $y(t)$. For general targets, $y(t)$ can be approximately expressed as

$$y(t) = x(t) * h_1(t) * h_2(t) * h_3(t) * h_4(t) + \xi(t) \tag{7.41}$$

where, $*$ represents the convolution, $x(t)$ is the pulses of transmitted laser beams, $h_1(t)$ is the shock response induced by atmospheric disturbance from the laser to the target, and $h_2(t)$ denotes the shock response induced by the target modulation. While $h_3(t)$ indicates the shock response induced by atmospheric disturbance from the target to the laser. $h_4(t)$ and $\xi(t)$ signify the response of the system for receiving and processing laser echoes and additive noise, respectively.

For relatively short ranges, the atmospheric disturbance can be neglected.

$$h_1(t) \approx \delta(t) \tag{7.42}$$

$$h_3(t) \approx \delta(t) \tag{7.43}$$

Then formula (7.41) can be simplified as

$$y(t) = x(t) * h_2(t) * h_4(t) + \xi(t) = (x(t) * h_4(t)) * h_2(t) + \xi(t) \tag{7.44}$$

When lasers vertically irradiate planar targets with diffuse reflection surfaces, the targets do not additionally modulate the laser echoes. Under such condition, the distribution of the reflected normalized laser pulses can be considered as identical to that of the incident laser pulses, then

$$h_2(t) \approx \delta(t) \tag{7.45}$$

Here, the laser pulse echo $y(t)$ is the echo pulse of targets with datum plane and represented as $y_2(t)$, that is

$$y_2(t) = x(t) * h_4(t) + \xi(t) \tag{7.46}$$

A reference echo pulse $y_0(t)$ can be constructed by designing $h_4(t)$ using the maximum distortionless principle. That is

$$y_0(t) = \widehat{y}_2(t) = x(t) * h_4(t) \tag{7.47}$$

Thus, it is revealed that $y_0(t)$ is merely related to the response of the system for receiving and processing the laser echoes.

For a complex non-planar target, the following formula is obtained by substituting formula (7.47) into formula (7.44):

$$y(t) = y_0(t) * h_2(t) + \xi(t) \tag{7.48}$$

Therefore, as long as obtaining $y_0(t)$, the shock response $h_2(t)$ induced by the target modulation can be estimated according to the laser echo waveform $y(t)$ of the actual target, which can inversely deduce the comprehensive features of the target.

Only the waveform of non-saturated laser echoes obtained by the laser vertically irradiating planar targets can be regarded as the reference echo pulse, which plays a very important role in the extraction of target features. The pulse waveforms of laser echoes, to certain extent, can be regarded as the translation, amplification, compression and combination of reference waveforms. Therefore, the reference echo pulses have to be representative. Because the reference echo pulse is only related to the system for receiving and processing laser echoes, it can be acquired by experimental measurement. In the measurement, the lasers need to vertically irradiate the selected planar target to make the are irradiated by the laser beam be completely located in the receiving field of view by considering the range between the target and laser components.

7.4.2 Steps for Constructing the Reference Echo Pulses

A series of waveforms of reference echo pulses is collected in the laboratory and then processed using the following algorithms: (1) calculating the root mean square of the noise; (2) filtering the waveforms; (3) eliminating the direct current component; (4) normalizing the waveforms; and (5) obtaining the reference echo pulses by counting and averaging the feature parameters of the series of waveforms.

By calculating the distribution of rise time and fall time of echo pulses, it can be known that, for the non-saturated laser echoes, their normalized waveforms show similar features, which is the basis of proposing the reference echo pulse. While for saturated laser echoes, the features of their normalized waveform are not similar.

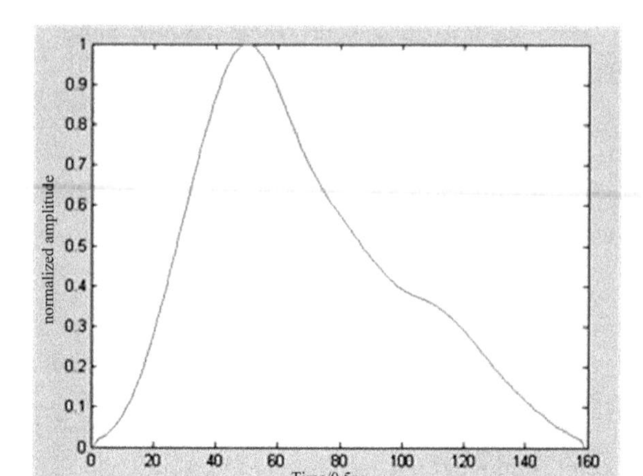

Fig. 7.23 The reference echo pulse of a typical target

Therefore, the saturated waveforms need to be avoided in practical applications. The obtained waveform of reference echo pulse is presented in Fig. 7.23, and it is the basis of analyzing other complex waveforms.

With different energies of emission lasers, for the same target, the similarity of the waveforms of laser echo pulses is the basis of acquiring the reference echo pulse. In this book, the sampled 16 groups of laser echo pulses cover the linear region of system response. By analyzing the data, it is found that the rising edge and falling edge of each group of echo pulses have a more stable relation compared with the energy of emission lasers, so that the waveforms of each group of echo pulses are similar. This is consistent with the case that the laser irradiates the ground surface with different backward scattering rates, and the back scattering waveforms are normalized through the reference echo pulse.

7.4.3 Features of Reference Echo Pulses

Conventional targets are diffuse scatters without a specific reflection direction. Therefore, the distribution features of the surface structures of targets in the spots with the variation of x and y can be represented by function $\xi(x, y)$ [7].

$$\xi(x, y) = \xi_0 + x \tan s_x + y \tan s_y + \Delta\xi(x, y) \tag{7.49}$$

where, ξ_0 is a constant. s_x and s_y denote the average slopes of targets within the laser spots in x and y directions, respectively. While $\Delta\xi(x, y)$ represents the random roughness of the target surface in the laser spots.

Suppose that a laser beam is transmitted to conduct one-dimensional linear scanning in the x direction, and the directional angle of the beam and the half divergence angle of the beam are ϕ and α, respectively. The numerical calculation shows that the root mean square width of the laser echo can be expressed as follows when the incident angle is within a certain range.

$$E\left(\sigma_p^2\right) = \left(\sigma_f^2 + \sigma_{OPT}^2\right) + \frac{4Var(\Delta\xi)}{c^2\cos^2\phi} + \frac{4z^2\tan^2\varpi\cos^2\phi}{c^2} \tag{7.50}$$

where, σ_f and σ_{OPT} are the root mean square widths of system response and laser pulses, separately.

According to the above formula (7.50), if the pulse width is at nanosecond scale and the divergence angle of the beam is at milli-radian scale, then the pulse width is mainly determined by the surface roughness of targets in the case of normal and small-angle incidence. Therefore, the surface structures (especially surface roughness) of different types of targets can be determined by measuring the broadening of laser echoes of various targets [8] to establish correlation between pulse broadening and target features. In addition, by researching on the waveform broadening of a same target with different incident angles, the corresponding relationship between the incident angle and the waveform broadening can be established.

For the reference waveform, $Var(\Delta\xi) = 0$ and $\phi = 0$, then formula (7.50) can be simplified as

$$E\left(\sigma_p^2\right) = \left(\sigma_f^2 + \sigma_{OPT}^2\right) \tag{7.51}$$

Therefore, the laser emission and the pulse width of system response are comprehensively considered in the acquisition of reference echo pulses. For complex waveforms of laser echoes, more attention is expected to be paid on the analysis of the influence of target fluctuation and incident angle using certain algorithms. From the theoretical perspective, the reference echo pulse belongs to Gaussian pulse, while the leading edge of reference echo pulses is close to Gaussian distribution but the lagging edge shows tailing phenomenon in this book. So the obtained reference echo pulse is not ideal Gaussian pulse, which shows the differences between the results obtained in ideal conditions and actual experiments. In the theoretical analysis, it is generally assumed that the echo pulses are the superposition of a series of Gaussian pulses, which is merely an approximate result compared with that obtained from actual measurements. Therefore, to obtain precise target features, it is necessary to explore the reference echo pulse of the laser detection system and then correct it on this basis.

In theory, the reference echo pulse is Gaussian pulse, however, in fact, the leading edge of the reference echo pulse is approximate to the leading edge of Gaussian pulse, but the lagging edge still has broadening and trailing compared with the lagging edge of Gaussian pulse. The laser echo pulses are generally considered as the superposition of a series of Gaussian pulses. Hence, it needs to correct the algorithms used to calculate the pulses which are decomposed into Gaussian pulses.

The reference echo pulse plays an important role in extracting target features. The reference echo pulse can be used to decompose complex echo waveforms of targets into the sum of the reference echo pulses and the waveforms of signals. After deducting the waveforms of the reference echo pulses, the target features including height distribution and hierarchical structure can be acquired by processing the remaining signal waveforms.

References

1. Hu Y, Shu R, Xue Y (2002) Experiment study on laser echo identities of terrain object. Infrared Laser Eng 31(2):105–108
2. Li L, Hu Y, Zhao N et al (2010) Experiment on the stretching characteristics of pulse width of laser remote sensing echo. Infrared Laser Eng 39(2):246–250
3. Hu Y (2009) Research on the retrieval method of target imaging based the laser remote sensing (research report of National Natural Science Foundation of China). College of Electronical Engineering, Hefei
4. Krawczyk R, Goretta O, Kassighian A (1993) Temporal pulse spreading of a return lidar signal. Appl Opt 32(33):6784–6788
5. Churoux P, Benson C (2000) Model of burst imagine lidar through the atmosphere. SPIE 4035:324–331
6. Zhang L, Chen Y, Zhang H et al (2004) Research on basic wave of laser imaging for earth observation. Infrared Laser Eng 33(3):260–263
7. Gardner CS (1982) Target signatures of laser altimeters: an analysis. Appl Opt 21(3):448–453
8. Abshire JB, Sun X, Afzal RS (2000) Mars orbitor laser altimeters: receiver model and performance analysis. Appl Opt 39(15):2449–2460

Chapter 8
Target Detection and Location Using a Laser Imaging System

Target detection and location based on a laser imaging system means discovering and locating targets by utilizing processing results of laser images. To find a target, the target needs to show obvious differences with the background in images. Laser images include gray, range, level and waveform images, so they are able to provide sufficient data for target detection. The modes of target location include direct location and indirect location, and the latter is not applicable to target detection. However, a laser imaging detection system can locate targets directly and accurately. The location of a target can directly be tracked by the detection system as long as range and direction of the target, as well as the coordinates and attitude of the system are measured in advance. This chapter mainly discusses the above aspects.

8.1 Target Detection Based on Range Images

Range images obtained through the detection based on a laser imaging system are distinct from general 2D infrared and visible images. One distinctive characteristic of gray images obtained through the detection based on a laser imaging system is that they contain 3D data of targets. Reflecting not only the position but also the elevation or distance of target, these data therefore provide effective information for target detection.

8.1.1 Analyzing Characteristics of Range Images

A range image shows the spatial depth, from which 3D information of targets can be directly obtained [1]. Figure 8.1 shows a true range image obtained by using a target detection system using a laser imaging technology and its corresponding visible image. In the range image, different gray values represent various range values.

Y. Hu, *Theory and Technology of Laser Imaging Based Target Detection*,
DOI 10.1007/978-981-10-3497-8_8

(a) the range image

(b) the corresponding visible image

Fig. 8.1 A range image obtained by using the detection system and its corresponding visible image

The main object in Fig. 8.1 is the car (car No. 1) in the center, and the background is the building. Besides, there are other objects such as motorcycles, a bike, another car (car No. 2) behind the motorcycles, shrubs, and trees.

There are great differences in the statistical characteristics of range images and images obtained through traditional passive imaging methods. It is commonly believed that there are two types of information: targets and backgrounds. For a regular optical gray image, its statistical histogram generally shows two peaks, which correspond to the gray values of targets and backgrounds, respectively. However, the statistical characteristics of range images show clear distinction. That is, points in a same target present little differences concerning the distance, while the average distances of points in different targets to the observation point is likely to exhibit great differences. In this case, the statistical histogram has multiple peaks, each of which is corresponding to one or more targets in the same distance [2]. Figure 8.2 demonstrates the histogram of the above range image, of which the abscissa is range with meter as the unit, instead of gray in general images. Besides, its ordinate is the number of pixels in the corresponding range. The statistical histogram of the range image presents distinct multiple peaks and targets in different ranges correspond to various peaks, as shown in Fig. 8.2.

Target detection is performed based on image analysis, and its flow is shown in Fig. 8.3. The most important steps of target detections are range separation and target fitting. Range separation means dividing a range image into several layers according to the multi-peak structure of the statistical histogram of the image, so that targets are separated from backgrounds. Then, separated image of each layer is morphologically filtered to reduce influences of noise and other small targets. Target fitting refers to obtaining the minimum bounding rectangle (MBR) in each independent region through the MBR fitting method and calculating the fitting degree of the rectangle.

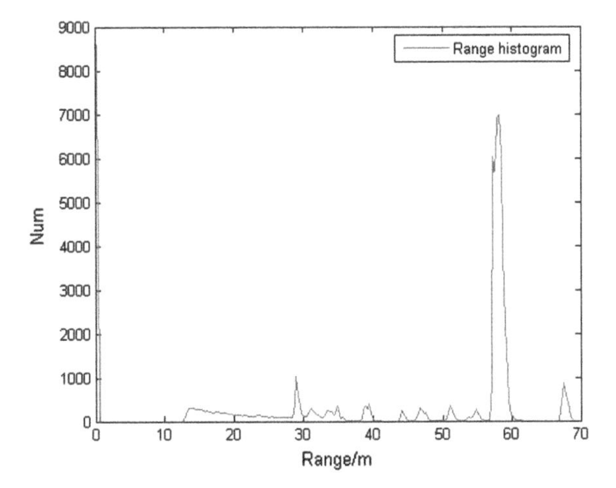

Fig. 8.2 The histogram of the range image

Combining the range information of a range image, information of the MBR is obtained such as natural size, range and centroid height. By matching the information with the natural size and shape of the target, the possible target is likely to be screened out from multiple targets. By doing so, the target is detected.

8.1.2 Range Separation and Layering of Images

The purpose of range separation is to divide targets at different distances to various layers. On the basis of peak evaluation function, an improved method for parting the function is proposed. In addition, by processing the histogram of range images, information of different objects is obtained, based on which the detected target at a certain distance is expected to be found.

Suppose that P_i denotes the frequency of a pixel whose distance value is *i*. By comparing P_i with P_{i+1} and P_{i-1}, if $P_i < P_{i-1}$ and $P_i < P_{i+1}$, then *i* is a valley point; if $P_i > P_{i-1}$ and $P_i > P_{i+1}$, then *i* is a peak point. The determination of peak and valley points provides useful parameters for target detections.

However, in actual operation, the simple method above is obviously rough because the peak points selected are possibly not rational as false peak points are likely to be included. In order to evaluate the rationality of peak points in the cases of multiple peaks and noise interference in histograms, an evaluation function for peaks is defined to guide peak extraction [3].

Based on the evaluation function for peaks above, the evaluation function is improved to be more applicable to peak extraction of histograms of range images. First, a range threshold is selected according to the size of the target to be extracted, and generally set as 1 m. Within the threshold, the maximum point $P_{\max}$ and the minimum point $P_{\min}$ are found. The threshold is slid towards the right side of the histogram, the

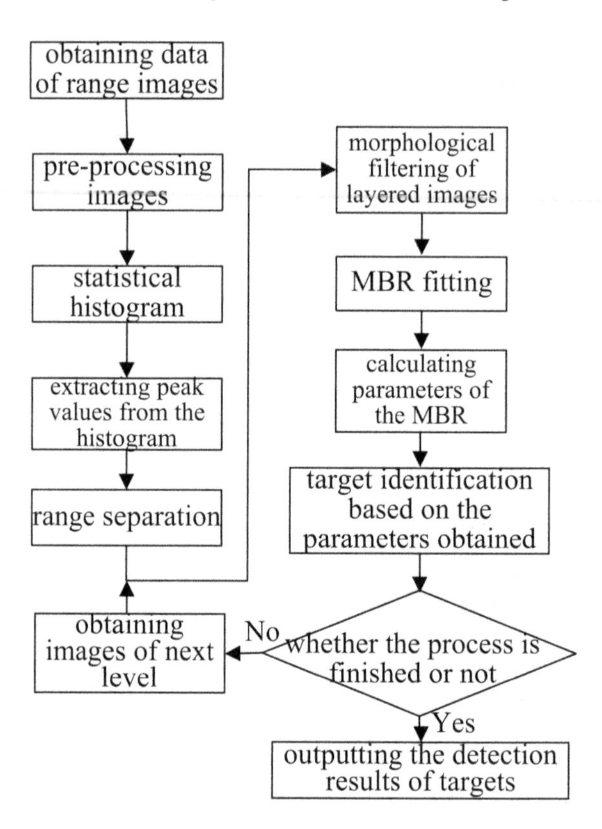

Fig. 8.3 Target detection and classification based on range images

newly covered datum which is less than P_{max} is considered as a peak point. Nevertheless, this method is sensitive to noise, so further processing is needed to offset the influence of the noise. Due to the small variation range of noise, whether P_{max} is a peak point or not can be determined by comparing the ratio of to P_{min}. It is generally believed that the datum is a peak point when P_{max} is 1.2–1.5 times larger than P_{min}. Therefore, $P(i)$, the improved evaluation function for peaks can be expressed as:

$$P(i) = \begin{cases} 1, & P_i = P_{\max} \text{ 且 } P_{\max} > aP_{\min}, \quad a \in (1.2-1.5) \\ 0, & else \end{cases} \tag{8.1}$$

where, $0 < i < L$, and L is the value of the maximum range in an image.

Figure 8.4 shows the detection result of Fig. 8.2 using the improved evaluation function for peaks.

After determining the position of peaks of the histogram, a range threshold is set in the front or back of each peak point to extract targets within the threshold.

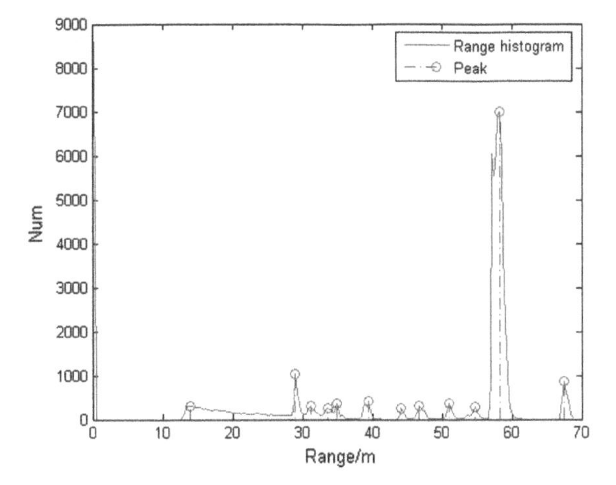

Fig. 8.4 The detection result of peaks of range histogram based on the improved evaluation function for peaks

Figure 8.5 demonstrates the outlines of the cars and the motorcycles at different distances extracted from Fig. 8.1.

As shown in Fig. 8.5, targets can be extracted from the range image through range separation and dividing image into layers. There into, Fig. 8.5a–c demonstrate the car No. 1, the motorcycles and the car No. 2, respectively. It reveals that by employing this method, car No. 1 and the motorcycles which are closer can be separated effectively. Besides, targets that are blocked out like No. 2 car also can be extracted. However, it also can be seen that although the targets are extracted, there are noise and other small targets influencing the identification for the required targets. To eliminate these influences, morphological filtering is performed on the obtained images to obtain target images with smooth edges, as shown in Fig. 8.6. It can be seen that through morphological filtering, influences of noise and other small targets are greatly reduced.

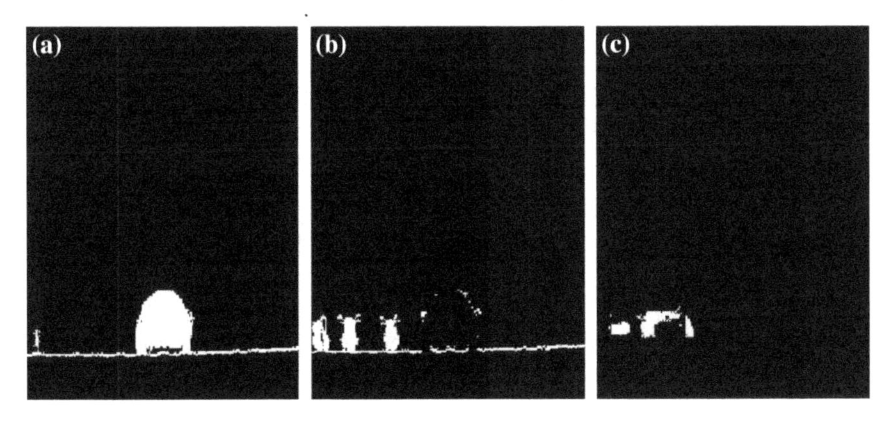

Fig. 8.5 The outlines of targets extracted from the range image

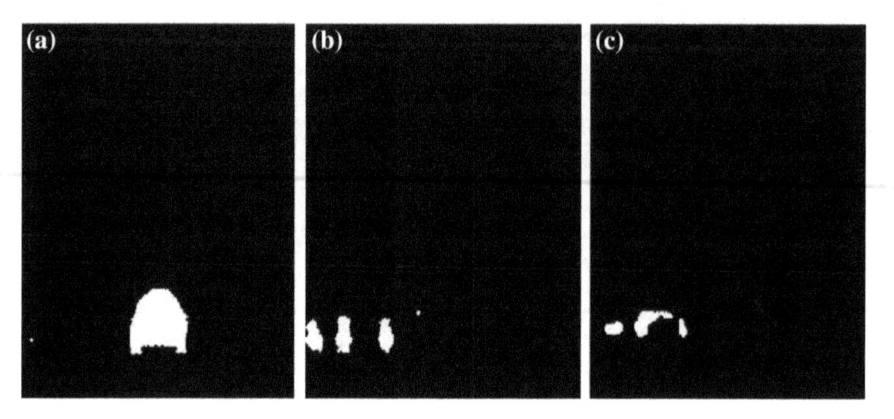

Fig. 8.6 The outlines of targets after morphology filtering

8.1.3 Target Fitting and Detection

After the above processing, range images of laser detection require further processing to find the required targets. The MBR fitting method is used to separate the outlines of targets, and the size of the MBR is the basis for determining size of a target. Additionally, the rectangle fitting degree of targets is used to analyze the shape of targets. Owing to the specific characteristics of range images, natural size, distance, centroid height, etc. of a target are expected to be obtained directly from the range images. For general artificial targets such as a car and an airplane with limited sizes in regular shapes, they can be extracted from range images through this method.

Bounding rectangles of a polygon can be classified into two types: one is MBR which is determined by the maximum and minimum coordinates of vertexes of the polygon; the other is minimum area bounding rectangle (MABR).

There are many methods for implementing MABR algorithm, and the most common method is to rotate an object within 90° by same intervals each time. The parameters of bounding rectangle of the object in the coordinate direction are recorded and the MBR is obtained by calculating the areas of bounding rectangles [4]. This method is easy to be performed with small computation but is less precise. Fortunately, the precision can meet the requirement of this application. In this research, the computing steps based on the MABR algorithm are as follows:

Step 1 by tracing the contours of binary images, a series of closed regions are obtained. Then, areas of bounding rectangles of each closed region are calculated, and the lengths, widths and areas of each bounding rectangle are recorded.

Step 2 After rotating the images anticlockwise by 3°, Step 1 is repeated; after thirty times of rotation, the process turns to Step 3.

Step 3 The areas of rectangles of the closed regions are calculated after each rotation to calculate the MBR, the length, width and fitting degree of which are recorded.

Step 4 The natural length and width of the MBR are calculated according to range information.

Table 8.1 shows the partial data of the range image in Fig. 8.2 calculated through the above steps.

Figure 8.7 shows the fitting results of partial MBRs.

In the above experimental data, serial number 3 represents car No. 1 in Fig. 8.1, and serial numbers 5, 6 and 7 demonstrate the motorcycles and the bike. It can be seen that their measured sizes accord with those of regular cars and motorcycles, which indicates high fitting degree of the rectangles. Besides, serial number 14 stands for shrubs, and it is characterized by low height, high fitting degree of

Table 8.1 The partial fitting results of the MBR

Serial number	Width of the target/m	Height of the target/m	Fitting degree of the rectangle	Range of the target/m	Centroid height of the target/m
3	1.873	2.061	0.739	29.02	1.432
5	0.561	1.153	0.696	31.35	0.612
6	0.608	1.247	0.686	30.78	0.587
7	0.514	1.153	0.717	31.07	0.627
14	0.629	0.873	0.777	35.48	0.352
26	0.711	1.777	0.558	39.50	1.824
28	0.859	0.661	0.792	44.95	0.453
29	1.884	1.146	0.486	44.87	0.612
30	0.397	0.925	0.714	44.73	0.408
38	0.727	2.40	0.426	51.06	2.023
50	17.73	16.02	0.778	58.31	8.244

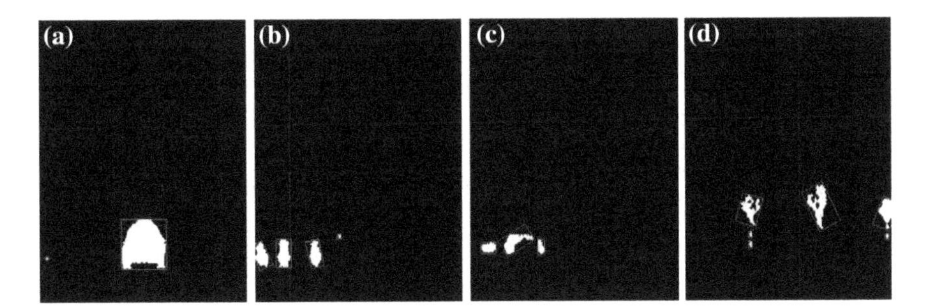

Fig. 8.7 The fitting results of partial MBRs

rectangles and regular shapes. In addition, serial numbers 26 and 38 represent roadside trees, which are high and narrow with a low fitting degree of rectangles. Serial numbers 28, 29 and 30 denote car No. 2. Owing to the outline of the car is separated by the motorcycles, etc., it is regarded as three separated targets in the classification. Additionally, serial number 50 shows the background building, which has a large size with a high fitting degree of the rectangle.

In the above analysis, car No. 2 is considered as three independent targets, while their distances to the detector differ slightly. Thus, the three targets can be regarded as one to process. According to the above data, various kinds of targets greatly differ in size, length-to-width ratio, rectangle fitting degree, etc. Therefore, these parameters can be used as the bases for discriminating the properties of targets to discover different targets and further to quickly find interested targets from mixed backgrounds. Moreover, the distance and centroid height of a target can be the foundation of target discrimination as well.

By processing the range image in Fig. 8.1 using the proposed method, 61 independent regions are extracted. Furthermore, by setting determining conditions of classification, seven areas probably containing targets are detected. The accuracy is 71.4% with two false targets. By using the MATLAB in a PC with CPU of 2.40 GHz and internal storage of 512 MB, data processing in this section consumes 28.420 s through this method. If a more efficient programming language and a better-performed hardware platform are used, the processing time is expected to shorten significantly. In this way, the real-time demands can be met.

8.2 Covered Targets Detection Based on Waveform Images

Covered targets are sheltered by covers or camouflages, which lead to the visible and infrared images too blurred to contain real properties of targets. In this case, covered targets are unlikely to be discovered by traditional imaging techniques. Waveform images can reflect waveform distribution of laser echoes in terms of time and amplitude. Furthermore, waveform distribution of laser echoes has a strong mapping relation to target features, so they can exhibit some characteristics of targets. Combining the above relationship, properties of synthesized echoes are analyzed according to the waveform images of targets. Besides, echo signals of concealed targets are expected to be detected from the hybrid echoes, in this way, real targets are likely to be found from coves or camouflages. This section discusses a method for detecting covered targets based on waveform images by taking targets hidden in forest for instance [5].

8.2.1 Principles of Target Detection

Trees are natural camouflages and covers. In the detection based on a laser imaging system for the targets hidden in forests, branches and leaves of each layer of trees and ground reflect laser echoes according to laser propagation paths. In the meantime, lasers irradiate the targets in the space between trees and generate echoes. The echoes reflected from trees and ground are superposed on those reflected from targets to form hybrid echoes in the detector, as shown in Fig. 8.8.

In the detection based on a laser imaging system for the targets hidden in forests, the beam size is the primary issue to consider. Undersize laser beams are likely to be entirely blocked out by branches and leaves of trees, thus they cannot irradiate the targets through spaces between trees. Conversely, oversize laser beams lead to energy dispersion, so that target echoes are mixed with those of backgrounds and hard to be separated. Therefore, while selecting laser beams, the sizes of laser footprints on ground are supposed to correspond to the sizes of targets, that is, laser spots on the ground are expected to cover the targets. In general, targets hidden in forest are mostly maneuvering targets such as an M1 tank of American army and an armored car of Taiwan, China. The length, width and height of the former are 9.77,

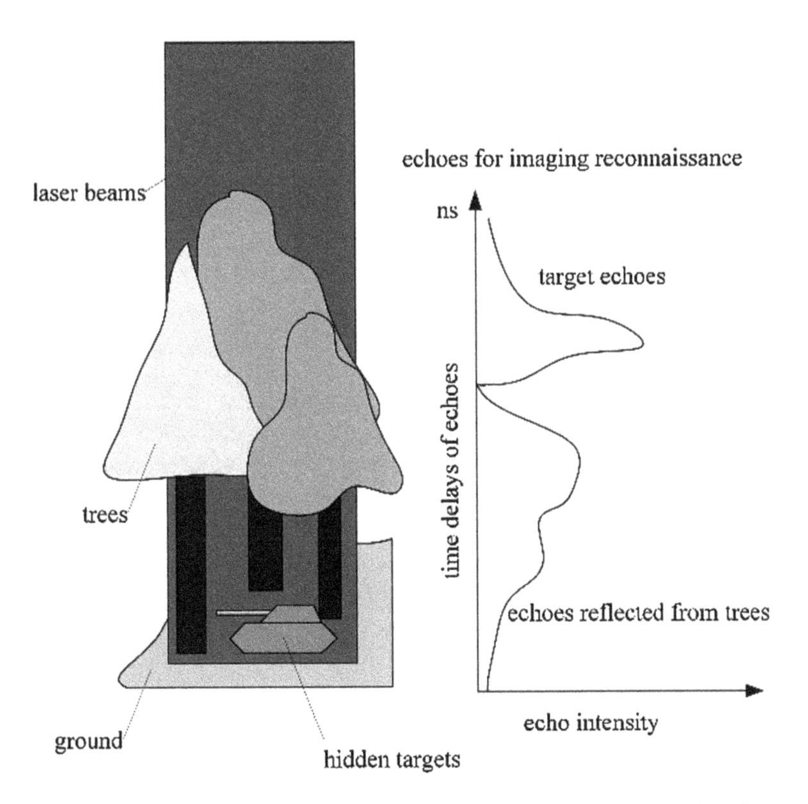

Fig. 8.8 The principles of detecting the targets hidden in forests based on a laser imaging system

3.65 and 2.4 m respectively while those of the latter are 7.2, 2.7 and 3 m (excluding a machine gun). For these targets, it is suggested to radiate laser beams showing a ground footprint whose diameter is 10 m. Under such condition, when the elevation of the detection platform is 3 km, the divergence angle of lasers is expected to be 3.4 mrad. Of course, the exact figures need to be adjusted based on practical situations of target areas such as density, average height and canopy closure of trees as well as possible sizes of hidden targets.

The key to the detection of hidden targets based on waveform images of targets is to make the best of the multi-valuedness of laser echoes. That is to separate the echoes reflected from targets and backgrounds from the hybrid echo signals and eliminate the influences of echoes of trees and ground.

8.2.2 Laser Echo Model for Hidden Targets

Geometric-optical and radioactive transfer (GORT) model [6–10] is simple as it is based on the relation between the statistical parameters of the trees in a particular area and waveforms of laser echoes. Furthermore, its parameters are easy to be obtained through echo waveforms. Therefore, it is suitable for describing the interaction between structures of trees and detecting lasers so as to obtain the relationship between laser echoes and the parameters of 3D structures of trees. In the GORT model, the characteristics of overall distribution, geometric structure (such as average size, shape and density of trees) and reflectivities of trees are the system parameters. In this case, echo features of hidden targets and trees can be analyzed as a whole regardless of the detailed features of the trees.

According to Fig. 8.1, detecting laser beams are continuously attenuated due to being blocked out by branches and leaves in the transmission process to the targets passing through trees [1]. Assume that the reflectivities of points of a tree are the same, void fraction is used to present the coverage degree of the detecting laser beams. Suppose that $P(z)$ represents the void fraction of laser beams at the height of z above ground, namely, the proportion of laser beams that can irradiate the lower height without being sheltered. $P(z-\Delta z)$ denotes the void fraction at the height of $z-\Delta z$ above ground, then $P(z)-P(z-\Delta z)$ stands for the proportion of the reflected energies of lasers in that of the whole trees at the thickness of Δz. Suppose that $R_v(z)$ refers to the total energies of the echoes reflected from the tops of the trees to the height of z received by a detector, while R_o and R_g are the received energies of echoes reflected by the hidden targets in the forest and the ground respectively. Additionally, J_0 and a represent the laser emission energy and the normalized attenuation factor, respectively. Then, it can be obtained through the GORT model that [1]:

$$-\frac{dR_v(z)}{dz} = aJ_0\rho_v\frac{dP(z)}{dz} \tag{8.2}$$

$$R_o = aJ_0\rho_o P(0)\frac{S_o}{S_g} \tag{8.3}$$

$$R_g = aJ_0\rho_g P(0)\left(1 - \frac{S_o}{S_g}\right) \tag{8.4}$$

Where ρ_v, ρ_o and ρ_g are the laser reflectivities of the trees, targets and ground, respectively. Besides, S_o and S_g are the sectional areas of the targets and the ground in the vertical direction of beams.

In the case that the reflecting surfaces of targets are at different heights, for example, as an armored car and its gun turret are not in the same height, their reflection echoes are not synchronous. In this case, $P(z)$ is expanded into $P'(z)$ which denotes the energy attenuation of detecting laser beams caused by the reflection of different parts of a target in forest and the occlusion of trees. $P'(z)$ can be expressed as:

$$\begin{cases} P'(z) = P(z) & z \geq z1 \\ P'(z) = P(0)\left(1 - \dfrac{\int_z^{z1} S_o(x)dx}{S_g}\right) & z < z1 \end{cases} \tag{8.5}$$

Where $z1$ is the minimum height of tree crowns, and $S_o(z)$ is the cross section areas of targets at the height of $z1$.

According to Eq. (8.5), Eqs. (8.3) and (8.4) can be rewritten as:

$$-\frac{dR_o(z)}{dz} = aJ_0\rho_o\frac{dP'(z)}{dz} \tag{8.6}$$

$$R_g = aJ_0\rho_g P'(0) \tag{8.7}$$

Equations (8.6) and (8.7) are the model for laser echoes of hidden targets. By solving $P'(z)$, hidden targets are expected to be discovered.

In the above equations, it is assumed that lasers are incident vertically and then horizontally irradiate upon the ground. Under such condition, the echoes reflected from all parts of grounds show same time delays. For inclined ground or non-vertical incidence of lasers, the heights of the ground in each part are varied, resulting in different echo delays. Moreover, the echoes from the ground are likely to be mixed with those of hidden targets and trees, which make it difficult to analyze the echoes. To facilitate the processing, both ground slope and oblique incidence of lasers are considered as the inclination of ground, which equals to adding a transformation of coordinates. That is to say, by rotating the incident angle of detecting laser beams for a certain angle, the lasers are regarded as vertically radiated in the new coordinate system while merely the slope angle of ground changes. Suppose that the total area of the reflecting section of ground is S_{gt}, and the actual reflecting section of ground at the height of z is $S_g(z)$. Besides, the highest point of the slope ground, as well as the highest and lowest points targets after inclination are z_{g1}, z_{o1} and z_{o2} respectively, then $P'(z)$ can be modified as:

$$\begin{cases} P'(z) = P(z) & z \geq z1 \\ P'(z) = P(0)\left(1 - \dfrac{\int_{z}^{z1}\left[S_o(x) + S_g(x)\right]dx}{S_{gt}}\right) & z < z1 \end{cases} \tag{8.8}$$

According to above equations, the improved function $P'(z)$ contains the information such as the spatial distribution of trees and the altitude distributions of targets and ground. Apart from being an important factor for the waveform modulation of laser echoes, the information is the bases for waveform analysis of laser echoes in the subsequent processing as well.

In this case, Eqs. (8.6) and (8.7) can be rewritten as:

$$\begin{cases} -\dfrac{dR_o(z)}{dz} = aJ_0\rho_o\dfrac{dP'(z)}{dz} & z_{g1} < z < z_{01} \\ -\dfrac{d\left[R_o(z) + R_g(z)\right]}{dz} = aJ_0\rho_{go}(z)\dfrac{dP'(z)}{dz} & z_{o2} \leq z \leq z_{g1} \\ -\dfrac{dR_g(z)}{dz} = aJ_0\rho_g\dfrac{dP'(z)}{dz} & z < z_{o2} \end{cases} \tag{8.9}$$

Where $\rho_{go}(z)$ refers to the average of the united reflectivities of ground and hidden targets at the height of z.

According to Eq. (8.9), for included ground, echoes reflected from targets and ground are likely to be overlapped in the time domain, which makes it hard to discover hidden targets. In general, owing to detecting laser beams are reflected by various layers of vegetation in the propagation process, the corresponding reflected echoes are mixed with those from targets and ground. The distribution of $P'(z)$ is expected to affect the energies of echoes received at different times.

8.2.3 *The Detecting Algorithm for Hidden Targets*

The echoes reflected from hidden targets in laser imaging based detections are mixed up with those from trees and ground. Therefore, echoes of vegetation need to be evaluated and eliminated to discover the hidden targets.

Based on the above analysis, the function $P'(z)$ contains the information such as the spatial distribution of trees and the altitude distributions of targets and ground. By analyzing and solving $P'(z)$, the information of echoes reflected from vegetation can be achieved. According to Eq. (8.9), $P'(z)$ can be defined as:

$$P'(z) = 1 - \frac{R(z)}{aJ_0\rho(z)} \tag{8.10}$$

Assume that a distribution model of trees is composed of trees at two different heights with the elevations of the crown centers being 8 m and 11 m, respectively. The variations of the waveform of laser echoes and the distribution of $P'(z)$ with the height z can be obtained according to the GORT model, as shown in Fig. 8.9.

As shown in Fig. 8.9, owing to $P'(z)$ represents the generalized void fraction within detecting laser beams, it changes continuously in the height range of trees (from the top to the bottom of a tree crown). While, it remains invariable in the height range from the bottom of a tree crown to the ground or hidden targets owing to the laser beams are not sheltered.

According to the characteristic of $P'(z)$, its distribution values varying with the height z can be gained by utilizing Eq. (8.10). Given that the attenuation coefficient α of laser energy is influenced by various external factors, its exact value is difficult to be achieved. Therefore, the obtained distribution value of $P'(z)$ is a relative value but its distribution characteristic remains unchanged. By analyzing the variation trend of $P'(z)$, the possible distribution range of the heights (time delays) of vegetation echoes, namely, the range where the value of $P'(z)$ continuously varies from the top of the vegetation, is obtained. Then, echoes in this range are eliminated from the waveforms of laser echoes. In this way, the echo signals of laser detection excluding vegetation echoes are obtained. However, the obtained echo signals contain the mixed information of targets and ground so target information requires to be extracted further.

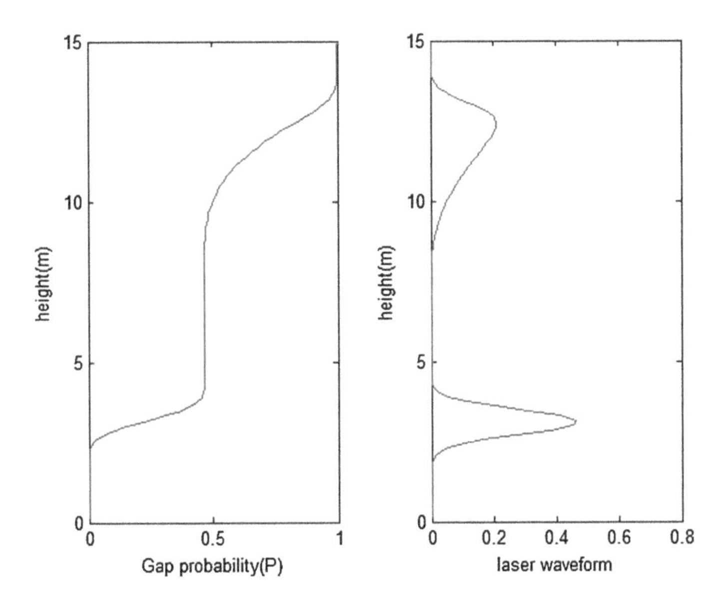

Fig. 8.9 The variations of the waveform of laser echoes and the function of void fraction with the height

After the echoes reflected from vegetation are removed, a hybrid echo signal of the ground and the hidden targets is obtained and denoted as $S(t)$. If the target echo is separated from the ground echo on the time axis and fractured into two wavelets, in this case, the target can be found directly. To meet this condition, the lowest surface of the target has to be higher than the highest surface of the ground. Suppose that δ, h and D indicate the inclination of the ground, the height of the target and the diameter of the laser footprint on the ground, respectively, the condition can be expressed as:

$$h/\cos\delta > D\tan\delta \tag{8.11}$$

Where $h/\cos\delta$ is the lowest surface of the target on an inclined ground, and $D\tan\delta$ represents the highest surface of the inclined ground. For example, when h =2 m and D = 10 m, the maximum value of δ is 11.5°. It can be seen that when the inclination of the ground is large (larger than 11.5°), echoes from hidden targets and the ground are overlapped on the time axis. In this case, it is hard to directly judge the existence of the target.

In the laser footprint (about 10 m in diameter) on the ground of hidden targets detection based on a laser imaging system, suppose that the distribution of laser reflective characteristics of the ground changes gently, then its echo signal can also be considered as a smooth signal which changes slowly. According to Eqs. (8.8) and (8.9), due to the abrupt variation of the height of targets and the differences between the reflectivities of targets and the ground, a target echo can be regarded as adding an unstable singular signal to a ground echo. Therefore, the methods relating to signal analysis can be applied to detect the existence of targets.

Triple-correlation peak detection [11, 12] is a method for detecting and analyzing signals by reusing useful signals and eliminating noise in the reutilization processing based on the correlation of signals and the irrelevance of noise. This algorithm can absorb the energies of useful signals and suppress the noise to highlight the peaks of signals at maximum. As this method can present the signal mutation in a certain time range, it can be used to detect the existence of targets.

The general form of the n-order correlation function of function $I(t)$ is

$$I^{(n)}(t_{1,}t_2, t_n) = 1/T\int I(t)I(t+t_1)I(t+t_2)\cdots I(t+t_n)dt \tag{8.12}$$

The triple-correlation function of $I(t)$ can be defined as

$$I^{(3)}(t_{1,}t_2) = 1/T\int I(t)I(t+t_1)I(t+t_2)dt \tag{8.13}$$

Suppose that noise is the additive stationary random process $N(t)$ that is independent of signals $I(t)$, and $J(t) = N(t) + I(t)$. By performing triple-correlation operation on $J(t)$, its ensemble average can be obtained as [13]:

$$\left\langle J^{(3)}(t_1,t_2)\right\rangle = I^{(3)}(t_1+t_2)+\left\langle N^{(3)}(t_1+t_2)\right\rangle + \langle N(t)\rangle\left[I^{(2)}(t_1)+I^{(2)}(t_2)+I^{(2)}(t_2-t_1)\right] + \langle I(t)\rangle\left[\left\langle N^{(2)}(t_1)\right\rangle+\left\langle N^{(2)}(t_2)\right\rangle+\left\langle N^{(2)}(t_2-t_1)\right\rangle\right] \quad (8.14)$$

The ensemble average of odd-order correlation functions of the zero-mean stationary Gaussian random process equals to 0. Therefore, if $N(t)$ is white noise, the second and third items in the right of Eq. (8.14) equal to 0, while the first and fourth items contain signal information. According to Eq. (8.14), The ensemble average triple correlation suppress almost all the noise, while, for the wide-band colored noise, its self-correlation function $N^{(2)}(t)$ reduces rapidly with the increase of t. In the case of noise with signals, the items containing signals are highlighted, which makes the ensemble average triple correlation of noise with signals is distinct from that of the noise without signals. Therefore, ensemble average triple correlation is used to suppress noise. In addition, under the conditions where noise contains signals, only the triple correlation of signals presents peaks. Thus, according to peaks and appropriate threshold discrimination, the existence of target signals can be determined.

Triple-correlation peak detection can highlight the signal information of $I(t)$, so a appropriate function $I(t)$ needs to be constructed to be applicable to the detection of echo signals of targets. As the echo signals of hidden targets singular signals in the slow-varying echo signals of grounds, they are different from the local signals around. Based on the difference, the discretized function $I(t)$ can be described as:

$$I^{(3)}(x) = \sum_{i=1}^{n}[s(x+i)-E_k][s(x+i+1)-E_k]^2[s(x+i-1)-E_k]^2 \quad (8.15)$$

Where n is the size of the window and E_k is the average signal in the kth local window.

$$E_k = \frac{1}{n}\sum_{i=0}^{n-1}s(x+i) \quad (8.16)$$

By using the signals in local windows and the average local deviation as characteristics, the corresponding peaks of varying signals in the local can be obtained through the triple-correlation peak detection on the signals in the window with a size of n [14]. As the size of the target is smaller than that of laser footprint on ground, the width of the echo signal of the target is narrower than that of the ground echo. Therefore, selecting a proper width of the local window can show the distinction between the peak of an echo signal of the target and that of ground.

By simulating using the GORT model, detecting laser echoes are obtained respectively in the conditions with and without hidden targets. The target is 7.5 m

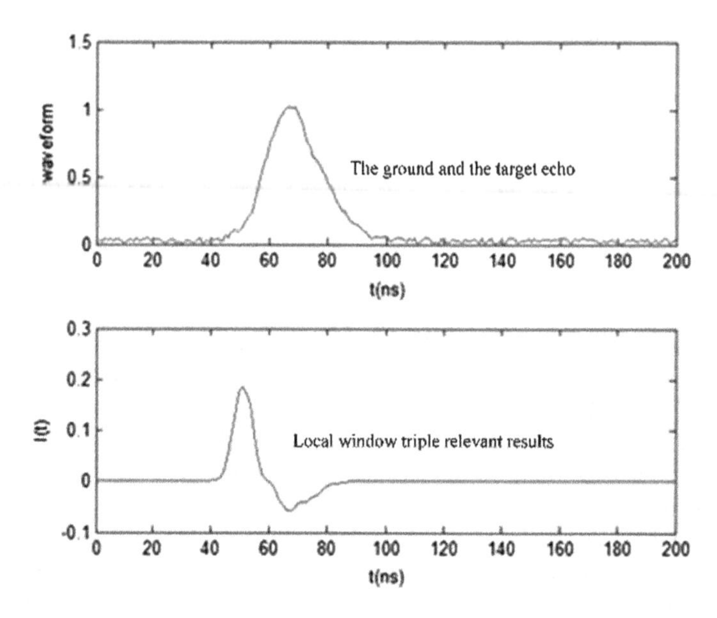

Fig. 8.10 The waveforms of laser echo signals of the target and the ground, together with the triple-correlation detection results of peaks when the target exists

long, 2.5 m wide and 2 m height. Besides, the diameter of the laser footprint and the inclination of the ground are 10 m and 30°, respectively. Based on the simulated echo signals obtained, Fig. 8.10 illustrates the waveforms of laser echo signals of the target and the ground, together with the triple-correlation detection results of peaks when there is the hidden target. While, Fig. 8.11 merely shows the waveforms of laser echo signals of the ground and the triple-correlation detection results of peaks in the condition without the target. Additionally, the size n of all the local windows is set as 11.

As shown in Figs. 8.10 and 8.11, when there is no target, the echo signal of the ground changes smoothly, and the absolute values of the positive and negative maximums of its triple-correlation peaks are basically the same. When there is a target, the target causes local singular signals. Under such condition, the triple-correlation peak when the target exists is larger than those in other period. Based on this, it can be determined whether there is a target.

8.2.4 Analyzing the Features of Hidden Targets

8.2.4.1 The Determination of Imaging Range of Hidden Targets

According to the separation of target and ground echoes on the time axis, the time delays of target echoes can be directly determined through echo waveforms, and

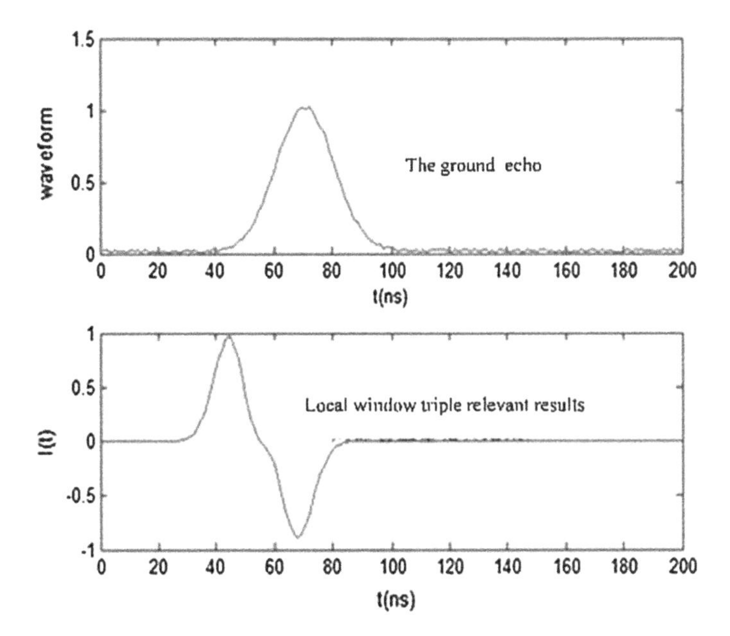

Fig. 8.11 The waveforms of laser echo signals of the ground and the triple-correlation detection results of peaks when there is no target

then the imaging range of the target is expected to be obtained. While, if the echoes of targets and ground are overlapped with each other, time delays of target echoes cannot be directly gained from echo waveforms. In this case, based on the above analysis, the echo signals of hidden targets are mixed with those of ground, which means that they present singularities in the time domain. By analyzing the singularities of echo signals of targets and ground, the distribution of the positions of the singular points of signals on the time axis can be obtained. In this way, the target signals can be located on the time axis.

In general, Lipschitz exponent is used to describe the local singularities of signals. Suppose that n is a non-negative integer $(n<a\leq n+1)$, if there are two constants A and $h_0(\geq 0)$ and a n-degree polynomial which make any $h\leq h_0$ satisfy Eq. (8.17), then $f(x)$ is called Lipschitz α at point x_0.

$$|f(x_0+h)-P_n(h)|\leq A|h|^a \tag{8.17}$$

If Eq. (8.17) is valid for any $x_0\in(a,b)$, then, in the range of (a,b), $f(x)$ is coincident Lipschitz α.

Lipschitz α function of $f(x)$ at point x_0 presents the regularity. The larger the Lipschitz α, the smoother the function is. If the function is continuous and differentiable at a certain point, then, Lipschitz α at this point is 1. Besides, if the function is differentiable but not continuous at a point, Lipschitz α at this point is 1 as well. If Lipschitz α of $f(x)$ at point x_0 is smaller than 1, then, $f(x)$ is singular at the point.

By employing the points of WTMMs, the singular points of signals can be detected. The theorems of WTMM and signal singularity are as following [15].

Theorem 1

Assume that n represents an integer and Θ *is the n-order vanishing moment. There are n-order continuously differentiable wavelets with compact support,* $f(t) \in L^1(c,d)$ *([c, d] is the range of real numbers). If there is a scale* $a_0 > 0$, *which satisfies* $\forall a < a_0$ *and* $t \in (c,d)$, *then, the wavelet transform coefficient* $|W_f(a,b)|$ *does not have a local maximum value point. In the case,* $f(t) \in L^1(c,d)$ *satisfies coincident Lipschitz* α *(*ε *is a small positive number).*

According the Theorem 1, by detecting the points of WTMMs, the singular points of signals can not only be detected but also can be located. For the laser echo signals of hidden targets obtained through simulation, wavelet analysis is performed. Daubechics3 wavelets with a three-order vanishing moment (suppose that the echo signals of vegetation are eliminated) are employed to decompose the echo signals into 4 levels. Figure 8.12 shows the decomposition coefficients and waveforms of echo signals in each scale.

In Fig. 8.12, if noise is regarded as rapidly changing and uncorrelated signals, scale 1 precisely matches with it. For the scales that match with the reflection signals of targets, they do not match with the influence of noise, which is therefore can be neglected. Owing to the reflection signals of targets vary faster than the echo signals of ground but slower than the noise, scales 2 and 3 are considered to match with these reflection signals. It can be seen in Fig. 8.12 that there is a distinct WTMM point at the target position ($x = 70$) in the second and third levels of wavelet transform, which verifies the feasibility of the WTMM-based method for target location. For other WTMMs in scales 2 and 3, they are probably caused by

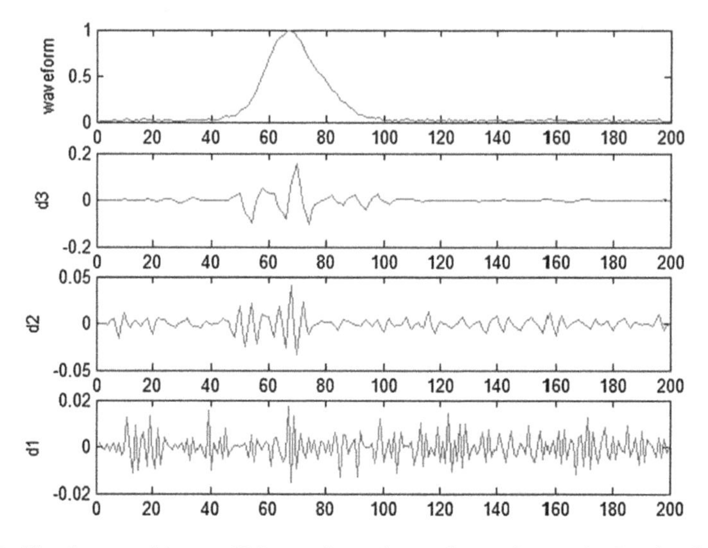

Fig. 8.12 The decomposition coefficients of wavelets and waveforms of echo signals in each scale

the reflection of uneven ground surface and nonuniform noise. Owing to the local mutation of these points is less obvious than the change of signals caused by targets, only the WTMM points in the same scale require to be considered. In addition, it is noteworthy that as a target has its spatial size, its echo signal presents a certain bandwidth instead of focusing on a point on the time axis. The positions of WTMMs stand for the locations where target signals change most violently with most concentrated energies. When the width of a target echo is not large, this point can represent the central position of the target.

8.2.4.2 Evaluation of the Features of Hidden Targets

In the hidden target detection based on a laser imaging system, after discovering and locating targets, other parameters of targets need to be obtained as much as possible for the further classification and identification. According to the characteristics of hidden target detection using a laser imaging system, the major information that requires to be obtained is dimension parameters of the targets. The targets hidden in forest are mainly vehicles, whose size varies for different types. Heavy vehicles like heavy-duty trucks are generally large. An American M977-heavy haulage truck, for example, is 8.2 m long and 2.4 m wide, with an overall height of 2.8 m. In general, a medium sized vehicle is 5–8 m long such as an armored car and a light tank. Besides, a small vehicle, for example, a jeep, is less than 5 m long. For a hidden target, its type can be preliminarily discriminated by obtaining its dimension information so as to make further threat assessment and decision.

(1) The separation between echoes of targets and ground

When the echoes reflected from targets and ground are separated on the time axis, their energies can be obtained by calculating the integrals of echo waveforms of targets and ground. Suppose that the echo energy, the reflectivity and the area of the reflecting section of a target are E_1, ρ_1 and S_1 respectively, and those of ground are E_2, ρ_2 and S_2 respectively. Besides, the reflectivities of the target and ground are assumed as uniformly distributed which is satisfied or approximately satisfied under a general condition. If the area of a laser footprint is S_0, $S_0 = S_1 + S_2$ is obtained. Then:

$$\frac{\rho_1 S_1}{\rho_2 S_2} = \frac{\rho_1 S_1}{\rho_2 (S_0 - S_1)} = \frac{E_1}{E_2} \tag{8.18}$$

$$S_0 = \frac{E_1 S_0}{\frac{\rho_1}{\rho_2} E_2 + E_1} \tag{8.19}$$

The area of the laser footprint is fixed and known, so only the ratio ρ_1/ρ_2 of reflectivity of the target to that of ground is needed so as to obtain the area of the reflecting section of the target. By means of experiments or data collection, the reflectivity of a general car for the laser with a certain wavelength can be obtained, based on which the ground reflectivity of a certain detection region is expected to be obtained from relevant data. In this way, the area S_1 of the reflecting section of the target can be obtained.

By simplifying the target as a cube model, the actual surface area S_1' (length × width) of the target can be obtained according to the area S_1 of the reflecting section and the equivalent inclination θ of the ground. Where, θ is the inclination of the ground when the incident angle is π/2 after equivalent coordinate transformation for the incident angle of lasers.

$$S_1' = S_1 / \cos\theta \tag{8.20}$$

The equivalent inclination of the ground can be estimated based on the width of echo waveform of the ground. While performing detection based on a laser imaging system on the ground with an incident angle of π/2, the echo width can be expressed as:

$$\tau_R = \tau_S + \tau_h + \frac{2Var(\Delta\xi)\cos\theta}{c} + \frac{2D\tan\theta}{c} \tag{8.21}$$

Where c, τ_S and τ_h are the velocity of light, the pulse width of emitted laser and the pulse broadening caused by the receiving circuit. Besides, $Var(\Delta\xi)$ and D are the surface roughness and the diameter of the laser footprint on the ground, respectively.

$Var(\Delta\xi)$ is negligible when the laser footprint on the ground is small. The values of τ_S and τ_h can be determined through experiments, and the diameter of the laser footprint can be calculated utilizing the values of imaging range and laser beam divergence. According to Eq. (8.20), the estimated value of the equivalent inclination θ of ground can be obtained, and then the estimated surface size of the target can be achieved by solving Eqs. (8.19) and (8.20).

(2) The overlap of echoes reflected from targets and ground

With a large inclination of ground, the echoes reflected from targets are overlapped with those of ground on the time axis. Under such circumstance, effective echo waveforms of targets are hard to be directly extracted, so other methods are employed to estimate the dimension information of the targets.

According to Eq. (8.19), the key to estimating the size of targets is to acquire the information relating the echo energy of targets. Echo energies of targets and ground are overlapped on the time axis, and therefore, exact value of echo energies of targets is difficult to be directly acquired based on a single echo. To solve this problem, information of adjacent echoes is utilized. The laser footprints on the ground of two adjacent detections are close so that the distribution characteristics of ground in the two detections can be considered as basically the same, so do the echo signals. Afterwards, the echo signals of ground obtained in the adjacent detections not containing target echoes can be used to represent the echo signal of ground obtained in the current detection to solve the above problem.

Suppose that, after eliminating the echoes of vegetation, the echo signal in the current detection and its adjacent detection are $R_2(t)$ and $R_1(t)$, respectively. In $R_2(t)$, the components of echo signals of the ground and the target are $R_g(t)$ and $R_o(t)$, respectively. According to Eqs. (8.18) and (8.19), it can be achieved that:

$$dS_1(t)\rho_1 + dS_2(t)\rho_2 = kdR_2(t) \tag{8.22}$$

$$dS_1(t)\rho_2 = kdR_1(t) \tag{8.23}$$

Where k is the normalized coefficient of energy attenuation. $S_1(t)$ and $S_2(t)$ are the functions describing the variation of the section areas of the target and the ground with the imaging range, separately. In addition, ρ_1 and ρ_2 represent the laser reflectivities of the target and the ground, respectively.

According to Eqs. (8.22) and (8.23), then

$$\int [S_1(t)\rho 1]dt = S_1\rho_1 = k\int [R_2(t) - R_1(t)]dt \quad R_2(t) - R_1(t) > 0 \tag{8.24}$$

$$\int [S_2(t)\rho 1]dt = S_0\rho_1 = k\int R_1(t)dt \tag{8.25}$$

$$S_1 = \frac{\rho_2}{\rho_1}\frac{\int [R_2(t) - R_1(t)]dt}{\int R_1(t)dt}S_0 \tag{8.26}$$

Where S_0 is the area of the laser footprint on the ground.

After obtaining the ratio of the reflectivity of the ground to that of the target using Eq. (8.26), the area of the reflecting section of the target can be calculated based on the echoes of the adjacent detection zone and this detection. Moreover, by using Eqs. (8.19) and (8.20), the estimated surface size of the target can be achieved.

It is assumed in all the above equations that the occlusion coefficients of vegetation in the adjacent detections are the same. However, when they are different, the attenuation ratio of incident energies of lasers varies, which causes the change of the ratio of echo energies in the detection of adjacent zones. Therefore, echo signals $R_1(t)$ and $R_2(t)$ have to be normalized. The following algorithm is adopted.

Step 1: Suppose that the overall echo energies in the two detections are equivalent. According to Eqs. (8.22) and (8.23), when the two detections show the same $P(z)$, the integral sum of $R_1(t)$ equals to that of $R_2(t)$. In this case, by adding $R_1(t)$ and $R_2(t)$ up, the overall echo energies E_1 and E_2 are acquired. Then, q which is the normalized coefficient, is calculated as $q = E_1/E_2$. Let $R_2(t) = qR_2(t)$.

Step 2: S_1 is computed by solving Eq. (8.26). Then, the actual value of E_2 is $E_2 = E_2(1 + (\rho_1 - \rho_2)/\rho_2)$ according to Eq. (8.22). The newly normalized coefficient satisfies $q' = q(1 + (\rho_1 - \rho_2)/\rho_2)$.

Table 8.2 The estimation error of target areas when the adjacent terrains change

Differences of the inclination	3°	−3°	5°	−5°
Variance	10.3409×10^{-5}	1.4987×10^{-4}	8.6205×10^{-4}	9.7084×10^{-4}
Standard deviation	0.0671	0.1031	0.2474	0.2625

Step 3: S_1 is calculated repeatedly according to the newly normalized coefficient. Then, the above processes are repeated until a stable S_1 is acquired.

Due to the possibly different surface distributions within laser footprints in the adjacent detections, the areas of targets obtained are estimated values. Table 8.2 illustrates the average variance and standard deviation of the areas of targets obtained by means of simulation with different inclinations of ground in the adjacent detection zones. The concrete parameters acquired through simulation are shown in Sect. 8.2.3. As shown in Table 8.2, with a small change of adjacent terrains, the error of target areas obtained employing this method is acceptable.

8.2.5 *Experiments and Analysis*

Based on the above models and algorithms, simulation experiments of laser imaging based target detection are carried out for three types of vehicle targets including a tank, a truck and a jeep, as displayed in Figs. 8.13, 8.14, and 8.15.

According to GORT model, a tree-based occlusion model is established, in which the trees are in the oval shape. Two types of trees are contained in this model and their average parameters are presented in Table 8.3.

On the basis of the above models, simulation experiments are performed using the laser imaging detection system at the elevation of 4000 m and with a laser divergence of 2.5 mrad. The distributions of simulated echo signals and functions $P'(z)$ of void fraction of the tank, the truck and the jeep hidden in forest on the ground with different inclinations are obtained, as shown in Figs. 8.16, 8.17, 8.18, 8.19, 8.20 and 8.21, respectively.

According to the distribution of functions $P'(z)$ of the void fraction, the echo signals of ground and the targets with the echoes of trees being eliminated are gained, as shown in Figs. 8.22, 8.23 and 8.24.

Triple-correlation peak detections are performed to the echo signals of ground and the targets after removing the echoes of trees. The results are shown in Figs. 8.25, 8.26 and 8.27, and the corresponding echoes reflected from the ground and the related results when there is no target are illustrated in Figs. 8.28, 8.29 and 8.30.

When the targets are hidden beneath trees, the correlation peaks of echo signals obtained based on triple-correlation peak detections are obviously asymmetric, and

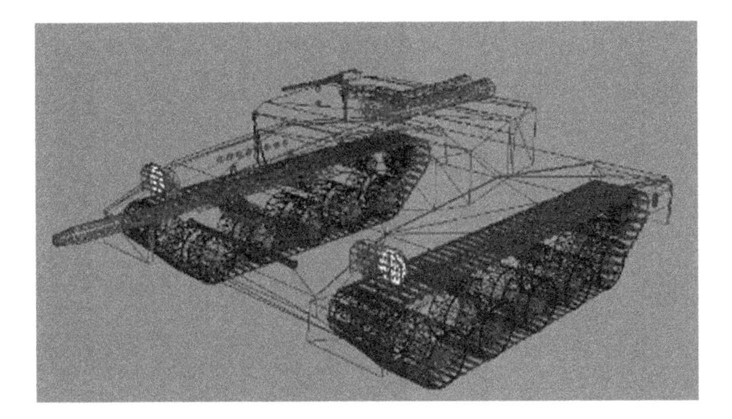

Fig. 8.13 The tank model

Fig. 8.14 The truck model

Fig. 8.15 The jeep model

Table 8.3 Parameters of trees

Type of trees	Horizontal diameter/m	Vertical diameter/m	Density of leaves and branches/(1/m)	Elevation of the center of tree crown/m
A	3.0	7.2	0.44	12
B	2.4	6.0	0.36	8

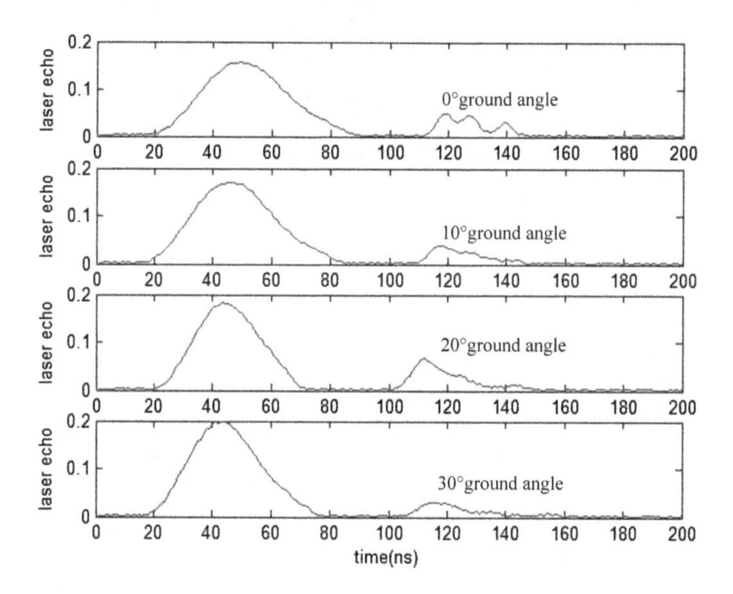

Fig. 8.16 The simulated echoes from the hidden tank

the correlation peaks obtained when there are targets are significantly larger than those at other times.

When there is no target, the absolute values of the positive and negative maximums of correlation peaks of the echo signals are almost the same. Therefore, it is verified that the method of triple-correlation peak detection is effective in discovering the targets hidden in forest.

Multiple simulation detections are performed on the above tank, truck and jeep under different terrain conditions. The probabilities of correct detection are 97%, 95% and 83% respectively, which proves that this method is applicable to target detections. By estimating the areas of the targets based on the simulated echo signals using the above method, the estimation errors are obtained at different inclinations of the ground, as shown in Table 8.4.

According to Table 8.4, it can be seen that the larger the inclination of the ground, the bigger the estimation error of the target area is. For the target with small areas, its estimation error is much bigger. For the above targets, the maximum evaluation error reaches 39.4%. As the targets show large differences in the areas (for example, the area of the tank is over three times larger than that of a miniature vehicle), the classification of targets can be basically realized.

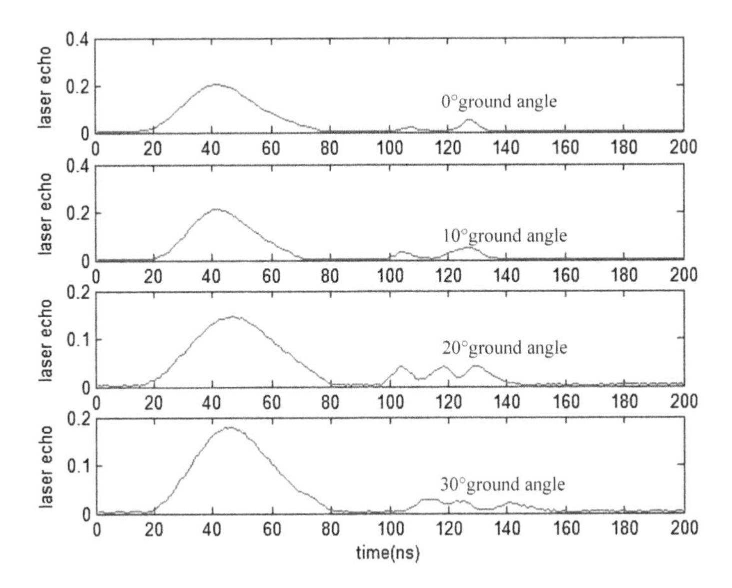

Fig. 8.17 The simulated echoes from the hidden truck

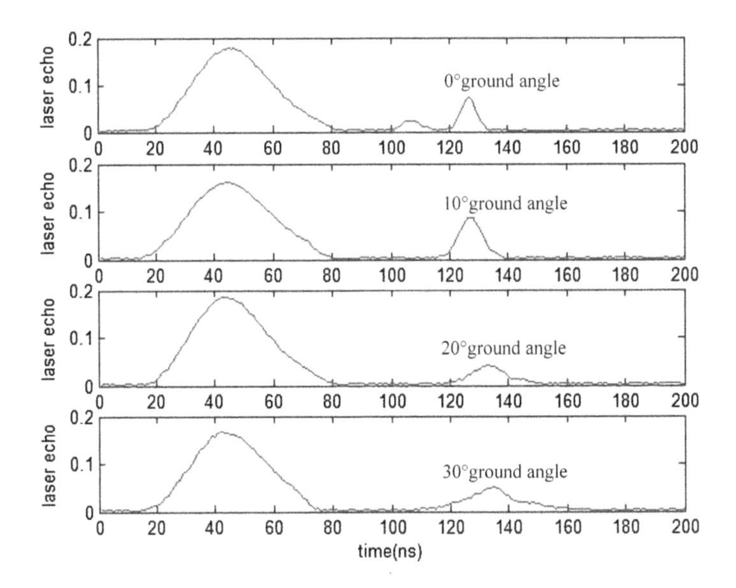

Fig. 8.18 The simulated echoes from the hidden jeep

Apart from waveform images, the differences of hidden targets and camouflages in the spatial level shown in the level images can also be utilized to detect hidden targets through data processing. Figure 8.31 shows the range image of a car covered by a camouflage net. Due to protection of the camouflage nets, it is hard to find the

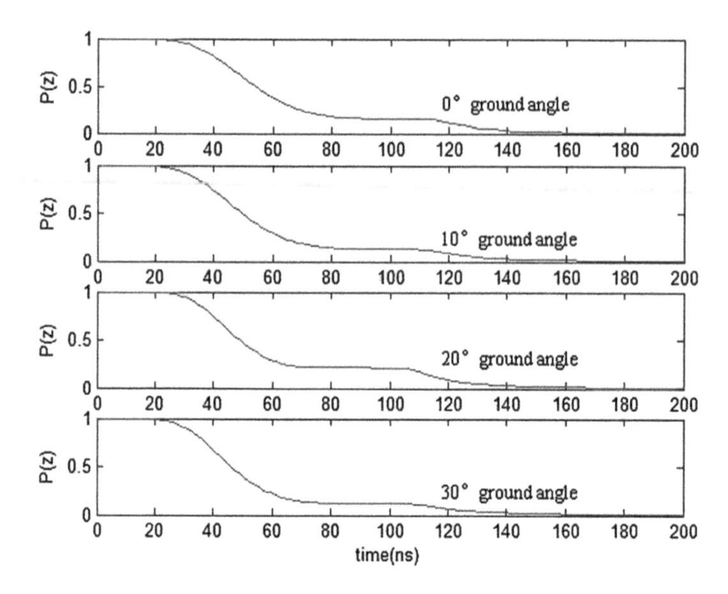

Fig. 8.19 The function of the void fraction of echoes from the hidden tank

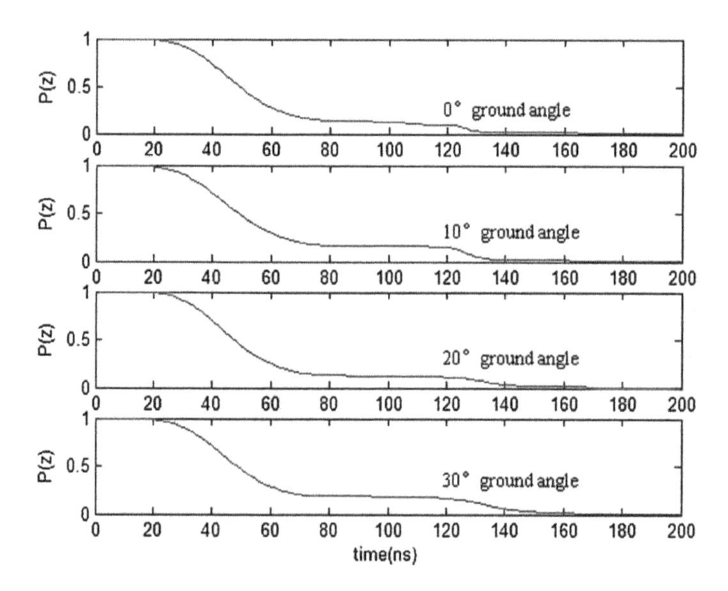

Fig. 8.20 The function of the gap probability of echoes from the hidden truck

car from the range image. Figure 8.32 illustrates the target imaging results acquired using level images, where the car covered by the camouflage net is easily to be discovered.

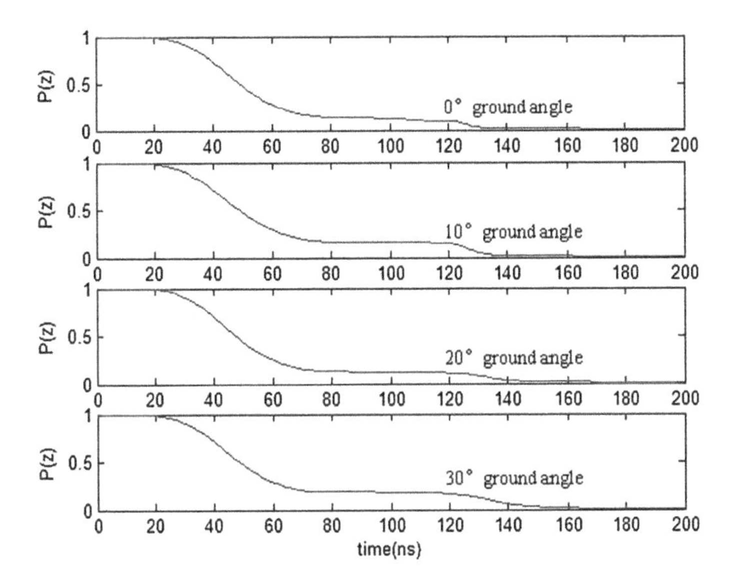

Fig. 8.21 The function of the void fraction of echoes from the hidden jeep

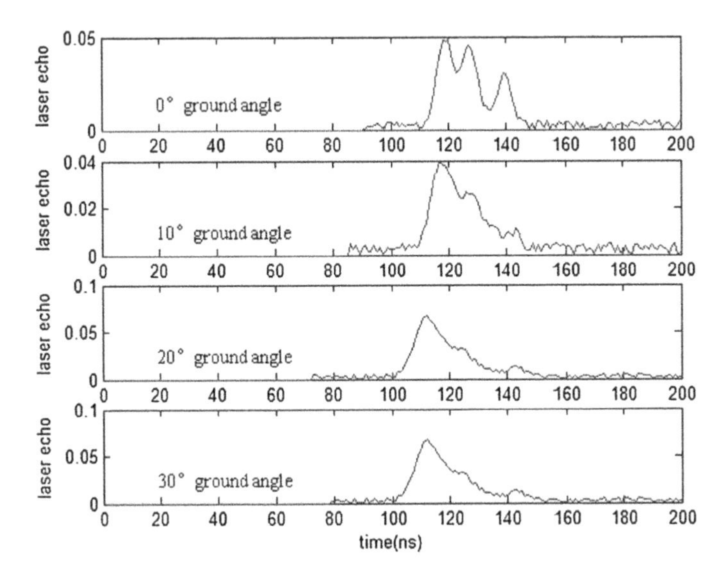

Fig. 8.22 The echo signals of ground and the tank with the echoes of trees being eliminated

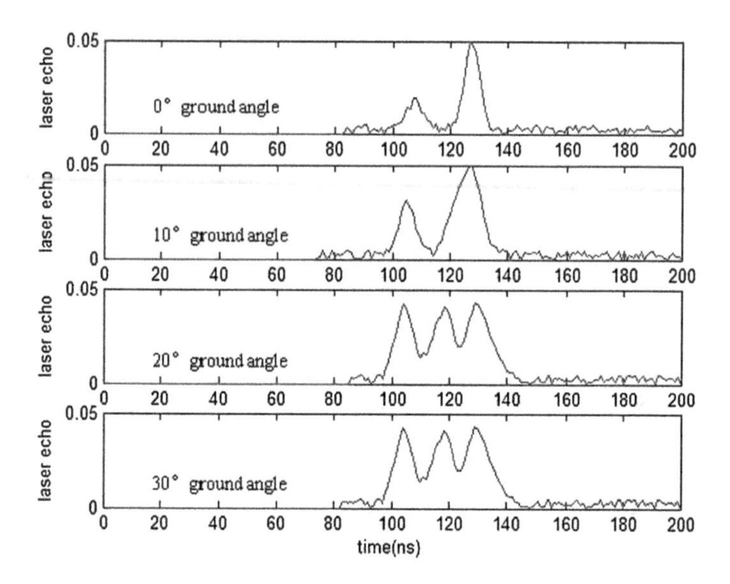

Fig. 8.23 The echo signals of ground and the truck after eliminating the echoes of trees

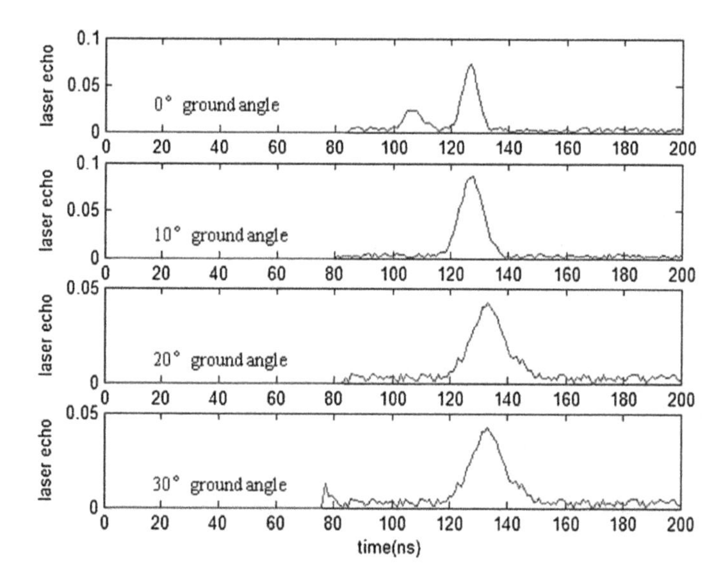

Fig. 8.24 The echo signals of ground and the jeep after eliminating the echoes of trees

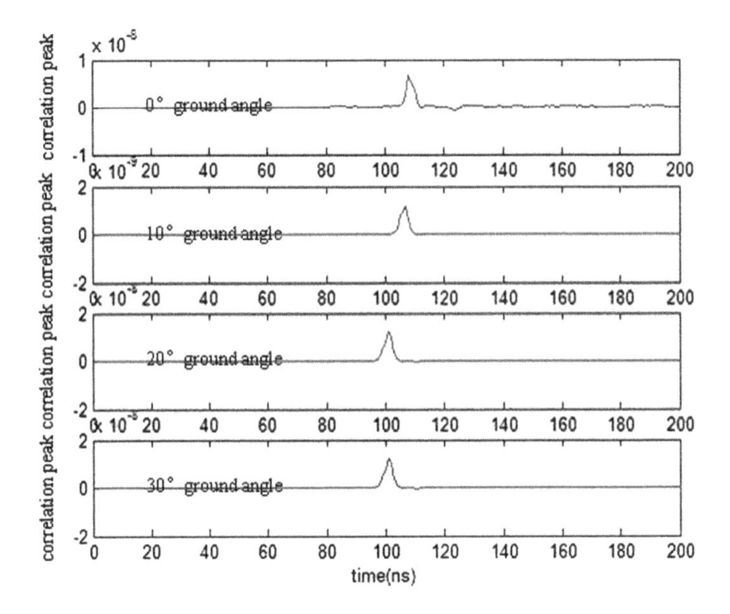

Fig. 8.25 The results of triple-correlation peak detections of echo signals of ground and the tank

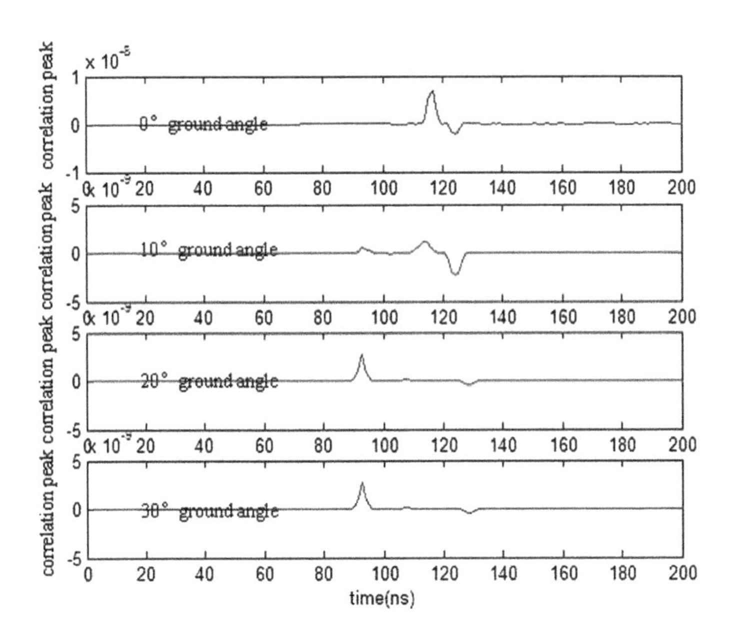

Fig. 8.26 The results of triple-correlation peak detections of echo signals of ground and the truck

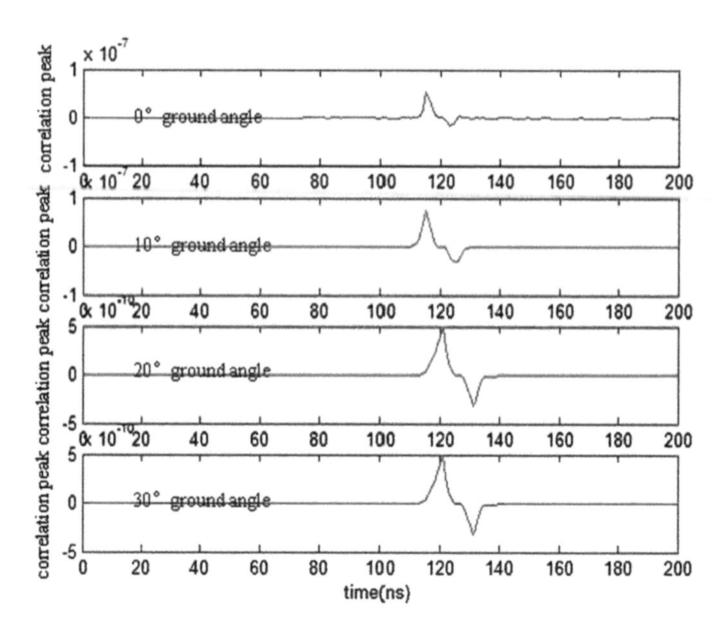

Fig. 8.27 The results of triple-correlation peak detections of echo signals of ground and the jeep

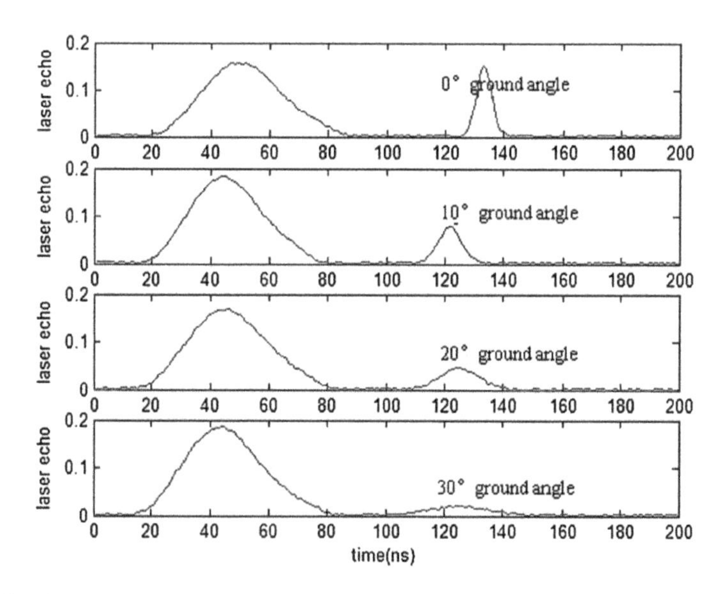

Fig. 8.28 The echo signals of ground and vegetation when there is no target

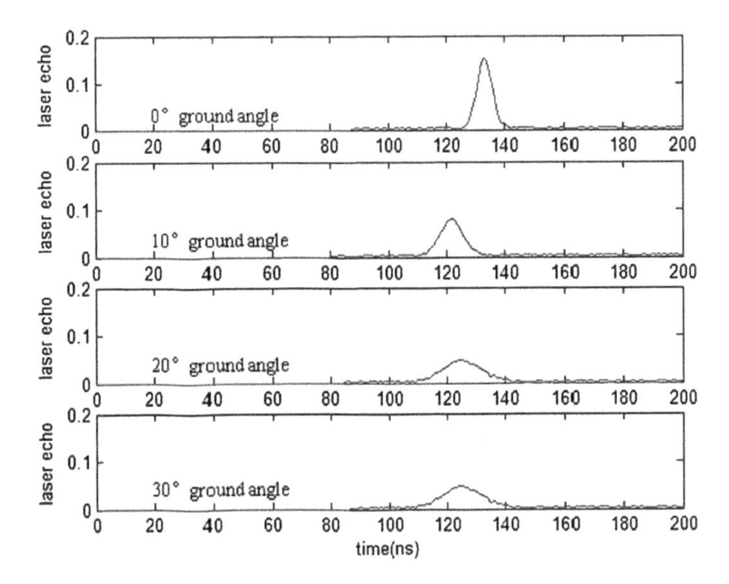

Fig. 8.29 The echo signals of ground after those of trees being eliminated

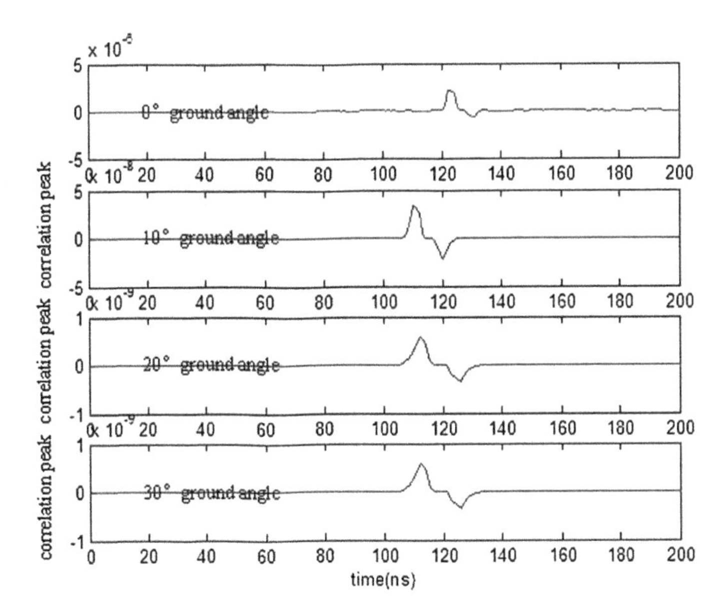

Fig. 8.30 The results of the triple-correlation peak detection of echoes reflected from ground when there is no target

Table 8.4 The estimation errors of target areas with different inclinations of ground

Inclination of ground/(°)	0	10	20	30
Tank/%	12.2	5.06	13.58	27.33
Truck/%	3.15	9.87	21.72	20.84
Jeep/%	10.76	30.57	26.05	39.4

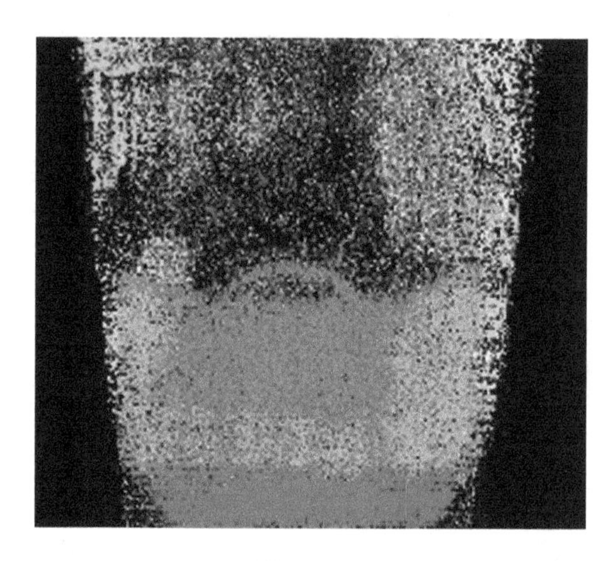

Fig. 8.31 The range image of the car covered by the camouflage net

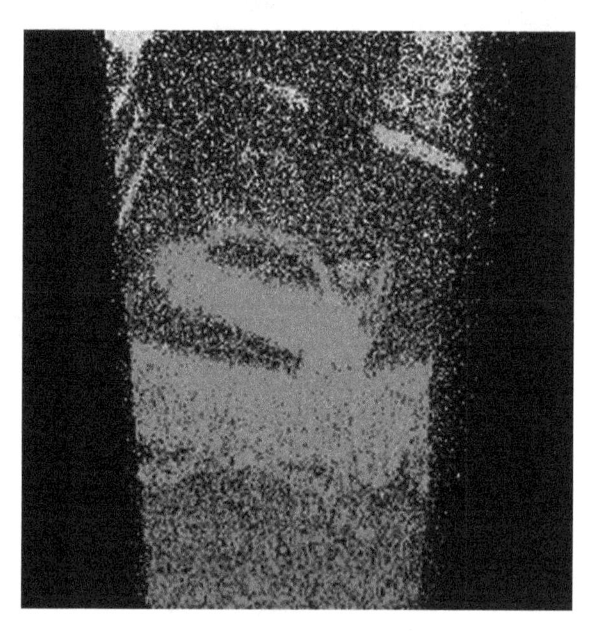

Fig. 8.32 The car covered by the camouflage net discovered using the range image and the level image

8.3 Target Location Based on a Laser Imaging System

Target location based on a laser imaging system is one of the important aspects of target detections. The 3D coordinates of targets can be obtained by comprehensively processing the data including the range and orientation of targets as well as the position and attitude of detection platforms.

8.3.1 The Principle of Target Location

The target location using a laser imaging system adopts positioning principle using space geometry. That is to say, by determining the parameters of geometrical vectors, the vertexes of the vectors can be directly located, as shown in Fig. 8.33. Assume that the laser imaging system is situate at the point G, and there is a vector $\vec{S}$ with the vector radius being S in the space pointing to the same direction of the laser beams and receiving field of view. The angle of direction ϕ represents the angle between $\vec{S}$ and the reference direction. $\vec{S}$ is intersected with the target at point P which is also called the footprint point of laser sampling. In addition, the direction of the vector is $(\varphi, \omega, \kappa, \phi)$, where φ, ω and κ are the angles of roll, pitch and yaw angles of the vector, respectively. While ϕ is the angle between the pixel corresponding to the sampling point P of laser detection and the reference pixel (the lower point of airplanes or sub-satellite point Q). If the coordinates (X_G, Y_G, Z_G) of the starting point of the vector are measured, then, the only coordinates (X_P, Y_P, Z_P) of the other vertex P can be determined. By recording the time delay of laser pulses from being emitted to being received after being reflected by the targets in target

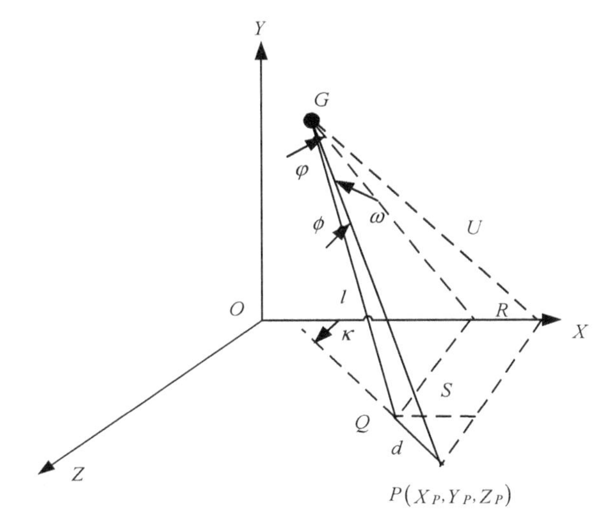

Fig. 8.33 The target location based on a laser imaging system

detection using a laser imaging system, the slope distance S from the projection center to the sampling points of targets (laser footprints) can be accurately measured. The GPS positioning system can be used to obtain the coordinates of point G, and then the high-precision attitude measuring device INS system is employed to measure the attitude (roll angle φ, pitch angle ω and yaw angle κ) of the projection center in the major optical axis. Then, the coordinates of the sampling point P of the target can be computed according to Eq. (8.27).

$$\begin{cases} X_P = X_G + \Delta X \\ Y_P = Y_G + \Delta Y \\ Z_P = Z_G + \Delta Z \end{cases} \tag{8.27}$$

Where ΔX, ΔY and ΔZ are the increments of coordinates of the target sampling point P relative to point G, the center of scanning projection, and they are closely related to $(\phi, \omega, \kappa, S)$.

Let $GQ = l$, $QP = d$, then

$$\begin{cases} \Delta X = l \cos \omega \sin \varphi + d \cos \kappa \\ \Delta Y = l \sin \omega + d \sin \kappa \\ \Delta Z = l \cos \omega \cos \varphi \end{cases} \tag{8.28}$$

Where φ, ω and κ are the roll, pitch and yaw angles of the platform carrying the detection system, respectively.

In ΔQPG, according to cosine law, it can be achieved that:

$$d^2 = l^2 + S^2 - 2lS \cos \phi \tag{8.29}$$

In the right angle ΔOQR, $QR = l \sin \omega$ and $PT = l \cos \omega \sin \varphi$.

In the right angle ΔOUR, $OU = l \cos \omega \sin \varphi + d \cos \kappa$ and $PU = d \sin \kappa + l \sin \omega$.

In the right angle ΔPGU, $GU^2 = S^2 - (d \sin \kappa + l \sin \omega)^2$.

In the right angle ΔGOU, $GU^2 = (l \cos \omega \cos \varphi)^2 + (l \cos \omega \sin \varphi + d \cos \kappa)^2$.

Then,

$$S^2 - (d \sin \kappa + l \sin \omega)^2 = (l \cos \omega \cos \varphi)^2 + (l \cos \omega \sin \varphi + d \cos \kappa)^2 \tag{8.30}$$

By combining Eqs. (8.29) and (8.30), the following equation can be obtained:

$$\begin{cases} l = S \cos \phi - \dfrac{S \sin \phi}{\sqrt{1 - b^2}} b \\ d = \dfrac{S \sin \phi}{\sqrt{1 - b^2}} \end{cases} \tag{8.31}$$

Where $b = \cos \omega \sin \varphi \cos \kappa + \sin \kappa \cos \omega$

By solving Eqs. (8.31), (8.28) and (8.29) simultaneously, the coordinates of point P can be obtained

$$\begin{cases} X_P = X_G + \left(S\cos\phi - \frac{S\sin\phi}{\sqrt{1-b^2}} b \right) \cos\omega \sin\varphi + \frac{S\sin\phi}{\sqrt{1-b^2}} \cos\kappa \\ Y_P = Y_G + \left(S\cos\phi - \frac{S\sin\phi}{\sqrt{1-b^2}} b \right) \sin\omega + \frac{S\sin\phi}{\sqrt{1-b^2}} \sin\kappa \\ Z_P = Z_G + \left(S\cos\phi - \frac{S\sin\phi}{\sqrt{1-b^2}} b \right) \cos\omega \cos\varphi \end{cases} \tag{8.32}$$

In general, after acquiring the distances of sampling points of targets while scanning and sampling targets using laser beams, the slope distance S of targets is obtained. In the meantime, ϕ is expected to be obtained by measuring or pre-setting the angle between the emitted laser beams and the reference point of the system. On this basis, the coordinates (X_G, Y_G, Z_G) of the optical center of the detection system are measured employing a kinematic positioning method based on a GPS or a differential GPS with higher precision. Then, an associated INS is utilized to determine the attitude information of the detection system including angles of roll, pitch and yaw denoted as φ, ω and κ, respectively. In this way, according to Eq. (8.32), the coordinates (X_P, Y_P, Z_P) of sampling points of lasers can be computed. Finally, by conducting traversal scanning of the target zone using laser beams emitted from the detection system, the coordinates of all the sampling points which are distributed according to a certain rule (relying on the scanning mode) are acquired. Through imagining and processing, the location results of targets in the detection region are expected to be obtained.

8.3.2 *The Positioning Error*

According to Eq. (8.28), the errors of GPS positioning, ranging, attitude, scanning angle, *etc.* affect the positioning accuracy to some extent.

8.3.2.1 Radial Error

As for the positioning error, especially that of planimetric position, it has been analyzed concretely in previous researches [16, 17]. Therefore, the section mainly focuses on the influences of each measured parameters on the positioning accuracy, particularly, the radial accuracy, and the concept of influence factors of error is introduced as well [18, 19].

According to Eq. (8.32), the positioning error of DGPS directly influences the three coordinates. At present, as the DGPS positioning technology has been well developed, the kinematic surveying can reach an accuracy at centimeter magnitude as long as GPS receiver shows high precision. Thereby, the influences of other

Table 8.5 The calculated influence factors of errors

Operating parameters					Influence factors of errors				
S	ϕ	φ	ω	κ	$\partial Z/\partial S$	$\partial Z/\partial \phi$	$\partial Z/\partial \varphi$	$\partial Z/\partial \omega$	$\partial Z/\partial \kappa$
600	0	2	2	2	1	−0.0002	−0.0001	−0.0001	0
600	5	2	2	2	0.990	−0.0005	−0.0001	−0.0001	−0.0003
600	10	2	2	2	0.971	−0.0007	−0.0006	−0.0001	−0.0005
600	20	2	2	2	0.915	−0.001	−0.001	−0.0001	−0.001
600	20	5	5	5	0.873	−0.0015	−0.0012	−0.0002	−0.001
600	20	10	10	10	0.792	−0.0019	−0.0015	−0.0005	−0.001
600	20	15	15	15	0.697	−0.0024	−0.0018	−0.0007	−0.0012
600	20	10	10	2	0.842	−0.0015	−0.0014	−0.0005	−0.0011
600	20	10	2	10	0.801	−0.002	−0.0016	−0.0001	−0.0011
600	20	2	10	10	0.864	−0.0015	−0.0011	−0.0005	−0.0011

parameters on radial precision are mainly analyzed here. By performing partial differential on the other five parameters using Z_P in Eq. (8.32), we can obtain

$$\begin{cases} \dfrac{\partial ZP}{\partial S} = \cos\omega\cos\varphi\left(\cos\phi - \dfrac{\sin\phi}{\sqrt{1-b^2}}b\right) \\ \dfrac{\partial ZP}{\partial \phi} = -S\cos\omega\cos\varphi\left(\sin\phi + \dfrac{\cos\phi}{\sqrt{1-b^2}}b\right) \\ \dfrac{\partial ZP}{\partial \varphi} = -S\sin\varphi\cos\phi\cos\omega + S\sin\phi\cos\omega\left[\dfrac{b\sin\varphi}{\sqrt{1-b^2}} - \dfrac{\cos^2\varphi\cos\omega\cos\kappa}{(\sqrt{1-b^2})^3}\right] \\ \dfrac{\partial ZP}{\partial \omega} = -S\sin\omega\cos\phi\cos\varphi + S\sin\phi\cos\varphi\sin\omega \\ \qquad \left[\dfrac{b}{\sqrt{1-b^2}} - \dfrac{\sin\varphi\cos\omega\cos\kappa + \cos\omega\sin\kappa}{(\sqrt{1-b^2})^3}\right] \\ \dfrac{\partial ZP}{\partial \kappa} = \dfrac{S\sin\phi\cos\omega\cos\varphi(\sin\varphi\cos\omega\sin\kappa - \cos\omega\cos\kappa)}{(\sqrt{1-b^2})^3} \end{cases} \tag{8.33}$$

Equation (8.33) shows the influences of errors of ranging, beam pointing angle as well as the pitch, roll and yaw angles of the attitude of the detection system, that is, the influence factors of error, on the radial accuracy. The computing results of examples adopting actual parameters are illustrated in Table 8.5. Among the parameters, the units of the range and the influence factor of slope distance error are m and that of the angle is (°). While, influence factors of errors in beam pointing angle and the three parameters relating the attitude are in the unit of m/(").

According to Eq. (8.33) and Table 8.5, conclusions can be obtained as follows:

(i) The influence factor $\partial Z/\partial S$ of slope distance error is not related to the range but to the slope distance error whose value is approximate to 1, which means that the slope distance error is almost directly transferred to the radial error. The errors of other parameters slightly influence the radial error; while with their increments, the influence of the slope distance error on the radial accuracy is reduced.

(ii) The influence factors of the four angles are proportional to the slope distance, and hence, with the rising working altitude, the errors of the four angles greatly affect the radial error. In the meantime, the lager the values of the four angles, the more significant the influences of their errors are.
(iii) Among the influence factors of the four angles, that of the error of beam pointing angle $\partial Z/\partial \phi$ is the biggest, followed by that of pitch angle $(\partial Z/\partial \omega)$ and then yaw angle $(\partial Z/\partial \kappa)$.
(iv) Among the four influence factors of errors concerning the four angles, that of yaw angle $(\partial Z/\partial \varphi)$, merely 1/10 that of other parameters, is always the smallest, indicating that the vibration of the side roll affects the radial accuracy slightly.
(v) The influence factor of error of yaw angle varies with the beam pointing angle while is unrelated to the other three attitude angles, which implies that the error of yaw angle shows a constant influence on the elevation presents.

8.3.2.2 Comprehensive Error

The above parts analyze the sources of individual errors affecting radial accuracy as well as their quantitative influences on the radial error in target location. However, these factors are hard to be avoided or eliminated, and they work as a whole instead of separately influencing the direct location of targets, which causes the errors affecting the 3D positioning of targets.

In order to estimate the final location error caused by a target detection system based on laser imaging based on the error propagation law, the equation of error propagation in 3D positioning can be obtained according to Eq. (8.28) [16].

$$\begin{cases} m_{X_P}^2 = m_{X_G}^2 + (\cos\omega\sin\varphi)^2 m_l^2 + (l\sin\omega\sin\varphi)^2 m_\omega^2 + (l\cos\omega\cos\varphi)^2 m_\varphi^2 \\ \qquad + (\cos\kappa)^2 m_d^2 + (d\sin\kappa)^2 m_\kappa^2 \\ m_{Y_P}^2 = m_{Y_G}^2 + (\sin\omega)^2 m_l^2 + (l\cos\omega)^2 m_\omega^2 + (\sin\kappa)^2 m_d^2 + (d\cos\kappa)^2 m_\kappa^2 \\ m_{Z_P}^2 = m_{Z_G}^2 + (\cos\omega\cos\varphi)^2 m_l^2 + (l\sin\omega\cos\varphi)^2 m_\omega^2 + (l\cos\omega\sin\varphi)^2 m_\varphi^2 \end{cases} \tag{8.34}$$

Where, $m_{X_P}^2$, $m_{Y_P}^2$ and $m_{Z_P}^2$ are the errors of direct positioning in three dimensions in target detections using a laser imaging system, while m_φ^2, m_ω^2 and m_κ^2 are those of the three angles relating the attitude of the detection system. In addition, $m_{X_G}^2$ and $m_{Y_G}^2$ are the errors of 3D location in GPS positioning. Then, m_l^2 and m_d^2 are respectively [17]:

$$\begin{aligned} m_l^2 = & \frac{\cos^2\phi - b^2\cos 2\phi}{1-b^2} m_S^2 + \frac{S^2 sin^2\phi + \cos 2\phi b^2 S^2}{1-b^2} m_\phi^2 \\ & + \frac{S^2\sin^2\phi}{(1-b^2)^3}(m_\varphi^2 + m_\omega^2 + m_\kappa^2) \end{aligned} \tag{8.35}$$

Table 8.6 The estimated positioning accuracies

	Range error/m	Errors of the three angles relating the attitude m/(″)	Error of scanning angle ϕ/(″)	Errors of components in GPS positioning/m	Errors of 3D location		
					X	Y	Z
1	0.5	30	60	3	3.18	3.18	4.0
2	0.5	60	60	3	3.49	3.49	4.0
3	0.5	120	60	3	4.77	4.77	4.0
4	0.5	240	60	3	9.83	9.83	4.0

$$m_d^2 = \frac{\cos^2 \phi}{1 - b^2} m_S^2 + \frac{S^2 \cos^2 \phi}{1 - b^2} m_\theta^2 + \frac{S^2 \sin^2 \phi b^2}{(1 - b^2)^3} (m_\varphi^2 + m_\omega^2 + m_\kappa^2) \tag{8.36}$$

According to the equation of error propagation above, the location accuracy of the targets to be imaged are estimated. Suppose that the target detection system based on laser imaging technique is 600 m above the target. As the altitude is an important factor influencing the positioning accuracy, φ, ω and κ vary from 2° to 5° with the scanning angle varying from −22.5° to 22.5°. Table 8.6 illustrates the final results of accuracy estimation.

In a word, to improve the accuracy of direct target location in detections based on a laser imaging system, the accuracies of attitude determination and beam pointing angle have to be improved.

References

1. Huang T, Hu Y, Zhao G et al (2011) Target extraction and classification base on imaging LADAR range image. J Infrared Millimeter Waves 30(2):179–183
2. Shen Y, Yang Z (2003) A hierarchical filtering algorithm for processing distance images. J Xidian Univ 30(1):136–140 (Natural sciences edition)
3. Ren S, Wang B, Luo B (1997) Method for automatic detection of threshold and peak point based on histogram exponential smoothing. J Image Graph 2(4):230–233
4. Rafael CG, Richard EW (2003) Digital image processing, 2nd edn. Publishing house of electronics industry
5. Zhao N (2006) Processing of target detection using a laser imaging system. Electronic Engineering College, Hefei
6. Ni-Meister W, Jupp DLB, Dubayah R (2001) Modeling lidar waveforms in heterogeneous and discrete canopies. IEEE Trans Geosci Remote Sens 39(9)
7. Li X, Strahler AH, Woodcock CE (1995) A hybrid geometric optical-radiative transfer approach for modeling albedo and directional reflectance of discontinuous canopies. IEEE Trans Geosci Remote Sens (33):466–480
8. Ni W, Li X, Woodcock CE, Roujean JL, Davis R (1997) Transmission of solar radiation in boreal conifer forest: measurements and models. J Geophy Res 102(D24):29555–29566
9. Ni W, Jupp DLB (2000) Spatial variance in directional remote sensing imagery—recent developments and future perspectives. Remote Sens Rev 18(2–4):441–479
10. Nilson M (1995) Estimation of tree heights and stand volume using an airborne lidar system. Remote Sens Environ 56:1–7

11. Lohmann AW, Wirnitzer B (1984) Triple correlations. IEEE 72(7):889–901
12. Lohmann AW, Wigelt G, Wirnitzer B (1983) Speckle masking in astronomy: triple correlation theory and applications. Appl Opt 22(24):4028–4037
13. Fuyuan Peng, Xinjie Zhou (2000) Rapid infrared detection of small targets based on multiple correlation analysis. J Infrared Millimeter Waves 19(6):454–456
14. Yang Y, Peng F, Xie Z et al (2003) Laser detection of underwater targets based on wavelet analysis. Comput Simul 20(5):17–19
15. Songjun S, Wang S, Wang X et al (1998) Laser detection based on wavelet analysis. J Naval Univ Eng 3:37–42
16. Liu S, Shao H, Xiang M et al (1999) Principle and error analysis of the airborne laser ranging and multispectral imaging mapping system. J Surveying Mapp 28(2):121–127
17. You H, Ma J, Liu T et al (1998) Direct geo-location based on 3D remote sensing including GPS, attitude and laser imaging. J Remote Sens 2(1):63–67
18. Hu Y (2000) Quantitative technology of airborne 3D imaging. The Shanghai Institute of Technical Physics of the Chinese Academy of Sciences, Shanghai
19. Hu Y (2009) Three-dimensional target detection based on laser staring imaging. Technical report. Electronic Engineering College, Hefei

Chapter 9
The Target Classification and Identification for the Target Detection Based on a Laser Imaging System

In the target detection based on a laser imaging system, target classification and identification means determining the types and identifying the features of targets by processing their laser images including range, gray, waveform and level images. The target classification and identification requires processing the above images individually to achieve a comprehensive result. As to the image processing, regular methods can be used to handle each type of the images before the use. There are two kinds of target classifications including unsupervised and supervised ones. The former means identifying different targets merely depending on their features so it cannot determine their attributes, while the latter refers to identifying the targets with the help of priori knowledge together with target features so that it can determine the attributes of targets. In addition, using the characteristics of laser images along with other features of targets is also one of the important methods for the target classification and identification in the target detection based on a laser imaging system. This chapter mainly discusses the methods applied in the target classification and identification.

9.1 Feature Transformation of Detection Images

In order to improve the aggregation of similar features, and the separability of heterogeneous features, so as to increase the correct rate of target classification and identification, a significant step needs to be performed. That is, the feature space transformation for the feature sets of the targets needs to be performed before the classification and identification and after the extraction of the waveform eigenvectors of targets [1]. The method of regular space transformation for the eigenvectors judges the separabilities of inner-class, between-class and global scattering matrixes using Fisher criterion function. The regular subspaces constituted by the eigenvectors are expected to aggregate similar classes and separate different ones [2]. The feature extraction based on Fisher criterion function is also called

Y. Hu, *Theory and Technology of Laser Imaging Based Target Detection*,
DOI 10.1007/978-981-10-3497-8_9

discriminant analysis feature extraction (DAFE) or orthogonal subspace projection (OSP). In the field of pattern recognition, the DAFE method, as an important method, has been applied extensively in the radar target identification, image classification and image identification of human faces [3, 4].

The principle of regular space transformation of features is analyzed as follows [5, 6]: Suppose that $\boldsymbol{X}$ is a $n \times p$-order matrix, the element x_{ij} in the ith row and the jth column is the value of the jth characteristic variable observed at the jth time $(i = 1, 2, \ldots n,\ j = 1, 2, \ldots p)$. Moreover, it is assumed that $\boldsymbol{X}$ can be classified into g classes and the samples in the kth class account for n_k rows $(\sum_{k=1}^{g} n_k = n)$. Then the matrix transformation is carried out on $\boldsymbol{X}$ to increase the differences of characteristic quantities in the different classes and reduce those of the same class.

Suppose that $\boldsymbol{X}$ is converted into $\boldsymbol{Y}$ through the transformation ($\boldsymbol{Y}$ is a $n \times R$ matrix). The transform is:

$$\boldsymbol{Y} = \boldsymbol{XA} \tag{9.1}$$

The key to the regular transformation is to obtain the transformation matrix $\boldsymbol{A}$. Suppose that the average column vector of $\boldsymbol{Y}$ is $\boldsymbol{y}$, then, the quadratic sum of the fluctuation of column elements in $\boldsymbol{Y}$ relative to their average can be expressed as:

$$\mathrm{SS} = \sum_{k=1}^{\mathrm{g}} \sum_{j=1}^{\mathrm{n_k}} (\boldsymbol{y}_{k,j} - \bar{\boldsymbol{y}})(\boldsymbol{y}_{k,j} - \bar{\boldsymbol{y}})^{\boldsymbol{T}} \tag{9.2}$$

where $\boldsymbol{y}_{k,j}$ is the jth row vector in the kth class of matrix $\boldsymbol{Y}$ and $\boldsymbol{T}$ represents the associate operation.

SS can be expanded into the between-class SSB and inner-class SSW components. If the column vector in the mth column of matrix $\boldsymbol{A}$ is $\boldsymbol{a}$, then, SSB and SSW are functions of $\boldsymbol{a}$, which can be defined as:

$$\mathrm{SSW(a)} = \sum_{\mathrm{k}=1}^{\mathrm{g}} \mathrm{n_k} (\boldsymbol{y}_{k,j} - \bar{\boldsymbol{y}})(\boldsymbol{y}_{k,j} - \bar{\boldsymbol{y}})^{\boldsymbol{T}} \tag{9.3}$$

$$\mathrm{SSW}(\boldsymbol{a}) = \sum_{\mathrm{k}=1}^{\mathrm{g}} \sum_{\mathrm{j}=1}^{\mathrm{n_k}} (\boldsymbol{y}_{\mathrm{k,j}} - \overline{\boldsymbol{y}_{\mathrm{k}}})(\boldsymbol{y}_{\mathrm{k,j}} - \overline{\boldsymbol{y}_{\mathrm{k}}})^{\boldsymbol{T}} \tag{9.4}$$

where $\overline{\boldsymbol{y}_{\mathrm{k}}}$ is the average row vector of the kth class.

SSB and SSW stand for the between-class and inner-class distinctions, respectively. To evaluate the between-class and inner-class distances after the feature transformation, Fisher evaluation function is introduced, as follow:

$$F(\boldsymbol{a}) = \left\{\frac{1}{g-1} SSB(\boldsymbol{a})\right\} \bigg/ \left\{\frac{1}{n-g} SSW(\boldsymbol{a})\right\} \tag{9.5}$$

If there is no average difference among target features of g classes, then the bigger the value of $F(\boldsymbol{a})$, the larger the between-class distinction is comparing with the inner-class distinction. That is, $F(\boldsymbol{a})$ describes the degree of the separability of target features. F complies with Fisher distribution with the degrees of freedom being $g-1$ and $n-g$.

According to equations from (9.1) to (9.5), it can be obtained that:

$$F(\boldsymbol{a}) = (\boldsymbol{a}^T\boldsymbol{B}\boldsymbol{a})/(\boldsymbol{a}^T\boldsymbol{W}\boldsymbol{a}) \tag{9.6}$$

where,

$$\boldsymbol{B} = \frac{1}{g-1}\boldsymbol{B_0} \tag{9.7}$$

$$\boldsymbol{W} = \frac{1}{n-g}\boldsymbol{W_0} \tag{9.8}$$

$$\boldsymbol{B_0} = \sum_{k=1}^{g} n_k(\overline{\boldsymbol{x_k}} - \overline{\boldsymbol{x}})^T(\overline{\boldsymbol{x_k}} - \overline{\boldsymbol{x}}) \tag{9.9}$$

$$\boldsymbol{W_0} = \sum_{k=1}^{g}\sum_{j=1}^{n_k} (x_{k,j} - \overline{\boldsymbol{x_k}})^T(x_{k,j} - \overline{\boldsymbol{x_k}}) \tag{9.10}$$

where $\boldsymbol{B}$ and $\boldsymbol{W}$ are the between-class and inner-class covariance matrixes, while $\overline{\boldsymbol{x}}$ and $\overline{\boldsymbol{x_k}}$ are the average row vectors of $\boldsymbol{X}$ and the kth class, respectively.

According to the above analysis, to achieve the high separability of target features, the maximum value of $F(\boldsymbol{a})$ needs to be solved. It can be obtained that:

$$\begin{aligned} \boldsymbol{Ba} &= \lambda \boldsymbol{Wa} \\ \boldsymbol{W}^{-1}\boldsymbol{Ba} &= \lambda \boldsymbol{a} \end{aligned} \tag{9.11}$$

where λ is the maximum value of $F(\boldsymbol{a})$.

According to Eq. (9.11), $\boldsymbol{a}$ is the eigenvector corresponding to the maximum eigenvalue of matrix $\boldsymbol{W}^{-1}\boldsymbol{B}$. Then, the eigenvalues of $\boldsymbol{W}^{-1}\boldsymbol{B}$ are solved and arranged in a decreasing order as $\sigma_1 > \sigma_2 > \cdots > \sigma_m > 0$. The corresponding eigenvectors obtained $\boldsymbol{a_1}$, $\boldsymbol{a_2}$, ... $\boldsymbol{a_m}$ are employed as the column vectors to constitute a transformation matrix $\boldsymbol{A}$. Based on $\boldsymbol{A}$, the eigenvector space is transformed into $\boldsymbol{Y}$ from $\boldsymbol{X}$.

The characteristic quantities extracted from laser images include the width τ of laser echoes, the energy e of laser echo waveforms, the distribution vector D of echo waveforms, and the invariant moment of images of echo waveforms $(\phi_1, \phi_2, \phi_6, \phi_7)$. These characteristic quantities are arranged successively to constitute a row vector

$\delta[\tau, e, D, \phi_1, \phi_2, \phi_6, \phi_7]$ of the characteristic space. Through the extraction from images of multiple laser echoes of different targets, characteristic quantities including δ_{11}, δ_{12}, …, δ_{21}, δ_{22}, …, δ_{k1}, δ_{k2}, …, δ_{kn} are obtained. Thereinto, δ_{11}, δ_{12}, … are acquired from target 1. While, δ_{21}, δ_{22}, … and δ_{k1}, δ_{k2}, …, δ_{kn} are obtained from targets 2 and k, respectively. The eigenvectors of training samples of all targets constitute a characteristic space $\boldsymbol{X} = [\delta_{11}, \delta_{12}, \ldots, \delta_{21}, \delta_{22}, \ldots, \delta_{k1}, \delta_{k2}, \ldots]^T$ which is transformed into $\boldsymbol{Y} = [\chi_{11}, \chi_{12}, \cdots, \chi_{21}, \chi_{22}, \cdots, \chi_{k1}, \chi_{k2}, \cdots]^T$ according to the above method. In the new characteristic space $\boldsymbol{Y}$, χ_{11}, χ_{12}, … are the new eigenvectors of training samples of targets.

9.2 Target Classification and Identification Based on the Radial Basis Function Neural Network

A radial basis function neural network (RBFNN) is a feed forward neural network model combining the parametric statistical distribution model and the non-parametric linear perceptron model [7]. According to the theory of pattern recognition, all non-linear separable problems in a low-dimensional space can be mapped into a high-dimensional one, where the problems become linearly separable. The mapping theory of RBFNN is to represent the non-statistical mixing density distribution in sparse sample spaces utilizing the decomposed statistical density distribution and then obtain the mappings of different classes by connecting the structures using neural networks. Compared with the traditional universal approximate feedforward networks (such as BP neural network), RBFNN has advantages in approximation capability, classification ability, learning rate, etc [8]. With these advantages, its application practice shows that it is applicable to target identification in the target detection using a laser imaging system.

9.2.1 The RBFNN-Based Sorting Algorithm

RBFNN, as a three-layer feed forward neural network, its structure is shown in Fig. 9.1.

The first layer is the input layer, which consists of the nodes of signal sources.

The second one is the hidden layer where the transformation function of hidden units is a local-distributed nonnegative and nonlinear function, which attenuates in a radially symmetric manner with the central point. The number of units in this layer is determined according to the problems it describes.

The third one is the output layer. Its network outputs the linear weightings generated by hidden units.

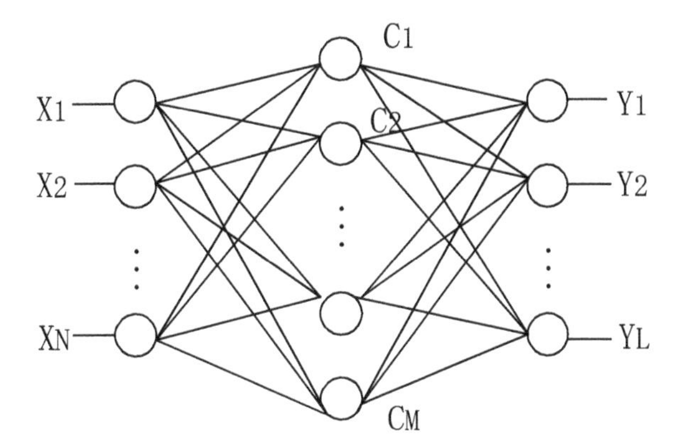

Fig. 9.1 The network structure of RBFNN

The RBFNN is transformed nonlinearly from the input to the hidden space while linearly from the hidden to the output space. The input-output relation can be expressed as:

$$Y_l = f(x) = \sum_{i=1}^{M} w_{il} g(\|\boldsymbol{X} - C_i\|) \tag{9.12}$$

where $\boldsymbol{X}$ is the input vector and $X \in R^n$, C_i represents the center of the units in the hidden layer ($C_i \in R^n$ and $i = 1, 2, \ldots, M$) and $\|\cdot\|$ refers to a norm. In addition, w_{il} is the link weight from the hidden to the output layer and g represents a radial basis function including Gaussian, multi-quadratic, inverse multi-quadric, spline, and thin plate spline functions.

When a Gaussian function is adopted, the input-output relation can be denoted as:

$$Y_l = f(x) = \sum_{i=1}^{M} w_{il} \exp\left(-\frac{\|X - C_i\|^2}{2\sigma_i^2}\right) \tag{9.13}$$

where σ_i is the response duration of the center of the ith hidden unit.

The key to the RBFNN-based classification and identification is to determine the clustering center C_i and select and train the parameters such as C_i, σ_i and w_{il} of the network structure.

9.2.2 *Classification and Identification Methods*

By employing the RBFNN, the eigenvectors extracted from the laser images of aerial targets are identified. To begin with, the features of the laser images are classified into k classes according to the types of the targets and ranges of attitude angles for same targets. Afterwards, the centers of the k classes are regarded as the clustering centers of the RBFNN. In this way, k unit centers are obtained in the hidden layer. The relation of each sample and its clustering center is represented by the membership degree μ_{ij}. If $\mu_{ij} = 1$, the sample X_i belongs to class j; while, if $\mu_{ij} = 0$, X_i does not belongs to the class.

The features of echo images of a target are classified into different classes by the ranges of attitude angles. In this case, 0 and 1 are not suitable for describing the membership degree of each sample and its clustering center. Therefore, fuzzy clustering is introduced here. Suppose that the laser images of an aerial target are classified into n clustering centers (C_{h1}, C_{h2}, … C_{hn}). Then, for a feature sample X_i containing a clustering center, its membership degree with the n clustering centers can be described as:

$$\begin{cases} \mu_{ij} = 1, & i = j \\ \mu_{ij} = d(i,j), & i \neq j \end{cases} \tag{9.14}$$

where $0 < d(i,j) < 1$. $d(i,j)$ represents the difference degree between the attitude angles of the sample X_i and the clustering center of class j. The smaller the difference, the higher the membership is. For a feature sample that does not belong to the same target, its membership degree with the clustering centers of other targets is 0.

According to Eq. (9.14), the value of the clustering center is:

$$C_k = \frac{\sum_{i=1}^{N} \mu_{ik} X_i}{\sum_{i=1}^{N} \mu_{ik}} \tag{9.15}$$

On the basis of the determined RBFNN structure and the sample distribution, the values of σ_i and w_{il} are selected randomly. By inputting the samples, the error between the output value and the correct identification result is obtained. Afterwards, C_i, σ_i and w_{il} are adjusted using the back-propagation model. By doing so, the correct result of classification and identification of laser images is achieved.

The algorithm for adjusting weights from the hidden layer to the output layer is:

$$w_{jk}(t+1) = w_{jk}(t) + \eta \frac{\partial \widehat{F}}{\partial w_{jk}} \tag{9.16}$$

where η and $\widehat{F}$ are the learning rate and the mean square error respectively.

Suppose that $d_i = \|X - C_i\|$, and $n_i = d_i (i = 1, 2, \ldots N)$ is input into the hidden layer. Then, according to the error back-propagation algorithm, the algorithms for adjusting C_i and σ_i are written as respectively:

$$C_i(t+1) = C_i(t) + \eta F(g) W \frac{\partial \widehat{F}}{\partial W} \frac{\partial d_i}{\partial C_i} \tag{9.17}$$

$$\sigma_i(t+1) = \sigma_i(t) + \eta \frac{\partial \widehat{F}}{\partial Y_i}\frac{\partial g_i}{\partial \sigma_i} \tag{9.18}$$

where $F(g) = \begin{bmatrix} \frac{\partial g_1(n_1)}{\partial n_1} & 0 & \cdots & 0 \\ 0 & \frac{\partial g_2(n_2)}{\partial n2} & \cdots & 0 \\ \vdots & \vdots & \ddots & \vdots \\ 0 & 0 & \cdots & \frac{\partial g_N(n_N)}{n_N} \end{bmatrix}$, g_i is the radial basis function, and Y represents the output of the ith node.

9.2.3 Experiments and Analysis

Four types of aerial targets, namely, the F-15 and F-16 fighters, and the 'Apache' and 'Cobra' helicopters are detected using a laser imaging system with the pitch angle varying from 0° to 30°. In the detection, the four targets are 4 km away from the detection system, and the lasers are emitted with the divergence angle and laser pulse width being 6 mrad and 8 ns, respectively. Through the simulation, a set of laser images are obtained, followed by the classification and identification applying the above algorithms.

The simulation models of the F-15 and F-16 fighters together with the 'Apache' and 'Cobra' helicopters are established by using 3DS MAX, respectively, as shown in Figs. 9.2, 9.3, 9.4 and 9.5. The waveforms of laser echoes obtained are illustrated in Figs. 9.6, 9.7, 9.8 and 9.9.

Stable transformation and feature extraction are performed for the echoes of the four targets to obtain partial eigenvectors of the four targets, as shown in Table 9.1.

According to the steps of classification methods for aerial targets based on RBFNN, 100 groups of echo waveforms of the four targets are obtained respectively at different distances and attitudes. Among which, 50 groups are selected randomly as the training templates. After the stable transformation, feature extraction and characteristic transformation, these templates are input into RBFNN

Fig. 9.2 The simulation model of the F-15 fighter

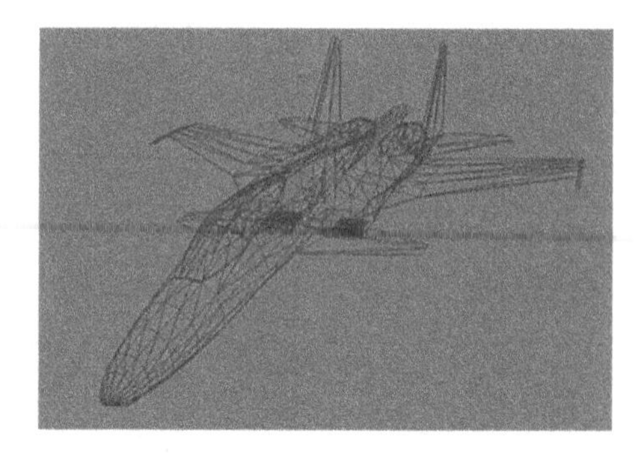

Fig. 9.3 The simulation model of the F-16 fighter

Fig. 9.4 The simulation model of the ‘Apache’ helicopter

Fig. 9.5 The simulation model of the ‘Cobra’ helicopter

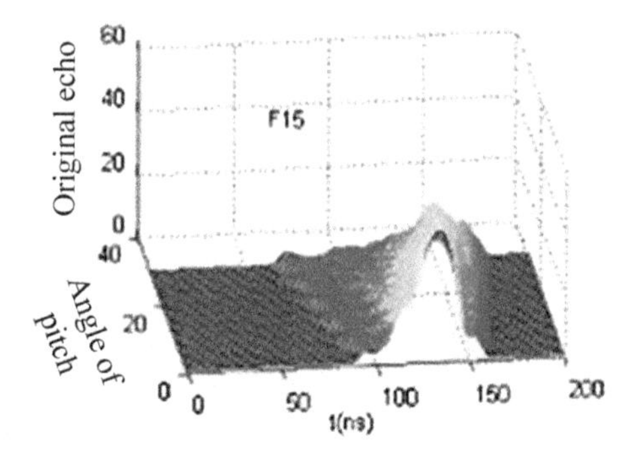

Fig. 9.6 The waveforms of laser echoes from the F-15 fighter

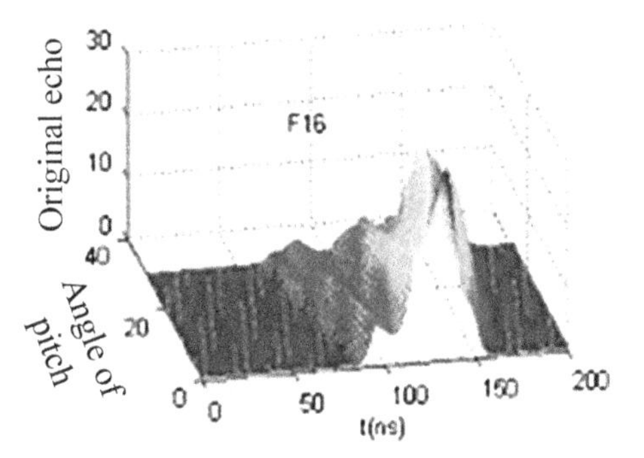

Fig. 9.7 The waveforms of laser echoes from the F-16 fighter

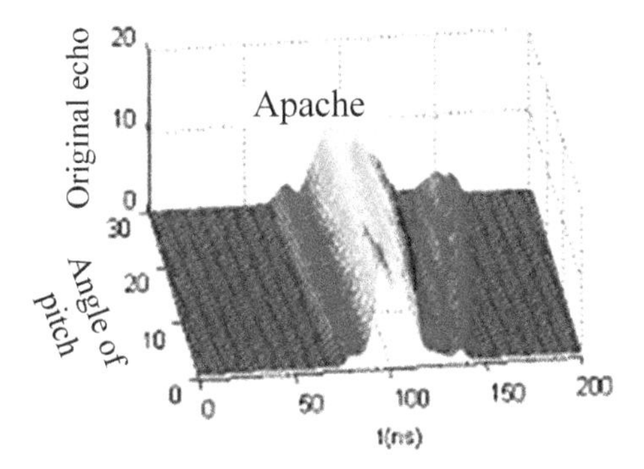

Fig. 9.8 The waveforms of laser echoes from the 'Apache' helicopter

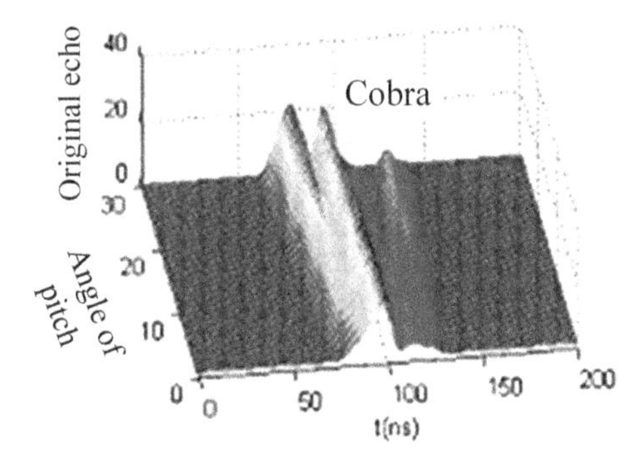

Fig. 9.9 The waveforms of laser echoes from the 'Cobra' helicopter

to be trained. While, the other 50 groups are considered as input data for identification and are identified through experiments. Then, Table 9.2 is obtained.

From the experiment results, it can be seen that by utilizing the RBFNN, a favorable result is obtained in the classification and identification for laser echoes reflected from the aerial targets.

Table 9.1 The results of feature extraction from laser images of the targets

Target	Serial number	τ	e	d1	t1	d2	t2	d3	t3	d4	t4	d5	t5	d6	t6
F-15	1	65	91.9	16.7	108	9.3	96	6.7	80	4.6	124	17.3	104	9.3	80
	2	66	91.8	16.9	108	9.1	96	6.4	80	4.7	124	18.1	104	9.4	80
	3	65	90.9	17.6	108	9.1	96	6.0	80	4.9	88	19.7	104	9.9	80
	4	64	90.1	16.5	108	9.5	96	6.6	80	5.3	124	17.2	104	9.4	80
	5	65	89.6	17.1	108	9.2	96	6.3	80	4.6	124	17.8	104	9.7	80
	6	64	88.8	19.3	108	8.8	96	5.9	80	5.1	88	20.2	104	10.2	80
F-16	7	64	67.3	11.2	108	7.3	96	6.9	80	6.2	116	24.8	104	8.9	120
	8	63	66.5	12.7	108	7.2	96	6.2	80	5.9	116	23.8	104	9.6	120
	9	64	66.7	11.0	108	8.1	96	7.6	80	7.2	116	24.9	104	10.2	120
	10	64	66.8	8.9	80	8.8	108	8.5	116	8.1	96	25.4	104	10.4	120
Apache	11	61	26.1	15.9	88	8.4	108	6.6	100	5.3	92	33.2	104	3.7	80
	12	65	25.4	11.4	88	9.2	108	6.8	80	5.3	100	32.5	104	3.1	80
	13	64	26.8	16.5	88	7.6	108	7.2	92	6.5	100	32.7	104	2.4	136
	14	64	27.2	16.9	88	11.6	100	6.4	92	6.0	108	32.3	104	2.7	136
Cobra	15	50	24.4	9.0	96	6.4	88	5.9	108	5.7	100	34.1	104	3.2	120
	16	52	24.4	10.9	100	5.4	80	4.1	108	3.9	88	33.7	104	3.4	80
	17	55	25.2	8.1	108	7.1	96	6.2	80	4.3	100	33.8	104	4.0	80
	18	50	26.3	6.7	108	6.7	88	5.8	96	3.9	80	33.8	104	2.9	120

(continued)

Table 9.1 (Continued)

Target	Serial number	d7	t7	d8	t8	d9	t9	ϕ_1	ϕ_2	ϕ_6	ϕ_7
F-15	1	6.1	120	33.6	112	1.8	64	8.1	2563	2.39e+005	2.71e+007
	2	6.6	120	32.9	112	1.9	64	8.1	2461	2.35e+005	2.74e+007
	3	7.5	120	32.3	112	2.0	64	8.0	2469	2.38e+005	2.79e+007
	4	7.9	120	32.8	112	1.9	64	8.0	2420	2.26e+005	2.65e+007
	5	8.9	120	32.4	112	2.1	64	7.9	2361	2.20e+005	2.60e+007
	6	10.1	120	31.4	112	2.2	64	8.0	2418	2.24e+005	2.64e+007
F-16	7	7.8	64	29.4	112	4.7	64	7.7	1803	1.81e+005	2.47e+007
	8	7.3	64	29.9	112	4.1	64	7.7	1794	1.77e+005	2.36e+007
	9	8.0	64	28.7	112	4.8	64	7.6	1676	1.71e+005	2.43e+007
	10	8.2	64	27.8	112	5.3	64	7.5	1525	1.52e+005	2.18e+007
Apache	11	1.3	136	29.5	96	22.7	112	7.3	2516	1.45e+005	8.77e+006
	12	2.3	88	28.4	96	23.1	112	7.1	2130	1.39e+005	9.03e+006
	13	2.1	80	30.6	96	21.6	112	7.5	2501	1.48e+005	1.04e+007
	14	2.5	80	29.3	96	19.5	112	7.6	2244	1.53e+005	1.36e+007
Cobra	15	1.5	80	20.8	96	17.8	112	6.9	2264	4.03e+005	8.88e+007
	16	3.0	120	20.8	96	16.4	112	6.7	1784	3.64e+005	8.82e+007
	17	2.4	136	18.9	96	17.9	112	6.7	1529	3.64e+005	9.92e+007
	18	2.1	80	21.9	96	17.7	112	6.4	1622	2.87e+005	5.50e+007

Table 9.2 The identification results of the four airplanes

	Target library	Targets to be identified			
		F-15	F-16	Apache	Cobra
Identification times	F-15	43	6	2	0
	F-16	6	41	0	0
	Apache	1	2	45	4
	Cobra	0	1	3	46
Correct rate of identification		86	82	90	92

9.3 Target Classification and Identification Based on the Support Vector Machine

The support vector machine (SVM) is a new machine learning method in the statistical learning theory. Developed based on VC theory and the principle of structural risk minimization, the SVM seeks the optimum compromise between the complexity of the model and the learning capacity by selecting proper discrimination functions according to the limited sample information. By doing so, the actual risk of the learning machine can be minimized so as to obtain a better generalization ability. The SVM is effective in solving problems such as small sample learning, high-dimensional features and nonlinear classification with high correct rates [9, 10], so it is suitable for the classification and identification of targets in the target detection based on a laser imaging system.

9.3.1 The Principle of the Classification Based on the SVM

For the binary-class linearly separable problems, the training data are assumed as:

$$D = \{(x_1, y_1), (x_2, y_2), \ldots, (x_l, y_l)\} \;\; x \in R^n, \;\; y_i \in \{+1, -1\}, \;\; i = 1, \ldots, l \tag{9.19}$$

The optimal classification hyperplane is the hyperplane that can separate the training data completely and correctly with the largest margins. The marginis defined as the sum of the distances from the nearest vectors to the hyperplane, where the nearest vectors to the hyperplane are named as the support vectors (SVs). Figure 9.10 illustrates the optimal classification hyperplanes. Although both the two hyperplanes can separate the training data completely and accurately, only the optimal hyperplane presents the maximum margin.

In general, the classification hyperplane can be expressed as $\omega \boldsymbol{x} + b = 0$, where ω is the direction vector and b is the bias. By normalizing the equation, the linearly separable sample set D is expected to satisfy:

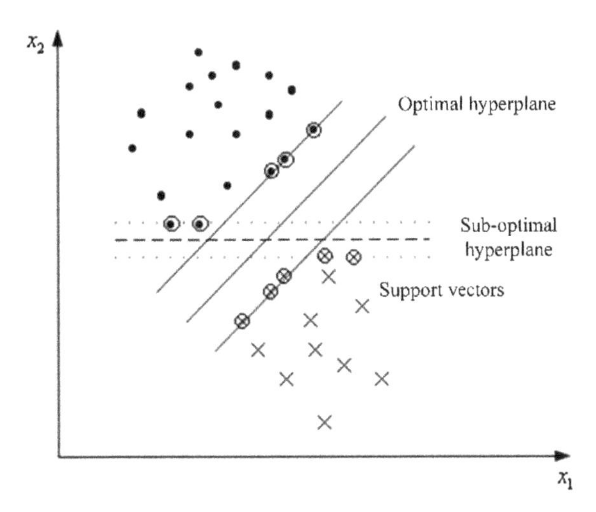

Fig. 9.10 The optimal linear classification hyperplanes

$$y_i[(\omega x_i) + b] - 1 \geq 0, \quad i = 1, \ldots, l \tag{9.20}$$

Under such condition, the margin equals to $2/\|\omega\|$, so the maximization of the margin is equivalent to minimizing the value of $\|\omega\|$. In order to find the optimal classification hyperplane, under the constraint of Eq. (9.20), the minimum functional is solved as

$$\Phi(\omega) = \frac{1}{2}(\omega \cdot \omega) \tag{9.21}$$

This optimization problem can be solved by utilizing Lagrange functional saddle points.

$$L(\omega, b, a) = \frac{1}{2}(\omega \cdot \omega) - \sum_{i=1}^{l} \alpha_i\{y_i[(x_i \cdot \omega) + b] - 1\} \tag{9.22}$$

where α_i is Lagrangian multipliers corresponding to each sample. At a saddle point, the solutions of ω, b and α have to meet the following conditions:

$$\frac{\partial L(\omega, b, a)}{\partial b} = 0 \tag{9.23}$$

$$\frac{\partial L(\omega, b, a)}{\partial \omega} = 0 \tag{9.24}$$

According to Eqs. (9.23) and (9.24), it can be obtained that:

$$\sum_{i=1}^{l} y_i \alpha_i = 0 \quad (9.25)$$

$$\omega = \sum_{i=1}^{l} y_i \alpha_i x_i \quad (9.26)$$

Equation (9.26) indicates that the weight coefficient vector of the optimal hyperplane is the linear combination of vectors of the training set. Then Eqs. (9.25) and (9.26) are substituted into Eq. (9.22) and the original problem is transformed into a dual problem, that is, α_i is solved under the constraint of Eq. (9.25) to maximize the function below.

$$Q(\alpha) = \sum_{i=1}^{l} \alpha_i - \frac{1}{2} \sum_{i,j=1}^{l} \alpha_i \alpha_j y_i y_j (\boldsymbol{x}_i \cdot \boldsymbol{x}_j) \quad (9.27)$$

In this way, the problem has transferred into the optimization of a quadratic function under the inequality constraints, which has a unique solution. If α_i^* is the optimal value, then $\omega^* = \sum_{i=1}^{l} y_i \alpha_i^* x_i$, and the optimal classification function is:

$$f(\boldsymbol{x}) = \mathrm{sgn}\{(\omega \cdot x) + b\} = \mathrm{sgn}\left\{ \sum_{i=1}^{l} \alpha_i^* y_i (\boldsymbol{x}_i \cdot \boldsymbol{x}) + b^* \right\} \quad (9.28)$$

where only a small part of α_i^* do not equal to 0 and their corresponding samples are the SVs. While b^* as the classification threshold, can be acquired through any SV or the mid-value of any pair of SVs in two classes.

The SVM-based target classification method means to serve the eigenvectors $[\delta_{11}, \delta_{12}, \ldots, \delta_{21}, \delta_{22}, \ldots, \delta_{k1}, \delta_{k2}, \ldots]^T$ after the feature transformations the input vectors which are mapped into the high dimensional feature space through the nonlinear mapping. Afterwards, the optimal classification hyperplane is constructed to maximize the distances from the original and the high dimensional sample spaces to the hyperplane. In a word, by using this method, the nonlinear problem is transformed into the linear one.

According to the related theories of functionals, as long as a kernel function satisfies the Mercer condition and corresponds to the inner product of a transformation space, the inner product operation in high-dimensional spaces can be realized using the function in the original space. Hence, in the optimal classification plane, the linear classification can be realized by using a proper inner product function $K(\boldsymbol{x}_i, \boldsymbol{x}_j)$ after the nonlinear transformation without increasing the computing complexity. In this case, the object Eq. (9.27) is converted into:

$$Q(\alpha) = \sum_{i=1}^{l} \alpha_i - \frac{1}{2} \sum_{i,j=1}^{l} \alpha_i \alpha_j y_i y_j K(\boldsymbol{x}_i, \boldsymbol{x}_j) \tag{9.29}$$

Then, the corresponding function for the classification is transformed into:

$$f(\boldsymbol{x}) = \operatorname{sgn}\left(\sum_{i=1}^{l} \alpha_i^* y_i K(\boldsymbol{x}_i, \boldsymbol{x}) + b^* \right) \tag{9.30}$$

Currently, the kernel functions commonly used are as follows:

1. The linear kernel function:

$$K(x, x_i) = (x, x_i) \tag{9.31}$$

2. The polynomial kernel function:

$$K(x \cdot x_i) = K(x \cdot x_i + c)^d \tag{9.32}$$

   where d is any positive integer. When $c > 0$ or $c = 0$, it is a non-homogeneous or homogeneous kernel function, respectively.
3. The radial basic function:

$$K(x \cdot x_i) = \exp\left(\frac{\|x - x_i\|}{2\delta^2} \right) \tag{9.33}$$

9.3.2 The Design of Classifiers

According to the above analysis, the key to the SVM-based target classification and identification is to select the sorting algorithm, kernel function and parameters of the kernel function.

The commonly used sorting algorithms in the SVM include C-SVC, V-SVC and LS-SVC. Taking the binary-class classification for instance, for the randomly generated 3D eigenvectors, the variations of training times of the three classifiers under different sample sizes are shown in Fig. 9.11.

According to Fig. 9.11, with the same sample size, the LS-SVC algorithm is more superior in the training time. For the target classification and identification in the target detection based on a laser imaging system, the sample sizes are different. When the sample size is large, the computation burden is expected to be reduced so as to accelerate the speed of classification and identification, which implies that realizing the real-time target classification and identification is the key index for the design of classifiers. Therefore, considering the characteristics of the target

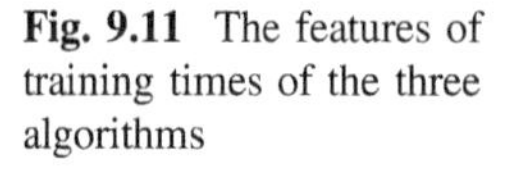

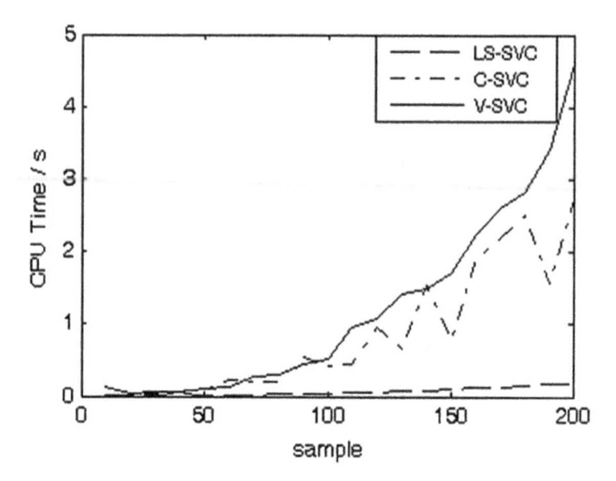

Fig. 9.11 The features of training times of the three algorithms

detection based on a laser imaging system, the LS-SVC algorithm is more applicable to the SVM-based target classification and identification.

Generally, there are mainly three types of kernel functions adopted in the design of SVM classifiers: the linear, polynomial and radial basis kernel functions. The type of kernel function determines the shape of the classification plane. For example, the linear kernel function generates linear classification planes, while the other two types of kernel functions generate non-linear ones, respectively. Therefore, in order to achieve a favorable classification result, appropriate kernel functions need to be selected for different problems. For most of the actual characteristic data, they are not always distributed linearly in the characteristic space. Moreover, the relationships among the features of laser echoes are not linear as well. For these reasons, the non-linear kernel functions are adopted, among which radial basis function, as a type of compact non-linear kernels, performs better than the other kernel functions in reducing the computing complexity of the classifier. In general, without knowing the spatial distribution of features, the application of the radial basis function is expected to achieve a better classification performance. Therefore, in the SVM-based target classification and identification, the radial basis function is adopted as the kernel function for the classifier.

The error penalty parameter C (the upper bound of the SV coefficient α_i) is used to compromise the proportion of the wrongly classified samples and the complexity of the algorithm. The selection of C is determined by the specific problems and the amount of noise in the data. In a certain characteristic subspace, a small C means that the empirical error penalty and the complexity of the learning machine are small, while the empirical risk is large. If the value of C is ∞, it requires meeting all the constraint conditions and all the training samples have to be correctly classified. In each characteristic subspace, there is at least one proper C which endows the SVM with the best generalization ability. The performance of the SVM is directly influenced by Gaussian kernel parameter σ as well. Owing to there is a one-to-one correspondence relationship among the kernel function, the mapping function and the characteristic space, the mapping function and the characteristic space can be determined through a known kernel function. For a specific problem, the kernel parameters need to be selected properly according to the particular case.

9.3.3 Experiments and Analysis

By adopting the constructed SVM classifier, the features of target images are classified and identified to verify the feasibility of the SVM-based target classification and identification [11]. In the experiments, two large civil airplanes B747 and DC10 together with two fighters F-16 and F-15are detected under the conditions of both the azimuth and pitch angles varying from −30° to 30°, and the SNR from 0 to 30 db. Then, 120 groups of laser images of each target are obtained, from which the energy features and the distributions of echo waveforms are extracted by applying the methods described in Chap. 6. The radial basis function is adopted as the SVM classifier, and the penalty factor and the kernel parameter are 20 and 3, respectively. Afterwards, 40 groups of data are selected from each target as training samples and the other 80 groups as testing samples. The results of classification and identification are listed in Tables 9.3 and 9.4.

As shown in Tables 9.3 and 9.4, in the classification and identification based on the distribution features of waveforms, except for the F-16, the correct rates of identification of the other three targets are above 80%. Among them, the correct rates of identification for both the B747 and DC10 are up to 100%, and the average correct rate of the identification is 87.81%. While, for the classification based on the energy features of echoes, the correct rates of the identification for the B747 and F-16 are rather high as 100 and 98.75%, respectively, while those of the other two types are rather low. These results indicate that for the civil airplanes like B747 and DC10, the classification and identification based on the distribution features of the waveform performs better than that based on energy features. However, for the fighters such as F-15 and F-16, both the two classification and identification

Table 9.3 The results of classification and identification with energy features as the input vectors

	Target library	Targets to be identified				Average correct rate of identification /%
		B747	DC10	F-16	F-15	
Identification times	B747	80	34	1	0	75
	DC10	0	46	0	0	
	F-16	0	0	79	37	
	F-15	0	0	0	43	
Correct rate of identification /%		100	57.5	98.75	53.75	

methods, either using the energy features or the distribution features of the wave forms the input vectors, fail to obtain dissatisfactory results. Therefore, the two features are supposed to be combined to improve the effect of classification and identification.

The two kinds of features constitute a combined eigenvector to be used in the classification and identification. Table 9.5 shows the classification results based on the combined eigenvector. The selected samples are the same as the above, that is, 40 groups of data for each target are selected as training samples and the rest as testing ones.

According to Table 9.5, all the correct rates of identification for the echoes of the four aircrafts are above 95%, among which those of B747 and F-15 are up to 100%. Therefore, with this set of parameters of the classifier, due to the significant increment of its dimensionality relative to the situation where separate features are used, the identification performance is improved greatly.

Apart from the input eigenvectors, the factors that affect the SVM-based target classification and identification include the kernel function, kernel parameter,

Table 9.4 The results of classification and identification with distribution features of waveforms as the input vectors

	Target library	Targets to be identified				Average correct rate of identification /%
		B747	DC10	F-16	F-15	
Identification times	B747	80	0	27	0	87.81
	DC10	0	80	0	11	
	F-16	0	0	53	0	
	F-15	0	0	0	69	
Correct rate of identification /%		100	100	66.25	86.25	

Table 9.5 The results of classification and identification with the combined eigenvector as the input vectors

	Target library	Targets to be identified				Average correct rate of identification /%
		B747	DC10	F-16	F-15	
Identification times	B747	80	0	0	0	97.81
	DC10	0	77	0	11	
	F-16	0	0	76	0	
	F-15	0	0	0	80	
Correct rate of identification /%		100	96.25	95	100	

Table 9.6 The results of classification and identification based on the three kernel functions (unit: %)

	Linear kernel function	Polynomial kernel function	Radial basis kernel function
B747	100	100	100
DC10	87.5	86	91.25
F-16	98.75	95	97.5
F-15	100	98.75	100
Average correct rate of identification	96.56	94.93	97.18

SNR, etc. The following discuss the influence of each factor on the results of classification and identification.

The combined eigenvector is selected as the input vector of the classifier and the samples are selected as the above. Then, the linear, polynomial and the radial basis kernel functions are adopted, respectively. The results of classification and identification based on the three types of kernel functions are illustrated in Table 9.6.

As seen in Table 9.6, by adopting the three kernel functions, the correct rates of target identification are above 90% with slight differences. Among which, the correct rates of identification by using the radial basis kernel function and the polynomial kernel function are the highest and the lowest, respectively. The above results accord with the theoretical analysis which indicates that the selection of kernel functions has little influences on the classification performance.

According the analysis in the previous sections, the parameters of the SVM mainly includes the penalty parameter C and the kernel parameter σ, the selections of which have a great effect on the performance of the SVM. In order to analyze the influences of the two types of parameters effectively, the penalty factor is fixed to be 30 to analyze the effect of the kernel parameter σ on the performance of the

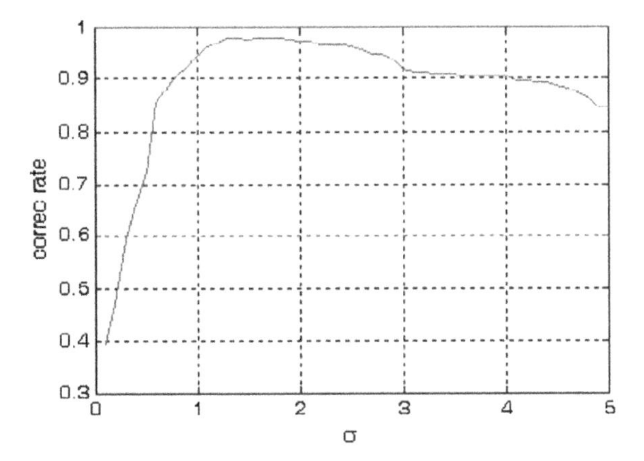

Fig. 9.12 The influence of the kernel parameter σ of the radial basis kernel function on the identification performance

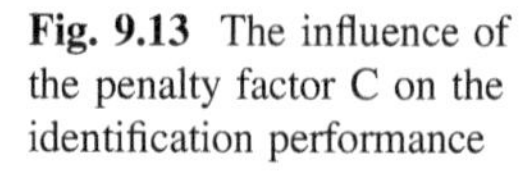

Fig. 9.13 The influence of the penalty factor C on the identification performance

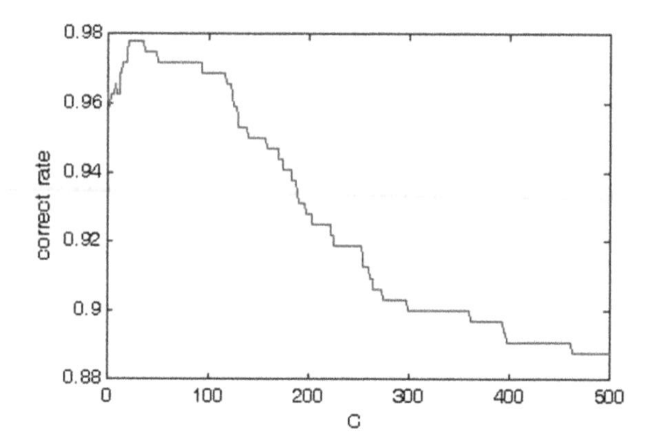

classifier. Figure 9.12 shows the variation of the correct rate of the identification with the kernel parameter σ.

As seen in the figure, the correct rate of identification rises at first and goes down latter with the increase of σ. Hence, there is an optimal σ that can generate the highest correct rate of the identification. In the above simulation experiments, the value of σ that leads to the highest correct rate of identification ranges from 1 to 2. Thus, the kernel parameter σ is supposed to be selected from the range.

To analyze the influences of the penalty parameter on the classifier, the kernel parameter σ requires to be fixed, being 1.3 for instance. Figure 9.13 illustrates the variation of the correct rates of identification with the penalty parameter C.

As shown in Fig. 9.13, with the increase of the penalty parameter C, the correct rate of target classification is shown to grow first and then drops down. Thereinto, the maximum correct rate of the identification is found to be with a penalty parameter C of 40. Therefore, in the design of the SVM classifier, after selecting the kernel parameter, the optimal kernel parameter and penalty factor can be determined by fixing the kernel parameter to analyze and compare the classification performances with different penalty factors.

With different SNRs, the discrepancies in the correct rate of the identification with the two varied eigenvectors are demonstrated in Fig. 9.14.

As seen from Fig. 9.14, as the distribution features of the waveforms are slightly influenced by the noise, the correct rate of identification remains above 85% when this type of features are used. However, as the noise has a significant influence on the energy features, the correct rate of identification applying energy features rises with the increment of the SNR. Moreover, the correct rate of identification based on the distribution features of the waveforms is higher than that based on the energy features as the SNR varies from 0 to 30 dB. The results are consistent with the evaluations for the two types of features in Chap. 8.

The target classification and identification for the target detection based on a laser imaging system is developed to sufficiently extract the information of the laser images and integrate the laser images with other types of images of targets. While extracting the laser images of targets for classification and identification, after

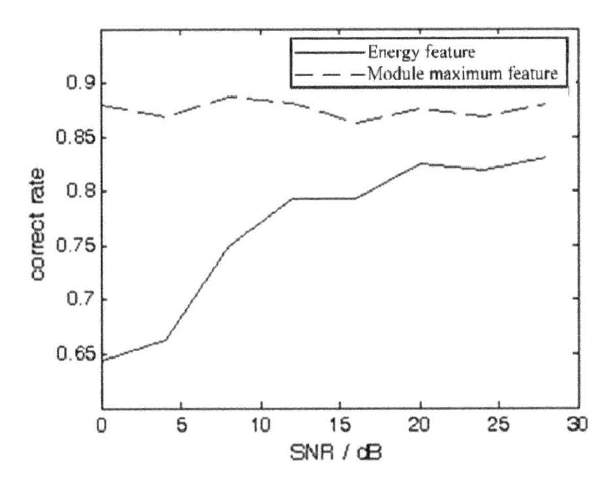

Fig. 9.14 The anti-noise performance of the SVM

performing 3D Fourier transform on the 3D matrix constituted of range and intensity images, Youman and Hart [12] used the technologies relating 2D templates to classify and identify the targets. Holmes et al. [13] studied the extraction of multi-scale geometric features from the range images obtained based on the laser imaging radar and identified three types of targets applying the combined classifiers based on the models and features. In addition, Pal et al. [14] extracted 13 features from the co-occurrence matrix of the intensity images obtained form laser radar imaging including correlation degree, moment, contrast ratio, entropy, mean value, standard deviation, etc. Besides, he utilized the k-nn classifier that can select features randomly and the improved conventional multi-layer perceptrons (CMLP) network to identify laser-imaging radar targets. By carrying out the feature level fusion between the geometric features and the reflection coefficient characteristics extracted from range and intensity images of targets, respectively, Soliday et al. [15] identified targets applying the fussy network classifier. Based on the statistical model, Koksal et al. [16] proposed a target identification method for laser imaging radars, which can be used to extract the contour edges of range images of targets through tracing contours. The extracted contour edges are considered as a kind of characteristics to establish a statistical model, which is then matched with the target model library to identify targets. For the target classification and identification based on the fusion of laser images and other types of images of targets, Selzer and Gutfinger [17] studied a method for the target classification and identification by combining the range images and the infrared emanation images obtained in laser radar imaging. By detecting the boundaries of the two types of images, this method can be used to extract the image edges and obtain the binary images, which are then matched with the binary templates. Then, based on the Dempster-Shafer (D-S) evidence theory, the results of target classification are achieved after the fusion. In addition, Li et al. [18] studied the target classification method in the decision level based on the fusion between the range images and the passive infrared radiation images obtained in laser radar imaging. Considering the correlation of the range images and the intensity images, the quality of the range images is improved first through the pixel level fusion of laser radar images. Then, the correlation coefficient is obtained by using the method of maximum close distance (MCD) and matched with the eight-value template. Finally, according to the D-S evidence theory, the classification results based on the fusion of the range images and the passive infrared radiation images acquired in laser radar imaging are achieved.

References

1. Zhao N (2006) Processing of the target detection based on a laser imaging system. Electronic Engineering College, Hefei
2. Hu R, Zhu ZD (1997) Researches on radar target classification based on high resolution range profiles. IEEE Conf 951–955
3. Belhumeur PN, Hespanha JP, Kriegman DJ (1997) Eigenfaces vs. fisherfaces. Recognition using class specific linear projection. IEEE Trans PAMI 19(7):711–720
4. Zyweck A, Bogner RE (1996) Radar target classification of commercial aircraft. IEEE Trans AES 32(2):589–606
5. Liu Y, Yang W (1999) Image identification of radar target based on canonical transformation. Syst Eng Electron 21(3):31–33
6. Zhou D, Yang W (2001) Range profile recognition of radar target based on modified canonical subspace. Syst Eng Electron 23(10):11–12
7. Girosi F (1994) Regulation theory, radial basis functions, and networks. Springer, Berlin, pp 166–187
8. Yang H, Ren Y, Li Y (2001) Aircraft target recognition method based on radial-basis-function neural network. J Tsinghua Univ (Sci & Tech) 41(7):36–38
9. Bian Z, Zhang X et al (2000) Pattern recognition. Tsinghua University Press, Beijing
10. Deng N, Tian Y (2009) Support vector machine, theory, algorithm and development. Science Press, Beijing
11. Li L (2010) Feature extraction of echoes and target classification based on laser remote sensing imaging. Hefei Electronic Engineering Institution, Hefei
12. Youmans DG, Hart GA (1999) Three-dimensional template correlations for direct-detection laser-radar target recognition. Schafer Corp, Chelmsford, MA, USA, ADA390244
13. Holmes QA, Zhang X, Zhao D (1997) Multi-resolution surface feature analysis for automatic target identification based on laser radar images. In: Proceedings of international conference on image processing, Sanfa Barbara, CA, USA, pp 468–471
14. Pal NR, Cahoon TC, Bezdek JC, Pal K (2001) A new approach to target recognition for LADAR data. IEEE Trans Fuzzy Syst 9(1):44–52
15. Soliday SW, Perona MT, McCauley DG (2001) Hybrid fuzzy-neural classifier for feature level data fusion in LADAR autonomous target recognition. In: Automatic target recognition XI, Orlando, FL, USA. SPIE 4379:66–77
16. Koksal AE, Shapiro JH, Wells WM (1999) Model-based object recognition using laser radar range imagery. In: Automatic target recognition IX, Orlando, FL, USA. SPIE 3718:256–266
17. Selzer F, Gutfinger D (1988) Ladar and FLIR based fusion for automatic target classification. SPIE 1003:236–245
18. Li Q, Dong G, Wang Q (2007) Target classification simulation for radar-passive-infrared imaging combination. Chin J Lasers 34(10):1347–1352

Chapter 10
Analysis of the Errors of the Target Detection Based on a Laser Imaging System

As a high-precision detection method, target detection based on a laser imaging system is conducted to obtain 3D and even higher dimensional information of targets. Therefore its data errors have an important influence on the positioning precision and the identification effect of target attributes. In order to ensure that the target meets practical requirements, it needs to comprehensively analyze error sources of the data and features of each error in the aforementioned detection. On this basis, it also needs to find error mechanisms and correction methods so as to guarantee the precision of target detection based on a laser imaging system. Therefore, the author mainly discusses the sources, features and relative methods for correcting the errors of such detection in this chapter.

10.1 Classification of Error Sources

Errors of target detection based on a laser imaging system are mainly reflected in the errors of obtained laser images, which directly lead to a decrease in the positioning precision, and reducing accuracy in the classification and identification of targets. The primary cause for generating the errors is mainly attributed to the errors including the errors in data acquisition, system, system integration detection and processing. The errors in data acquisition mainly include the errors of timing discrimination, measurement of time delays, directional angle of beams, position measurement of detection systems, attitude measurement of detection systems and sampling errors of A/D. The system errors primarily contain those existing in spot dispersion, atmospheric modulation, timing synchronization, sampling distribution and imbalance of multi-channel detections. As for the errors concerning system integration, they are generally found in the coaxial transceiving of lasers, azimuth errors of the detection system (registration errors between the system for measuring position and attitude and optical measurement) and optics-machines assembling errors. The errors relating to detection and processing mainly include errors in

Y. Hu, *Theory and Technology of Laser Imaging Based Target Detection*,
DOI 10.1007/978-981-10-3497-8_10

generating laser images, extracting image features and target features, detecting and positioning, and classification and identification. The error sources and influence of target detection based on a laser imaging system are shown in Fig. 10.1.

Generally, these errors are divided into gross errors, system errors and random errors. The gross errors which are caused by the less standard measurement and the influence of targets and environments can be avoided. The system errors are induced by some fixed factors in the system operation and data acquisition, which are expected to produce fixed deviations. The random errors are caused by certain random factors such as the changes of system performances in the data acquisition. From the perspective of the error types, the above errors in the target detection based on a laser imaging system are mainly random errors and partial deviations which can be eliminated by different means according to their different natures of

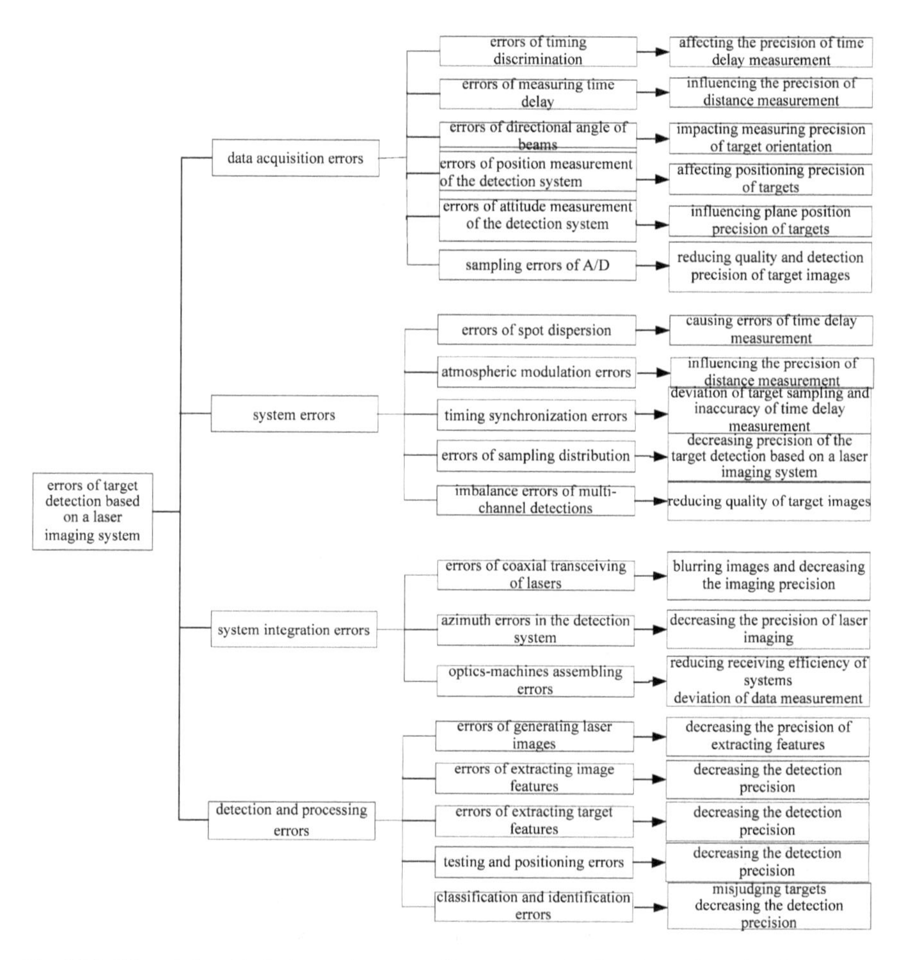

Fig. 10.1 The relationship between the errors of target detection based on a laser imaging system and their influences

various types. For example, for the gross errors caused merely by losing the laser detection information, they can be reduced by testing using detection data and marked correspondingly. While the system errors can be eliminated or weakened by means of strictly testing or changing working modes. As for the random errors, they need to be eliminated by taking corresponding measures.

The detection and processing errors are mainly shown in the uncertainty of data processing results, and can be eliminated by using methods including providing accurate data sources, optimizing processing methods and improving processing procedures. These methods are not supposed to be discussed in this chapter. As the imbalance errors of multi-channel detections of the system errors have been discussed in the previous chapters, it is not explained neither in this chapter. Other errors are mainly caused by all kinds of uncertainty generated in the design, implementation and operation of the system, and have to be addressed by the designers by taking various measures. This chapter is supposed to focus on analyzing data acquisition errors, system errors and system integration errors.

10.2 Data Acquisition Errors

The calculation of laser imaging and target positioning calls for 8 parameters, namely, target range, directional angle of beams, and 3D position and attitude coordinates of sensors. The acquisition errors of these data directly determine the precision of laser imaging. The target range corresponds to the time delay of target echoes, which are measured by firstly discriminating the arrival time of echoes.

1. **Errors of timing discrimination**

In the process of laser detection, in order to measure time delays, it needs to discriminate the arrival time of laser echoes. However, laser echoes are influenced by the receiver, transmission path and irregularity of targets. As the influences of various factors are input into the input port of timing discrimination circuit, the circuit shows the randomly irregular signal shape of laser echoes. No matter performing the timing discrimination for these irregular signals through A/D conversion or comparison using the simulating threshold, errors are difficult to be avoided, thus leading to the errors of time delay measurement and the errors in distance measurement. In the section, the author mainly analyzes the reasons causing timing discrimination errors using the latter method [1, 2].

Timing discrimination is generally to compare the echo signals with a threshold. As the signal exceeds the threshold, it is supposed to be an echo pulse at a logic level. Figure 10.2 presents the process of timing discrimination. After comparing with the threshold, the leading edge of the obtained echo pulse represents the time delay of the target. However, due to the irregularity of echoes, the leading edge of a same echo is not always linear. Moreover, waveform, amplitude and leading edge often change randomly in a string of echo signals. Therefore, the echo pulses obtained by comparing with a threshold move forward and backward in the time

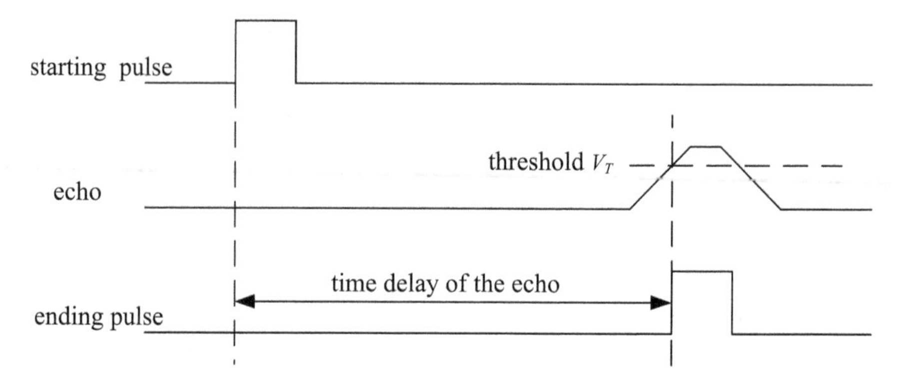

Fig. 10.2 The timing discrimination

axis, that is, the range walk phenomenon, which brings difficulties for the precise extraction of time delay information of laser echoes and easily generates errors of time delay measurement. In the air-to-ground target detection experiment based on a laser imaging system, by detecting fixed targets, it can be found that if the energy of laser emission is changed by 3 times, there is an error of 0.5 m around in the measured results of distance. In order to solve this problem, the author has come up with some solutions in the previous chapters. Here, the author mainly analyzes the error condition in this case.

The error sources of timing discrimination are mainly noise and changes of signal amplitude and signal width (or waveform). Suppose that the echo signal is a bell-shaped pulse, its voltage can be expressed as [2].

$$v_S(t) = V_{SO} \exp\left\{-4\,ln\,2\left[\frac{t}{\tau_{RV}}\right]^2\right\} + v_n \tag{10.1}$$

where, v_n is the instantaneous noise level (whose influence can be ignored with a large SNR). V_{SO} and τ_{RV} represent the amplitude of the echo signal and the pulse width of the processed echo signal, respectively. t is the time interval between any point in the signal waveform and the peak.

By comparing echo signals with the threshold, V_T the obtained timing discrimination errors caused by the changes of signal amplitude and width can be expressed as the following formulas, respectively [2].

$$\Delta t_{vr} = \frac{\tau_{RV}}{4\sqrt{ln\,2\,ln(V_{SO}/V_T)}} \cdot \frac{\Delta V_{SO}}{V_{SO}} \tag{10.2}$$

$$\Delta t_{rr} = \sqrt{\frac{\ln(V_{SO}/V_T)}{4\,ln\,2}} \cdot \Delta\tau_{RV} \tag{10.3}$$

The above results are related to the ratio of the signal to the threshold, namely, V_{SO}/V_T, while show opposite effects. Thus, it is very important to select a reasonable ratio of the signal to the threshold for reducing the errors of timing discrimination.

The existence of noise increases the uncertainty of timing discrimination, so the estimation value induced by noise can merely be calculated by using the statistical method, that is

$$\Delta t_{nr} \approx \frac{\tau_{RV}}{2\sqrt{2}\, ln(V_{SO}/\sigma_0)} \tag{10.4}$$

where, σ_0 is the effective value of the noise voltage which is expressed by the mean square root of noise.

According to formula (10.4), the bigger the SNR, the more significantly the precision of timing discrimination can be improved. Thus, in order to improve the precision of timing discrimination, the SNR needs to be improved at first as much as possible to reduce the errors induced by noise in addition to raise the reliability of distance measurement. For the errors caused by the variation of signal amplitude and width, it needs to select a reasonable signal and threshold based on the results of formulas (10.2) and (10.3). Therefore, it is necessary to process echo signals self-adaptively according to the change rules of signal features. For instance, a constant-fraction trigger or a self-adaptive constant-fraction trigger is adopted to decrease the influence of amplitude variation so as to keep the sampling pulse at the central point in the leading edge of pulses. However, this method needs to show the function of optical automatic gain control or it is likely to be out of action because of the too small dynamic range. In addition, the method for sampling and correcting waveforms also can be used to process echo signals. The specific process is as follows: the waveforms of the stimulated echo signals are sampled at high speed while forming echo pulses through comparing with the threshold. Then the results of time delay measurement are corrected by processing signals and referring to the prior data. The acquisition of prior data relating the relation between the results of time delay measurement and the characteristics of echo signals is the key to improving the precision using the correction method based on waveform sampling. In addition, the design of the circuit for sampling and recording waveform of echo signals is critical for the precision improvement as well.

The method for sampling and correcting waveforms is to compare and correct the built database of echo waveforms, and the specific ways and steps are as follows:

(i) In the calibration and experiment of the system, a number of typical ground targets are selected to perform the data acquisition of time delay measurement. Meanwhile, it needs to acquire and record data including the amplitude, waveform features and broadening of the echo signals to establish a relationship of these features of echo signals with the results of time delay measurement.

(ii) The exact range and time delay of the selected targets need to be measured by a laser range finder or a measuring equipment for time delay of higher precision in the static condition.

(iii) After comparing the two time delays above obtained, the correction rules of time delay measurement is found according to the relative echo signal features of each target. These rules include the relations of the signal amplitude, the pulse width, and the waveform fluctuation with the corrected time delay.

(iv) In the practical measurement, the waveform of echo signals of each target point needs to be recorded.

(v) In the processing of ground information, the amplitude, width and waveform features of echo signals are calculated. By correcting the results of time delay measurement using the rules obtained from the analysis, the result of timing discrimination which is less influenced by the variation of signal amplitude and width can be acquired.

2. ***Errors of time delay measurement***

The commonly used method for measuring time delays is pulse counting. In order to improve the measurement precision, the time delay can be measured by the following methods: high-resolution measurements of time delays based on A/D conversion and delay line interpolation as well as the measurement through waveform sampling. The measurement principle of the above methods is to convert the analog quantity into the digital quantity and then obtain the results of time delay measurement. Under such circumstance, it is inevitable to produce digital quantization errors no matter using any methods. These errors can be divided into the quantization errors of counting, time-voltage conversion and insertion. As the three methods of time delay measurement have been discussed in Chap. 4, the author just makes a simple conclusion.

The quantization errors of counting mainly include three influence factors: ① quantization errors caused by the interval of counting clock, namely, counting resolution; ② quantization errors induced by the period of counting clock, that is, the difference existing between T_0 used for calculating time delays and the actual period of counting clock; and ③ the difference existing between the actual speed of light in the atmosphere and the light speed C used for computing time delays in the theory. The mean square root errors of the range in the calculation of targets' distance based on the errors of time measurement are written as

$$\varDelta Z_N = cT_0/2\sqrt{3} \tag{10.5}$$

$$\varDelta Z_{T_0} = \frac{Z}{3} \cdot \frac{\varDelta T_0}{T_0} \tag{10.6}$$

$$\varDelta Z_N = NT_0 \varDelta C/2 \tag{10.7}$$

when $T_0 = 6.7$ ns, $\Delta Z_N = 0.58$ m. While, when $Z = 1000$ m and $\Delta T_0/T_0 = 10^{-5}$, $\Delta Z_{T_0} = 3.3$ mm. While, ΔC is related to the atmospheric condition, so the most important error is distance measurement error ΔZ_N in the time measurement error produced in the distance measurement of the target detection, and the error is caused by the distance measurement with limited resolution.

Therefore, the precision of distance measurement can be improved by raising the resolution of range counting. However, due to the limitation of electronic circuits, it is hard to realize the distance measurement with higher resolution, so it needs to take some improvement measures to make up the shortage of distance measurement.

The quantization errors of time-voltage conversion are the errors generated in the process of measuring time by performing digital quantization to the charging voltage of capacitors. The charging voltage is converted from the simulated time interval existing in the high-resolution measurement of time delays based on A/D conversion. This quantization error can be as small as that at picosecond level in the theory.

The quantization errors of insertion are caused by the limited levels of the tapped delay line in the time delay measurement using the delay line interpolation method, whose highest resolution of time interval reaches only one LSB. Meanwhile, due to the influences of comprehensive factors including the dispersive process of realizing circuits, the nonlinearity of delay lines, the random variation of environment and the time jitter, the measurement errors are likely to be larger.

3. ***Errors of directional angle of beams***

In the target detection based on a laser imaging system, the positioning of target sampling points needs to use the angle ϕ between the laser beams and optical reference direction (see relevant content in chapter eight), that is, the directional angle of beams. The optical reference direction is usually the central optical axis of the receiving telescope and changes with the attitude variation of the detection system, so it can be given by an attitude measurement device. When a target is detected, the direction of the laser beams changes dynamically, no matter employing any scanning mode to sample the target. This change can be realized either by mechanical scanning or splitting of optical beams. The directional angle of laser beams is generally detected by angle encoders of scanning mechanisms or the calibration in laboratories. Therefore, the real-time directional angle and the determined directional angle show a difference, which is the error of directional angle and changes randomly. The error of directional angle of beams is the main factor affecting the measurement errors of target orientation. Simple calculation reveals that if an error of 10 μrad exists in the directional angle of beams, which is very likely to happen, then the error of 100 mm around appears in the plane position in the detection range of 10 km. This is a considerable error.

For scanning and sampling in the target detection, the error of scanning angle directly determines that of directional angle of beams, which is mainly caused by many factors. These factors include the control errors due to non-uniform rotation of the scanning motor, the micro vibration in the rotation of scanning mirrors and the plane angle error of the scanning mirror surface. In the sampling of laser array, the splitting ways such as grating splitting are commonly used. Under such circumstance, the errors of directional angle of beams are mostly closely related to the incident light beam, followed by the calibration error of splitting angles and the stability of beam splitters.

4. ***Errors of position measurement of the detection system***

In the target detection based on a laser imaging system, GPS is mainly used to obtain precise position of detection platforms and generate high-precision 3D laser images of targets using a laser range finder through dynamic differential positioning. As target detection systems are always moved, they are difficult to be positioned along with easy occurrence of positioning errors. Moreover, positioning errors are inevitable while using existing positioning modes and signal types. Generally, positioning errors directly influence the precision of target position parameters. In order to guarantee that errors of target detection are as small as at sub-meter level, GPS positioning errors need to reach meter and even smaller levels.

In the GPS positioning, there are many errors, which can be divided into three types according to error sources: ① the errors related to GPS and mainly include the errors of satellite orbits, the ephemeris errors, the clock errors of satellites, and the relativistic effect; ② the errors related to signal propagation, including the errors of atmospheric delay, the ionosphere refraction, the troposphere refraction, and the multi-path effect; and ③ the errors pertinent to receivers, such as the clock errors of receivers, the channel deviation, the observation errors, and the miscalculation of integer ambiguity [3]. To reduce the influence of GPS positioning errors, while performing the dynamic differential GPS positioning, multiple base stations which are distributed uniformly in measuring regions can be established to ensure GPS is not far from base stations. Meanwhile, the precise single-point positioning technology also can be used to reduce the influence of GPS positioning errors.

5. ***Errors of attitude measurement of the detection system***

The generation of high-precision 3D laser images of targets also needs to measure attitude parameters of detection platforms. Therefore, errors of attitude measurement also are one of main factors influencing the precision of the data obtained in the detection based on a laser imaging system.

Errors of attitude measurement of detection systems are mainly shown as the errors in the three attitude directions, namely, the errors of the yaw, roll, and pitch angles. Because the gyroscope of attitude measuring devices points north linearly, the yaw angle errors are large errors that relatively hard to be eliminated among the errors of attitude measurement. In addition, the errors are likely to influence the motion direction and side of system platforms whose 3D coordinates need to be positioned. While the roll angle errors mainly affect the motion side and height direction of system platforms whose 3D coordinates require to be positioned.

The pitch angle errors mainly impact the motion direction and height direction. Generally speaking, errors of attitude measurement mainly result in precision reduction of plane position, which therefore deviates from the orientations of sampling points obtained by target detection and processing from the actual orientations. When errors of attitude measurement cannot be avoided, their influence on positioning can be weakened by lowering the flight height.

Errors of attitude measurement main include the errors of device installation, the constant errors of accelerometers, the proportional errors of accelerometers, the gyroscope drift, the measurement noise, the nonorthogonality of bearings, the errors of the gravity model, and the geoid errors [4]. At present, the precision of attitude measurement can be improved by using GPS calibration technology, through which the highest course precision and level precision can reach arc-second level in the inertial navigation of GPS.

6. ***Errors of waveform digitalization***

The waveform digitalization of laser echoes is an important content of high-precision target detection based on a laser imaging system. It includes the high-fidelity acquisition of laser echoes and the sampling, maintaining, and quantization (A/D) of analog signals of the laser echoes.

The high-fidelity acquisition is the basis of waveform digitalization of laser echoes. Only when a receiver has an appropriately high bandwidth can the maximum distortionless acquisition of echo signals be achieved. Approximately, when the product of the receiver bandwidth and pulse width of laser echoes is equal to 4, this high-fidelity acquisition effect can be achieved. For example, when the echo width is 10 ns, bandwidth can be selected as 400 MHz. Under such circumstance, the inappropriate selection of the receiver bandwidth is the root for generating errors. If the receiver bandwidth is too large, large noise is apt to be introduced. Then, once the noise is superposed onto pulse signals, echoes are supposed to be distorted. While if the receiver bandwidth is too small, then the high-frequency components of echo signals are filtered, which results in that true laser echo signals modulated by targets cannot be acquired.

The sampling of analog signals of laser echoes needs to obtain values of echo signals in a short time in a sampling pulse to precisely reflect waveform features of echo signals. Meanwhile, sampling frequency needs to be at least 3–5 times of the echo bandwidth. Even so, due to the limitation of sampling frequency, sampling errors are also produced, which finally results in the loss of high-frequency components of the echo pulses. Moreover, the instability of sampling clocks and the noise of sampling circuits are also the factors generating sampling errors. In this case, the non-instantaneity and jitter of sampling and the limitation of sampling frequency are the primary causes of errors.

The maintaining of analog signals is an important content of achieving high-fidelity requirement of waveform digitalization. The sampled analog signals are generally stored in a memory of analog signals and then undergo A/D conversion through quantization circuits. Here, the maintaining circuit needs to have minimum discharge characteristic, so the instability of the maintaining circuit is the source of generating errors. If the maintaining circuit decreases by 1% in a sampling period, it means that the intensity of laser echo signals declines by 1% as well.

The final result of waveform digitalization is A/D conversion, and quantization bits determine the precision of waveform digitalization. Because the quantization error of 1 LSB is inevitable in the waveform digitalization, it needs to select an A/D converter with high bits. Moreover, noise of A/D sampling circuits also can generate quantization errors. The quantization errors are expected to directly affect the quality and detection precision of target images, because the generation of target images and the target detection based on the waveform images depend on the high-precision digitalization of laser echoes.

The influences of waveform digitalization errors are mainly reflected in the following aspects: ① the waveforms of digitized laser echoes are incomplete or deformed, which influences the width of equivalent pulses, the echo position within the sampling gate, and the calculation precision of the echo power and back scattering rate. Furthermore, it leads to the decline of the quality and image detection precision of target images; and ② external noise such as flicker noise and thermal noise are supposed to be introduced into the sampling circuit, which can change waveforms of laser echoes to a certain extent and cause distortion of digitized echo waveforms. As a result, the digitized echo waveforms show a certain degree difference with the actual ones, which further affects the precision of waveform images and gray images of targets.

10.3 System Errors

1. ***Errors of spot dispersion***

The ground is supposed to present a laser spot when lasers irradiate the ground. The error of spot dispersion refers to the uncertainty of the laser spot for plane positioning and the uncertainty of the distance measurement due to the target fluctuation within the spot in the presence of the divergence angle of laser beams. This phenomenon is similar to the dispersion of optical imaging. Target fluctuation and the distortion of echo signals of complex targets within the laser spot lead to the generation of errors between the measured time delay of targets and the actual values, and ultimately affect the precision of 3D laser imaging.

Because of the existence of the divergence angle of laser beams, terrain fluctuation (including slopes, sags, and swells) in the irradiated spot is expected to lead to a relatively big error of distance measurement and plane position errors which cannot be ignored, and finally cause 3D positioning errors. The smaller the divergence angle of beams is, the smaller the directional angle of beams and then the smaller the 3D positioning error. However, as the directional angle of beams is an operating parameter of systems, it cannot be too small. Obviously, the errors caused by conical scanning with a small dip angle are smaller than those caused by linear scanning. While, the divergence angle of beams can be decreased. Meanwhile, the decreasing of the angle not only can reduce the errors of distance measurement and plane position, but also improves the dynamic detection effect [5], so decreasing the divergence angle of beams is a way of reducing errors. Moreover, with the diminution of laser spots, the possibility of ground object fluctuation within the spots is likely to be reduced and therefore the ground objects within the laser spots show relatively single geometrical features. In addition, the waveforms of laser echoes are slightly deformed small. Under such condition, the measurement of the leading edge of laser echoes can basically reflect the actual range of a single target. However, laser beams cannot be too small, otherwise, targets are not able to completely modulate the echo waveforms within a laser beam, and therefore it is hard to obtain preferable waveform images of targets.

2. ***Atmospheric modulation errors***

The atmosphere produces the absorption, scattering and turbulence effects on transmitted lasers to change the intensity, energy distribution and incident direction of lasers irradiating targets, thus resulting in atmospheric modulation errors. Owing to the laser transmission energy and the received signals are reduced by atmosphere, the SNR of the optical receiving system in the detection system based on laser imaging technique is reduced, which thus affects the precision of distance measurement. Atmospheric turbulence changes the distribution of laser beams, leading to the energy distribution of laser beams which irradiate targets deviating from original rules: (1) the energy distribution is not consistent with Gaussian distribution which is applied as the basis of theoretical analysis; (2) the spot is likely to be distributed as several irregular spots, which dynamically change all the time. All these make the atmospheric modulation greatly influence laser echoes and restrain the transmission of target modulation information. Under such circumstance, it is unfavorable for implementing precisely laser imaging as well as classifying and identifying for targets. As atmospheric density varies with the height, the refractive indexes of atmosphere for lasers are not the same in different heights, thus causing refractive effect. The refractive effect is supposed to change the transmission path of lasers and impact the precision of time delay, therefore producing the errors of distance measurement. When lasers pass through atmosphere, the influence of atmospheric modulation on lasers depends on the wavelength of laser pulses. For the same laser pulse signal, errors generated by atmospheric modulation are mainly related to the velocity, composition, temperature, pressure and humidity of atmosphere.

3. ***Timing synchronization errors***

In the target detection based on a laser imaging system, all data are acquired in the same timing synchronization, and all data are stored after being labeled with time stamps. Therefore, the synchronization degree of time determines the acquisition stability and the mutual integration of various data. However, in practice, the existence of some factors is likely to lead to results of instable and incongruous synchronization, that is, timing synchronization errors. These factors include the jitter of synchronous pulse sources, the difference of various timing drivers, the delay and jitter of Q-switched lasers, the instable rotation speed of scanning mechanisms and the delay of timing circuits [6]. Timing synchronization errors affect the target sampling, the data acquisition, the attitude and position measurement and the direction of instantaneous field of view in the timing synchronization relation. Besides, they mainly result in the deviation of target sampling and inaccurate time delay measurement. In the short-range target detection based on a laser imaging system, timing synchronization errors have less significant influence. While in the long-range detection, these errors need to be considered.

Timing synchronization errors are directly expressed as the random drift and jitter of the timing synchronization relation in each channel. Meanwhile, different from other errors, timing synchronization errors are likely to be accumulated. In order to solve this problem, the commonly timing reset method is generally used, thereinto, the pulsed reset of GPS collection is an effective method. A GPS receiver can transmit a precise pulse per second, the leading edge of which resets the timer in the laser detection process and affects the time delay of the pulses from being generated to sending data from GPS receivers to INS. Under such circumstance, the range from the detection platform to target sampling points and the direction of laser beams can be measured precisely. Meanwhile, the time for transmitting pulses by the laser can be triggered precisely as well.

4. ***Errors of sampling distribution***

In the target detection based on a laser imaging system, laser beams need to irradiate target regions for sampling according to certain rules, and a receiving telescope receives laser echoes scattered by targets. Obviously, only the echoes reflected from the irradiated points can be detected by a detection sensor, so the sampling irradiation condition of lasers for targets determines the sensing ability of detection systems for targets. Targets are sampled by lasers complying with Nyquist sampling theorem, and the sampling frequency is more than twice of the spatial frequency of targets. That is, the spatial fluctuation interval of targets is more than twice of the distribution interval of laser sampling points in the target region. If a flat target is detected, the sampling points can be distributed relatively sparsely; while if the target fluctuates significantly, then the sampling points need to be arranged densely. However, the requirement is difficult to be met absolutely in the real target detection. For this reason, the existence of the errors of sampling distribution is difficult to be avoided.

Errors of sampling distribution are mainly caused by the unmatched sampling intervals, which give rise to the loss of high-frequency fluctuation components of targets. Meanwhile, as detailed information of targets is hard to be detected, the precision of laser imaging for target detection is reduced, which finally affects the result of target identification. In order to solve this problem, the laser sampling points need to be distributed as densely as possible with the laser beams being as narrow as possible. However, in practical applications, the above conditions are difficult to be guaranteed, so that the detection efficiency declines inevitably. According to previous researches, two methods are adopted to decrease the errors of target sampling distribution: the first is the sampling with variable resolution based on self-adaptively scanning which is mentioned in Chap. 4. By using the method, it is able to perform the high-density sampling in fluctuating target regions, while the low-density sampling in flat regions. The second method is super-resolution laser imaging based detection technology applying overlapped sampling [7]. It is a kind of detection method used to perform the overlapped sampling of laser beams with a large divergence angle based on the target features so as to obtain the detail features of targets and achieve the super-resolution laser imaging. By employing the partial overlap among imaging spots to process the echo signals from the overlapped spots, the target information within the imaging spots can be obtained to improve the imaging resolution and imaging precision of targets. In the process of implementation, it needs to change the overlapped size according to target features. If targets fluctuate significantly, large areas are overlapped; otherwise, small areas are overlapped.

10.4 System Integration Errors

1. ***Errors of coaxial transceiving of lasers***

For target detection based on a laser imaging system systems which work actively, their basic requirement is that transmitted lasers can be detected, of course, the more the better, by the receiving system after experiencing target scattering. Therefore, in the implementation of the detection system, laser sampling spots need to be located in the laser receiving footprints. To realize this, the following requirements need to be satisfied: (1) the receiving instantaneous field of view is slightly larger than the laser divergence angle; (2) the axis of the receiving instantaneous field of view coincides or nearly coincides with the central axis of transmitted beams, that is, the coaxiality of transmitting laser and receiving echoes. The first requirement is easily to be met by selecting appropriate parameters of transmitting, receiving and detectors. In generally, the laser transmitting and the laser receiving systems can be arranged in two structure types with coaxial structure or parallel axes. In the design of the coaxial structure, an optical emitting device is equipped in the center of the receiving telescope. In this way, it is easy to achieve the coaxial transceiving of lasers. However, the optical emitting device is likely to block the reflected echoes, which diminishes the effective receiving aperture. In the design of parallel axis structure, an optical emitting device is placed outside a receiving telescope, whose central axis is parallel to the axis of the receiving optical device. This structure is equivalent to the coaxial transmitting and receiving in long-range detections.

However, if the coaxial transmitting and receiving is destroyed due to the changes of design, assembling, calibration or work environment, the laser sampling points cannot coincide with the receiving footprints. Under such circumstance, the following situations are likely to occur: (1) reducing the effective receiving energy or even no laser echo can be received; (2) not all the received target echoes are those reflected from laser sampling points, which are likely to make mistakes in the laser imaging, resulting in the blur target images and affecting the imaging precision; (3) in the target positioning, using the directional angle of beams to calculate the position coordinates inevitably causes the positioning errors. For example, if the divergence angle of beams and the receiving field of view are 1 and 1.2 mrad, respectively and the error of coaxial transceiving is 100 μrad, laser sampling points are located in the laser footprints. Under such condition, almost no receiving energy is lost, but the fuzzy images and positioning errors are expected to appear. Suppose that the operating range is 10 km at this moment, then the plane positioning error obtained by simple calculation reaches 1 m. So, for targets whose structure fluctuates within 1 m are supposed to lose detailed information, so that blur images are generated.

2. ***Azimuth errors in the detection system***

In the target detection based on a laser imaging system, the laser detection system, the GPS positioning system and the inertial measurement unit (IMU) are integrated synthetically. While conducting laser imaging and target positioning, it firstly needs to measure data relating the external azimuth elements including the directional angle of the main optical axis, the pitch, roll and yaw angles of sensors, the coordinates of sensors, and the time delay information of lasers. Secondly, it also has to know accurate spatial relations of GPS/IMU positioning and attitude determination with the laser detection system, namely, the relations of azimuth elements within the system. These relation data are factors need to be considered in the generation of high-precision laser images, which mainly include: (1) the offset between the central point of the laser detection system and the GPS phase center; (2) the offsets of the GPS phase center and the central point of attitude measurement; (3) the offset between the central points of attitude measurement and the laser detection; (4) the angle of main axis, that is, the angle between the three main axes of the coordinate system of the laser detection and those of the coordinate system of the platform carrier. The angle is caused by the fact that the three main axes of the two coordinate systems are not parallel in the assembling and calibration process.

The above three offsets and the angle are inevitable in the system integration. The offsets can be obtained by measuring range and angle using the off-line detection method based on the conventional theodolite and range finder or be measured directly. The angle difference of coordinate axes is firstly detected in the laboratory and then estimated directly from the data acquired in the target detection based on a laser imaging system using the on-line detection technique. However, if the measurement is not precise or the drift occurs in the work process, azimuth errors in the detection system are supposed to be generated, which affect the high-precision laser imaging.

3. ***Optics-machines assembling and calibration errors***

Optics-machines assembling and calibration errors refer to the angle difference and position difference induced by the optics-machines assembling precision in assembling of system components and the installation of the system on the detection platform. In the high-precision optics-machines system, the assembling and calibration errors are difficult to be avoided. It is shown in the following aspects: (1) the position errors and angle difference cannot be adjusted to zero because of the limitation of equipments in the process of assembling and calibration; (2) the offset of position and angle appears due to the influences of motion, deformation and environment in the assembled and calibrated optics-machines systems. Optics-machines assembling and calibration errors are likely to reduce the receiving efficiency of the system or produce errors of measurement data.

In the target detection based on a laser imaging system, the main optics-machines assembling and calibration errors include the followings: the position errors between the light points of laser emission and the photosensitive surface of laser echo detectors, the coaxiality errors between the central axes of laser emission and receiving systems, and the assembling errors among optical components. Apart from these, the alignment errors of the beam-splitting components, the assembling errors among multi-pixel detectors, and the position and azimuth errors among the positioning, attitude determination and laser detection systems installed on the same rigid substrate belong to the assembling and calibration errors as well.

References

1. Jiang Y (2001) A precise positioning equation and error analysis and precision evaluation for integrated positioning system of airborne GPS, INS and laser scanning distance measurement. Journal of Remote Sensing 5(4):241–247
2. Hu Y, Xue Y, He Y (1999) The research into the precision of distance measurement for airborne laser scanning laser. Chin J Quantum Electron 16(3):193–197
3. Liu J, Zhang X, Li Z (2002) Analysis of systematic error influencing the accuracy of airborne laser scanning altimetry. Geomatics Inf Sci Wuhan Univ 27(2):111–116
4. Zhang X (2007) The principle and method of light laser detection and distance measurement technology. Wuhan University Press, Wuhan
5. Li S, Xue Y (2001) The integrated technology system of high efficiency three-dimensional remote sensing. Science Press, Beijing
6. Hu Y, Shu R, Zhao S (2002) The positioning technology and precision analysis of remotely-sensed image. The memoir of seminar of imaging spectrum technology and application sponsored by specialized committee of spatial remote sensing of Chinese society of space research, Wenzhou, pp 7–9
7. Zhao N, Hu Y (2006) Relation of laser remote imaging signal and object detail Identities. Infrared Laser Eng 35(2):226–229

CPSIA information can be obtained
at www.ICGtesting.com
Printed in the USA
LVHW011611160619
621375LV00002B/100/P